Fundamentals International Aviation

MW01092340

International aviation is a massive and complex industry that is crucial to our global economy and way of life.

Fundamentals of International Aviation, designed for the next generation of aviation professionals, flips the traditional approach to aviation education. Instead of focusing on one career in one country, it has been designed to introduce the aviation industry on a global scale with a broad view of all the interconnected professional groups. Therefore, this is an appropriate introductory book for any aviation career (including aviation regulators, maintenance engineers, pilots, flight attendants, airline managers, dispatchers, air traffic controllers, and airport managers among many others).

Each chapter of this text introduces a different cross-section of the industry, from air law to operations, security to remotely-piloted aircraft (drones). A variety of learning tools are built into each section, including case studies that describe an aviation accident related to the content of each chapter.

This book provides a foundation of aviation industry awareness that will support the next generation as they choose a career path that best aligns with their interests and ambitions. It also offers current professionals an enriched understanding of the practices and challenges between the many interconnected professional groups that make up the rich fabric of international aviation.

Dr Suzanne K. Kearns is an Associate Professor within the aviation program at the University of Waterloo in Canada. She is a former aeroplane and helicopter pilot, the author of three other books, has developed and delivered 'Introduction to Aviation' courses to hundreds of students over the past 15 years.

Fundamentals of International Aviation

Suzanne K. Kearns

Routledge
Taylor & Francis Group

LONDON AND NEW YORK

First published 2018
by Routledge
2 Park Square, Milton Park, Abingdon, Oxon OX14 4RN

and by Routledge
711 Third Avenue, New York, NY 10017

Routledge is an imprint of the Taylor & Francis Group, an informa business

British Library Cataloguing-in-Publication Data
A catalogue record for this book is available from the British Library

Library of Congress Cataloging-in-Publication Data
A catalog record for this book has been requested

ISBN: 978-1-138-70894-5 (hbk)
ISBN: 978-1-138-70897-6 (pbk)
ISBN: 978-1-315-20099-6 (ebk)

Typeset in Minion Pro
by Apex CoVantage, LLC

Visit our additional e-resources here:

www.routledge.com/9781138708976

Dedication

This book is for my husband Mike, who is my partner in every aspect of life. Thank you for supporting me, challenging me, and lifting me higher. It is also for my children, Katie, Sam, and Andy, who bring sunshine and meaning into every day of my life.

This book is dedicated to all students and teachers whose dreams and aspirations cause them to move through life with their eyes turned skywards.

Contents

Figures

Tables

TABLES

Boxes

Case Studies

Preface

International aviation represents a massive and complex industry that is crucial to our global economy and way of life. This book is an introduction to international aviation for *you* – the next generation of aviation professional.

Traditionally aviation education began with the study of a specific profession within a particular country (to be a pilot or air traffic controller in Canada, for example). After several years of study, students would transition to professionals and over the course of their careers, they would learn about other aviation professions as well as how operations are conducted in other countries.

You are joining the aviation industry at a time of dramatic international growth. This means your career path may be quite different than that of your predecessors. You may find job opportunities in foreign countries, and perhaps even experience several different job roles. For that reason, this textbook flips the traditional approach. Instead of focusing on one career in one country, it has been designed to introduce you to the aviation industry from an international perspective and with a broad view of all the interconnected professional groups.

Each chapter of this text introduces a different aspect of the industry. You will discover interesting careers in aviation that you might not have previously considered. This book will give you a foundation of industry awareness that will help you make an informed career choice, one that best aligns with your interests and ambitions.

After you have chosen a career path, and progress in your studies, you will discover that entire books have been written about the topics that make up each individual chapter of this text. There are many fantastic advanced resources available to support your transition from a learner to a professional.

Welcome to the exciting world of international aviation!

Dr Suzanne K. Kearns
Associate Professor, University of Waterloo
200 University Avenue West, Waterloo, Canada N2L3G1

Acknowledgements

I am sincerely thankful to colleagues and representatives from professional associations who volunteered their time and expertise to review drafts of every chapter in this book. In particular, members of the Next Generation of Aviation Professionals (NGAP) programme at ICAO were tremendously helpful in providing photography, reviewing edits, and answering many questions about 'how aviation works' in their part of the world. Without these volunteer efforts, it would have been impossible to represent the scope of international aviation.

This work was also tirelessly supported by Guy Loft, editor at Routledge, who was my partner in this project from proposal to completion.

My thanks also go out to Nadine Coderre, a talented editor who helped ensure clarity and standardization throughout the chapters. Finally, I appreciate the talented student digital artists who created the line art within this book: Jeremy Prapavessis, Mimi Wang, Jeffery Li, and Yifei Ren.

Learning Tools

This book has been designed to introduce you to key concepts and terminology used in the aviation industry. Each chapter incorporates recurring features that serve as an orientation to the topic area. These features include:

Opening Quizzes – Learning science suggests that attempting to answer questions before studying new material results in improved learning. Give it a try before beginning each chapter.

Chapter Outcomes – A list of the key points covered within the chapter.

'Did you Know?' Textboxes – Short facts or stories that relate to the chapter content.

Quick Reference Tables – In the cockpit, pilots will use quick reference handbooks (QRHs), which provide a listing of normal and abnormal flight procedures. In this book, quick reference tables (QRTs) offer detailed information, which might not be immediately needed, but may serve as a reference for future studies.

Examples from ICAO Council States – There are 192 countries that participate in international aviation. As it wouldn't be feasible to include examples from all these countries in each chapter, examples from the 'States of chief importance' of the International Civil Aviation Organization (ICAO) Council are included throughout the chapters. These States are Australia, Brazil, Canada, China, France, Germany, Italy, Japan, the Russian Federation, the United Kingdom, and the United States.

Language of Aviation – Rather than formal definitions, each chapter includes a Language of Aviation feature, with key terms that will help you understand chapter content.

Acronym Rundown – Acronyms are heavily used within the aviation industry, oftentimes so frequently that it is difficult to remember what the acronym stands for. Each chapter's acronym rundown presents key acronyms to help you learn to speak the language of aviation.

Summary of Key Points Chapter Review Questions and Case Study Questions – To help you review content and check your understanding, the key points of each chapter are summarized at the end of the chapter. Each chapter also includes questions to encourage reflection on the content. Questions include knowledge checks, opportunities for personal reflection, and independent research questions for you to seek out examples specific to your country.

Case Studies – Mid-chapter and end-of-chapter case studies describe aviation accidents associated with the chapter content, and allow you to apply what you have learned to a real-world example.

A note for readers: As an international textbook, this book uses the grammar, vocabulary, spelling, and style of British English. To ensure understanding for readers from all countries, the units of measurement have, for the most part, been provided using both the metric and imperial systems.

1. Annexes contain standards and recommended practices (SARPs) to the Chicago Convention, describing regulations states must abide by and those they should comply with. Currently, the Convention has ____ Annexes.

   a. 9
   b. 13
   c. 19
   d. 26

2. The Wright brothers are famous because they were the inventors of aviation.

   a. True
   b. False

5. International organizations, such as the International Air Transport Association (IATA) and the Airports Council International (ACI) do not play important roles within international air law.

   a. True
   b. False

3. Under international air law, all member countries that have signed on to the Chicago Convention have complete freedom to each other's airspace.

   a. True
   b. False

4. The sovereign body of the International Civil Aviation Organization (ICAO) is its Assembly, with representatives from 191 member States. ICAO's Assembly meets at least once every:

   a. Month
   b. Year
   c. 3 years
   d. 5 years

Learning science suggests that thinking through a few questions before you begin studying new material, even if you answer incorrectly, results in improved learning and retention.
Give it a try!

CHAPTER 1

International Air Law

CHAPTER OUTCOMES

At the end of this chapter, you will be able to . . .

- Discuss the origins of aviation and the history of international aviation law.

- Describe the 1944 Chicago Conference, as well as the structure and function of the International Civil Aviation Organization, which was created as a result of the Conference.

- Differentiate between multilateral, bilateral, and national aviation regulations.

- Identify the various international organizations that influence international aviation regulation.

- Use your understanding of international air law to discuss a case study on the shooting down of KAL 007, a civilian aircraft, by a military aircraft.

Introduction

The sheer complexity of international aviation law can be overwhelming to those first studying it. Questions are raised about how so many countries, with varying laws and cultural values and practices, can agree to follow the same rules and policies. Yet without large-scale international agreement, a safe and efficient aviation system would be impossible.

Consider for a moment what air travel might be like if each individual country designed its own unique methods for certifying aircraft as fit to fly, communicating and navigating, planning airports, or establishing pilot licensing standards. Whenever a flight crossed international borders, the flight crew would have to follow a new set of rules, which would be very confusing and probably unsafe! In addition, an aviation professional trained in one country might never be able to work in another country, as the standards and practices would vary.

For the aviation community to function safely and efficiently, international regulations must be standardized. *International air law* refers to the rules and regulations that impact global air transport, and is a unifying element of civil aviation. To gain a solid understanding of how the aviation industry functions on a global scale, it is necessary to develop a familiarity with international air law – for this reason, aviation law is the foundation upon which the following chapters in this text are built.

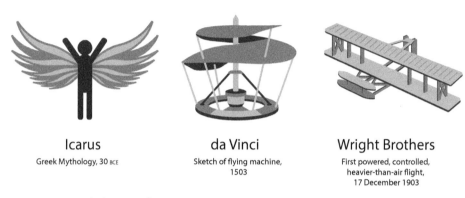

Icarus	da Vinci	Wright Brothers
Greek Mythology, 30 BCE	Sketch of flying machine, 1503	First powered, controlled, heavier-than-air flight, 17 December 1903

Figure 1.1 Early history of aviation

The Origins of Aviation

Flight has fascinated humankind for probably as long as people have walked the earth. Early recorded history is filled with legends of flight and early experiments – from the tale of Icarus who flew too close to the Sun, which melted the wax holding his wings together causing him to fall to Earth (Greek mythology, approximately 30 BCE) to the work of Leonardo da Vinci, who was the first to apply scientific principles to aviation in his sketches of flying machines (late 1400s). As long as humans have observed the flight of birds, there have been those who dreamed of joining them. In fact, the term *aviation*, which refers to the operation of aircraft, is derived from the Latin *avis* meaning 'bird'.

Did You Know?

Early attempts at flight involved the use of ornithopters. An ornithopter is a device designed to achieve flight by flapping wings (powered by either an engine or the pilot's muscles). Unfortunately, in the era of da Vinci, several monks lost their lives jumping off buildings or bridges with unsuccessful ornithopters.

The aviation industry, as we understand it today, is still relatively young. On 17 December 1903, brothers Wilbur and Orville Wright made history when Orville successfully flew their flyer about 37 metres (120 feet), earning the distinction of being the first to accomplish powered, controlled, heavier-than-air flight. In the years that followed the Wright brothers' success, several other functional aircraft were developed.

Did You Know?

The Wright brothers were not the 'inventors of aviation'. The Wright brothers built their success upon the contributions of earlier aviation pioneers. Otto Lilienthal, considered the Father of Gliding, performed many controlled glides in the late 1890s. Lighter-than-air flight – using hot-air balloons and airships – began with the French Montgolfier brothers in the late 1700s and evolved to include dirigibles (blimps) offering scheduled passenger service during the time of the Wright brothers. Several other pioneers were experimenting with powered flight, including Clément Ader, Hiram Maxim, Karl Jatho, and Augustus Moore Herring. Therefore, to be accurate, the Wright brothers were the first to accomplish *powered, controlled, heavier-than-air* flight.

World War I, which began in 1914, was the catalyst that dramatically accelerated the use of aircraft. During the war, aircraft were initially used for reconnaissance, and then eventually for bombing and air-to-air combat. When World War I ended in 1918, a surplus of trained pilots and aircraft were available for the first time in human history and the market for civil aviation began to develop. Air shows, flight training businesses, chartered passenger services, aerial surveying, firefighting, and advertising outfits that scattered printed material over towns grew in availability and popularity in the interwar era.

From the end of the First World War through the mid-1930s, airlines began springing up, including KLM from the Netherlands and Avianca from Colombia (both in 1919), Qantas from Australia (1920), Czech Airlines from Czechoslovakia (1923), Lufthansa from Germany (1926), Iberia from Spain, and Pan American World Airways from the United States (both in 1927), among others. For the first time in civil aviation, negotiations between countries over landing rights and privileges were important issues.

In World War II (1939–1945), aviation no longer served a supporting role in military combat; air power was crucial. The German air force, called the *Luftwaffe*, and the Imperial Japanese Army Air Service battled the Allies' British Royal Air Force (which included the Royal Canadian Air Force and Royal Australian Air Force), the United States Army Air Forces, and the Soviet Air Force in brutal bombing and fighter missions throughout the war. Incredible operational and technical aviation advancements occurred during this period. By the time the Allies won the war in 1945, both military and civil aviation had grown exponentially and the public perception had shifted – aviation was no longer seen as something mystical, but rather as a part of everyday life.

History of Multilateral Aviation Regulation

International air law falls into three categories: 1) multilateral agreements between three or more States, 2) bilateral agreements between two States, and 3) national regulations within a single State. This first section will explore the development of multilateral agreements within international civil aviation.

Reviewing the early history of aviation helps in understanding when and how international regulations were established. In 1908, before World War I, at least 10 German balloons crossed the border and landed in France. This raised concerns and caused the French government to propose an international conference to determine regulations for flight over and into foreign countries.[1]

This first important conference on international air law, called the International Air Navigation Conference, was held in Paris in 1910 and attended by 19 European States – it was the first effort to diplomatically create multilateral legal principles related to air navigation. States from other continents were not invited because it seemed unrealistic at the time that their aircraft could travel the great distance to Europe. Discussions broke down as States disagreed about the ownership of airspace – whether there should be freedom of the air, similar to the freedom of international waters in the ocean, or if a nation's sovereignty included control over airspace above its territory.[2]

1919 Paris Convention

World War I interrupted the progress of diplomatic negotiations on civil aviation. After the war ended, the true beginning of aviation regulation was marked with the 1919 Paris Convention, which grew out of the Paris Peace Conference. The war had demonstrated that aircraft had tremendous, yet possibly devastating, potential and therefore required international attention. The *Convention Relating to the Regulation of Aerial Navigation* was signed by 37 States on 13 October 1919 and included 43 legal articles that outlined agreements on technical, operational, and organizational aspects of civil aviation. The International Commission for Air Navigation (ICAN) was also created, under the direction of the League of Nations, as an organization with responsibility for managing and creating new aviation regulations as necessary.[3] The work of ICAN and its subcommissions was a tremendous help in drafting the annexes of the Chicago Convention, which was to come in 1944.

Did You Know?

It is generally accepted that the international air transport industry was born in 1919, with the signing of the Paris Convention. The year 1919 is also notable for the establishment of the precursor to the International Air Transport Association (IATA), the group that represents the world's scheduled airlines.

1928 Havana Convention

The United States and 20 other countries from North, South, and Central America met in Havana in 1928 for the Havana Convention. The goal was to establish a foundation of international cooperation specific to the Americas; however, the resulting convention weakened ICAN's international position. Building from the Paris Convention, several important modifications were made. The Havana convention applied to civil aircraft (excluding government/military aircraft) and established basic rules for air traffic, determining that every State had exclusive authority to the airspace above its land and connected territorial waters.

Yet there were some weaknesses to this convention, as no uniform technical standards were included, no permanent organization was established to manage or create new regulations, no provisions were made for annexes, and aircraft regulation was completely under the laws of each country – in short, the convention lacked uniformity.[4]

Although the Paris and Havana Conventions moved discussions forward, they also led to confusion as they created two separate sets of rules. At a time when there were about 50 States involved in aviation, the Paris Convention's ICAN represented 33 of them and the Havana Convention was ratified by 11.[5] With the dramatic growth of air travel following World War II, a single unifying convention was required.

1.1   The Language of Air Law

Regulation is the delivery of authoritative direction to create a desired degree of order.

A **convention** is a type of a treaty – an international agreement between States governed by international law – that is sponsored by an international organization; it is typically signed by many States. A **conference** is a meeting of representatives from States, during which the specifications of a convention are discussed.

The term **State** refers to a country. This differs from the common use of the word *state* in North America, which typically references the United States of America.

Signatory to the convention and **contracting State** both refer to a country that has signed on to a convention. Currently, 192 of the 195 States in the world are party to the Chicago Convention.

Ratified means that a convention has been signed by enough States for it to become valid. The Chicago Convention was ratified on April 4 1947 when the 26th State signed on.

Annexes are add-ons to a convention. When new issues are identified that require international regulation, new annexes are added. Annexes allow the original Chicago Convention to remain valid so that additions can be made without requiring all 192 States to sign on to a new Convention.

A **standard** is a specification that contracting States must adopt.

A **recommended practice** is a specification that States should adopt.

ICAO (International Civil Aviation Organization) is pronounced 'I-K-O' among aviation professionals, not 'I-see-A-O' or 'I-cow'.

Cabotage is a freedom of the air that allows an air carrier to transport passengers between two points within a foreign country.

1944 Chicago Convention

In 1942, three years before the end of World War II, it was clear that civil aviation was a critical international issue. Political and diplomatic discussions about international aviation arrangements began in Canada, the United Kingdom, and the United States. In 1943, at the Anglo-American Conference in Quebec City, Roosevelt and Churchill began discussions about post-war aviation policies under a United Nations (UN) organization.[6] As 1944 began, the war seemed to be coming to an end, and it was apparent that commercial aviation would be growing internationally.

On 11 September 1944, the United States invited 53 governments to an international civil aviation conference. The Chicago Conference was convened on November 1 1944, and lasted 37 days.[7] Fifty-two States attended the conference with a total of 955 people, including delegates, advisors, secretaries, stenographers, and members of the press.[8]

The result of the conference was the drafting of the *Convention on International Civil Aviation* (commonly called the Chicago Convention). The Chicago Convention established the International Civil Aviation Organization (ICAO) as the sole international organization responsible for civil aviation, replacing ICAN that had been created with the Paris Convention. The primary objective of international civil aviation was determined to be air transport in a safe and orderly manner that was economically sound and offered States equal opportunities.[9] The Convention defined universal rules associated with sovereignty of airspace, navigation, aircraft airworthiness and registration, and global standards and recommended practices (SARPs) for operational harmonization between States.

The Chicago Convention begins with the following preamble:

> WHEREAS the future development of international civil aviation can greatly help to create and preserve friendship and understanding among the nations and peoples of the world, yet its abuse can become a threat to the general security; and
>
> WHEREAS it is desirable to avoid friction and to promote that co-operation between nations and peoples upon which the peace of the world depends;
>
> THEREFORE, the undersigned governments having agreed on certain principles and arrangements in order that international civil aviation may be developed in a safe and orderly manner and that international air transport services may be established on the basis of equality of opportunity and operated soundly and economically;
>
> Have accordingly concluded this Convention to that end.[10]

The *Convention on International Civil Aviation* (the Chicago Convention), which provided the basis for international law of the air, included several instruments.

1. *The Interim Agreement on International Civil Aviation*, which allowed the global effort to begin before the Convention was ratified. The interim agreement created the Provisional International Civil Aviation Organization (PICAO); ICAO became a permanent organization on 4 April 1947.

2. *The International Air Services Transit Agreement* (called the Two Freedoms Agreement), which allowed aircraft of contracting States to

   • fly over each other's territory without landing (First Freedom of the Air); and

- land in another's territory for non-traffic purposes (such as refuelling), without picking up or dropping off passengers, cargo, or mail (Second Freedom of the Air).

3. *The International Air Transport Agreement* (called the Five Freedoms Agreement) added three freedoms associated with commercial transport. These included the rights to

   - carry passengers from an air carrier's home country to a foreign destination (Third Freedom of the Air);

   - carry passengers from a foreign destination to an air carrier's home country (Fourth Freedom of the Air); and

   - carry passengers between two foreign countries, when the flight begins or ends in the air carrier's home country (Fifth Freedom of the Air).

4. The drafts of 12 technical annexes (organized A to L) to cover operational and technical aspects of international civil aviation, such as airworthiness of aircraft, air traffic control, communications, and so on. Today, the 12 annexes have grown to 19 and are organized by number rather than letter.

5. A standard form of bilateral agreement for the exchange of air routes between two countries. Bilateral agreements are discussed in more detail later in this chapter.

1.2 Freedoms of the Air[1]

Beyond the five freedoms of the air that are incorporated within the Chicago Convention, there are additional freedoms with respect to scheduled international air service. These freedoms are not included in the multilateral Chicago Convention, but are occasionally agreed to on a State-by-State basis through bilateral agreements. These additional freedoms are

- carrying passengers between two foreign countries, passing through the air carrier's home country (Sixth Freedom);

- carrying passengers between two foreign countries, operating entirely outside of the air carrier's home country (Seventh Freedom);

- carrying passengers between two points in a foreign country, serving as a domestic airline within a foreign state, yet originating within the air carrier's home country (called *consecutive cabotage*) (Eighth Freedom); and

- carrying passengers between two points within a foreign country, operating entirely separately from the air carrier's home country (called *standalone cabotage*) (Ninth Freedom).

Note

1 Adapted from ICAO, 2004

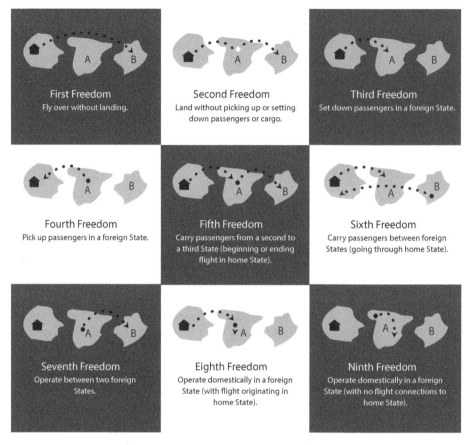

Figure 1.2 Freedoms of the air

The Chicago Convention was the result of one of the most successful and influential conferences ever held. The conference established key principles that made international flying much safer and less complex than it had been prior to WWII. It terminated the 1919 Paris and 1928 Havana Conventions, and meant that for the first time in history, a single international organization (ICAO) would standardize technical issues in aviation and harmonize practices between States.

Annexes to the Chicago Convention

When the Chicago Convention was created in 1944, 12 annexes were included with technical standards and recommended practices (SARPs). In the years since, as issues have arisen that were not considered in 1944, it has not generally been practical to revise the Convention (as this would require all 192 States to sign on to the new Convention). Therefore, annexes are added or modified to address these new issues.

The number of annexes has grown to 19. The protection of the environment is a good example of an international concern that was not considered in 1944 but has become increasingly important in modern times; therefore, Annex 16: Environmental Protection was added to the Convention. An overview of the topics covered in each of the annexes is included in Table 1.1. Many of these topic areas are explored in detail in later chapters of this textbook.

Table 1.1 Annexes 1 to 19 of the Chicago Convention

Annex	Title	Description	Textbook chapter
1	Personnel Licensing	Relates to the training and licensing of flight crew members (pilots, flight engineers, and navigators), air traffic controllers, aeronautical station operators, maintenance technicians, and flight dispatchers in order to ensure sufficient skill among professionals and to foster international licensing standards.	Chapters 2, 3, 4
2	Rules of the Air	Covers general rules of the air, including specific requirements for visual flight rules and instrument flight rules operations, right-of-way rules, flight plan requirements, and collision avoidance principles.	Chapter 4
3	Meteorological Service for International Air Navigation	Relates to the provision of meteorological information to aviation users, including weather reports, forecasts, landing forecasts, weather briefings for operators, meteorological watch offices to monitor changes in the weather, world area forecast systems, and the International Airways Volcano Watch.	Chapter 7
4	Aeronautical Charts	Sets standards for aeronautical charts to facilitate international navigation, including chart coverage, format, standardized symbols, and colour use. Defines 21 types of charts for specialized purposes.	
5	Units of Measurement to Be Used in Air and Ground Operations	Establishes the metric system as the international standard for civil aviation, recognizing that a consistent measurement unit is crucial for safety and efficiency.	
6	Operation of Aircraft	Standardizes safe operating practices for aircraft within international air transport operations. Defines standards for aircraft operations and performance, communications, navigation equipment, maintenance, flight documents, security, and responsibilities of flight personnel.	Chapter 2, 3

(Continued)

Table 1.1 (Continued)

Annex	Title	Description	Textbook chapter
7	Aircraft Nationality and Registration Marks	Requires aircraft to be registered in its contracting State and marked, with letters, numbers, or other graphic symbols, to indicate its nationality and registration.	Chapter 2
8	Airworthiness of Aircraft	Relates to the specifications of an aircraft's airworthiness and includes requirements that describe how aircraft must be designed, built, and operated. When requirements are met, aircraft are issued a certificate of airworthiness, which indicates that it is fit to fly.	Chapter 2
9	Facilitation	Refers to international practices to expedite customs, immigration, quarantine, and clearance to prevent unnecessary delay of aircraft, crews, passengers, or cargo.	Chapter 5
10	Aeronautical Telecommunications	Includes standards and recommended practices and procedures for air navigation services as well as guidance material for aviation communication, navigation, and surveillance.	Chapter 4
11	Air Traffic Services	Describes ground-based air traffic services (flight information centres and air traffic control units), which are designed to prevent collisions between aircraft (during taxi, take-off, cruise, or approach to land) within the global flight information regions.	Chapter 4
12	Search and Rescue	Includes the structure and cooperative principles needed for search and rescue operations, preparatory measures, and operating procedures for actual emergencies, recognizing that the international response to aircraft accidents must be quick and efficient.	Chapter 8
13	Aircraft Accident and Incident Investigation	States that the objective of an investigation is the prevention of future occurrences and contains international requirements for investigation, such as which States have a right to participate in the investigation and the rights and responsibilities of these States. Describes the investigation process and the organization of the final report.	Chapter 8

Annex	Title	Description	Textbook chapter
14	Aerodromes	Describes the planning of airports and heliports, including design of movement areas (runways and taxiways), airspace considerations to ensure safe arrivals and departures, lighting, operation and maintenance, rescue and firefighting resources, as well as many other considerations.	Chapter 5
15	Aeronautical Information Services	Ensures immediate flow of quality flight and terrain data required for the operation of international civil aviation. Makes reference to *international notices to airmen* (now simply *notices to airmen* or NOTAM), which are used to alert pilots to important issues; today, data is often fed directly into on-board navigation systems.	Chapter 7
16	Environmental Protection	Publishes standards designed to limit the environmental impact of aircraft engine emissions and aircraft noise.	Chapter 7
17	Security: Safeguarding International Civil Aviation Against Acts of Unlawful Interference	Incorporates global standards to safeguard international civil aviation against unlawful acts. Requires each State to develop its own civil aviation security programme.	Chapter 6
18	The Safe Transport of Dangerous Goods by Air	Establishes standards for safe transport of potentially dangerous cargo (anything radioactive, toxic, flammable, explosive, or corrosive) and provides a limited list of substances identified by ICAO as unsafe to carry on an aircraft.	Chapter 3
19	Safety Management	Describes how the aviation industry must proactively identify safety risks and reduce them, rather than take a reactive approach after an accident. Requires States to develop a state safety programme based on a strong safety management system and incorporating safety oversight, data collection, analysis, and sharing of safety information for the benefit of the international civil aviation system.	Chapter 9

Structure of ICAO

The International Civil Aviation Organization (ICAO) was created through the Chicago Convention as a special agency of the UN. The vision of ICAO is to achieve the sustainable growth of the global civil aviation system through the development of SARPs for supporting air navigation and developing air transport.

ICAO is managed by a *Secretariat*, which is a grouping of permanent administrative offices led by a *Secretary General* who functions as the chief executive officer of ICAO.

The high-level policies and work programme of ICAO are developed by the *Assembly* (the sovereign body of the organization with representatives from all 192 contracting States), and overseen by a permanent governing body called the *Council*, which is responsible to the Assembly. While the Assembly includes representatives from all member States and meets at least once every three years, the Council is composed of elected representatives from 36 contracting States, who work full time over their three-year terms. Note that in the year 2016 the Assembly voted to increase the Council to 40 elected representatives, which will take effect when the Resolution is ratified.

Working from the high-level objectives, the *Air Navigation Commission* (ANC) and panels of experts are then responsible for completing the items within the work programme. The ANC creates the detailed SARPs included within the annexes to the Convention.

Secretariat

The Secretariat of ICAO is a grouping of permanent administrative offices, based in Montreal, Quebec, with staff recruited from the member States of ICAO. The Secretariat is led by the Secretary General and includes five main divisions:

1. the Air Navigation Bureau;

2. the Air Transport Bureau;

3. the Legal Affairs and External Relations Bureau;

4. the Bureau of Administration and Services; and

5. the Technical Cooperation Bureau.

It also includes seven regional offices around the world.

The *Secretary General* serves as the chief executive officer of ICAO and has responsibility for the direction and work of the Secretariat. The Secretary General is the secretary of the ICAO Council and is responsible to the Council. The Secretary General also oversees the seven regional offices and any work assigned to the Office of the Secretary General, including communications, finance, and internal audits.[11]

The *Air Navigation Bureau* (ANB), in partnership with industry associations and stakeholders, manages a range of ICAO policies related to air navigation safety and infrastructure. The ANB creates and maintains the Global Air Navigation Plan and the Global Aviation Safety Plan.[12]

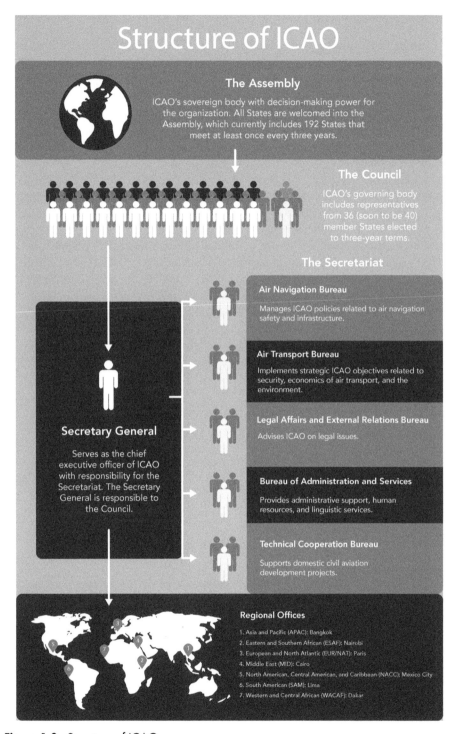

Figure 1.3 Structure of ICAO

The *Air Transport Bureau* supports the implementation of strategic ICAO objectives related to security, the economics of air transport, and protection of the environment.[13]

The legal functions of the *Legal Affairs and External Relations Bureau* include advising the Secretary General, Council, and other bodies within ICAO on issues related to air, commercial, labour, and international law. The external relations activities include reviewing and advising on relations with States, the UN, and other international organizations.[14]

The *Bureau of Administration and Services* provides ICAO with administrative support and management through human resources, information management, and linguistic services covering the six working languages of ICAO (English, Spanish, French, Arabic, Chinese, and Russian).[15]

The *Technical Cooperation Bureau* works with member States by supporting civil aviation development projects in their home countries. These may include the implementation of ICAO SARPs, strengthening civil aviation authorities, and providing training and infrastructure, among other activities.

Following the Chicago Conference in 1944 the Interim Council determined that, to address region-specific issues, the globe should be subdivided into air navigation regions. This would allow for operational and technical issues and air navigation facility planning to be completed in cooperation with the key States involved, rather than involving all 192 member States included in the Chicago Convention.

In 1945, the Interim Council established the air navigation regions, divided roughly along oceanic and continental borders, which have evolved over the years to become the following (head office for each region is provided in parentheses):

1. Asia and Pacific (APAC): (Bangkok)
2. Eastern and Southern African (ESAF): (Nairobi)
3. European and North Atlantic (EUR/NAT): (Paris)
4. Middle East (MID): (Cairo)
5. North American, Central American, and Caribbean (NACC): (Mexico City)
6. South American (SAM): (Lima)
7. Western and Central African (WACAF): (Dakar).[16]

The Council

The Council is made of up representatives from 36 member States who serve as the governing body of ICAO. Council members work at ICAO headquarters in Montreal on a full-time basis. States that are of particular importance to civil aviation are chosen to have seats on the Council. The 36 Council States are elected by the 192 member States within sessions of the Assembly for three-year terms. The Council then elects its president for a three-year term and three vice-presidents for one-year terms. Council States are organized into three groups or *parts*. The 2016–2019 Council is made up of representatives from these States:

- Part 1 – States of chief importance in air transport:
  - Australia, Brazil, Canada, China, France, Germany, Italy, Japan, Russian Federation, United Kingdom, and the United States

1.3 Council States

Throughout this textbook, look for tables that highlight chapter-specific examples from certain States. The 11 Council States that ICAO has designated as being of 'chief importance to air transport' are referred to in these tables with domestic examples of their aviation agencies or organizations.

- Part 2 – States that make the greatest contribution to the provision of facilities for international civil air navigation:

  - Argentina, Colombia, Egypt, India, Ireland, Mexico, Nigeria, Saudi Arabia, Singapore, South Africa, Spain, and Sweden

- Part 3 – States ensuring geographic representation:

  - Algeria, Cabo Verde, Congo, Cuba, Ecuador, Kenya, Malaysia, Panama, Republic of Korea, Turkey, United Arab Emirates, United Republic of Tanzania, and Uruguay.[17]

The Council documents and submits its work to the Assembly in annual reports. The Council is responsible for implementing resolutions of the Assembly, adopting policy, and setting the tasks and priorities for the ICAO work programme.[18]

The Assembly

The Assembly is ICAO's sovereign body, meaning that it is the group with decision-making power for the organization. It is in everyone's best interest for all States to be welcomed into ICAO, as the universal application of ICAO SARPs promotes the safety and efficiency of international aviation. Therefore, ICAO welcomes all States to participate.

The Assembly includes representatives from all 192 contracting States who come together at least once every three years. Assembly sessions are convened by the Council and opened by the President of the Council. Many international organizations (e.g. the International Air Transport Association, Airports Council International, and so on) are also invited to the Assembly.

During Assembly sessions, a variety of decisions are made, relating to taking action on Council reports, approving budgets, reviewing and approving ICAO work programmes, and electing new Council States. When the Assembly is underway, each State has one vote and decisions are based on support from the majority.

Occasionally an extraordinary meeting of the Assembly will be called, by the Council or by at least one-fifth of the contracting States, in response to a timely issue of high importance. For example, in 1970 an extraordinary session of the Assembly was held in response to the alarming increase in hijacking incidents in the late 1960s.

Air Navigation Commission

The Air Navigation Commission (ANC) can be considered an independent advisory board to the ICAO Council on air navigation matters.[19] The ANC is made up of 21 members with 'suitable qualifications and experience in the science and practice of aeronautics' as specified by the Chicago Convention, as well as industry and State observers. ANC members are nominated by ICAO member States and appointed by the Council to one-year terms, although they act independently to promote what they believe is in the best interest of international civil aviation (rather than supporting the interests of their particular State). Typically, the work of the ANC is accomplished in three sessions per year, with each lasting nine weeks (including a three-week recess). The ANC is responsible for the technical work programme of ICAO, which has high-level objectives linked to the safety, efficiency, and capacity of air navigation.[20]

ICAO Work Programme

When new issues are identified – resulting from an accident investigation, presented by an industry group, or arising from discussions at an Assembly meeting – they are added to the ICAO work programme. This will cause the issue to be reviewed by the ICAO Secretariat, the ANC, and the Council.

Did You Know?

At Assembly meetings there have, on occasion, been challenges keeping States focused on technical issues rather than political differences. To understand international law, it is crucial to remember that States are primarily concerned with supporting their own interests.

Louis Henkin, Columbia law professor, is famously quoted as saying 'almost all nations observe almost all principles of international law and almost all of their obligations almost all of the time'.[1] He notes that nations will comply with international law if it is in their interest to do so, but that they may ignore the law if the advantages of violating it outweigh the advantages of obeying it.

There have been only a few cases when States have faced the possibility of expulsion from ICAO – generally, as it is in everyone's best interest for all States to participate, ICAO chooses compromise and accommodation when problems arise.[2] To ensure global safety and security, ICAO has the power to audit compliance, enforce regulations, and sanction nations that choose not to comply.

Notes

1 Henkin, 1979, p. 47   2 Mackenzie, 2010

Standards and Other Requirements

A typical response to an issue is the creation or modification of SARPs or *procedures for air navigation services* (PANS). SARPs are contained within the annexes and define the standardized process by which aviation activities are carried out – they are crucial for achieving safe and efficient air travel around the world.

Of the standards contained within the 19 annexes, the majority (17 annexes) are the responsibility of ICAO's Air Navigation Bureau. The other two (Annex 9: Facilitation and Annex 17: Security) are the responsibility of the Air Transport Bureau. ICAO standards and other requirements are organized into:

- *Standards and Recommended Practices* (SARPs) – Standards are specifications that *must* be applied consistently around the globe to ensure safe and efficient air travel. States that cannot abide by a standard must notify the ICAO Council. Recommended practices are specifications that promote safety and efficiency that States *should* abide by, as possible. The universal adoption of ICAO international SARPs is one of the foremost objectives of the organization.[21]

- *Procedures for Air Navigation Services* (PANS) – PANS are operating practices that are too detailed for inclusion within SARPs, but are applicable on a global scale.

- *Regional Supplementary Procedures* (SUPPs) – SUPPs apply to certain ICAO regions. In terms of content, they are similar to PANS, but they do not have worldwide applicability.

- *Guidance materials* – supplement SARPs and PANS to help professionals facilitate their implementation.[22]

Since the Chicago Convention, ICAO has incorporated over 12 000 SARPs within the 19 annexes and five PANS, along with supplementary and guidance materials.

The Standards-Making Process

The creation of SARPs and PANS is accomplished through the *standards-making process* by which technical and non-technical groups (typically ICAO working groups, panels, or committees containing representatives from States and subject matter experts from industry) write recommendations that include impact assessments and implementation plans. Once developed, these recommendations are brought to the ANC for preliminary review. If the recommendations are not considered sufficient, they may be sent back for further work; if they are acceptable to the ANC, they are sent to States and international organizations for comment. The Secretariat compiles all comments and brings the recommendation back to the ANC for final review. After final review, the ANC sends a draft report to Council recommending adoption. Through a two-thirds majority vote, the Council can adopt the recommendation and States are informed of new provisions, which normally become applicable in November of the following year. This entire process takes about two years from initial proposal to formal adoption of a SARP within an annex or PANS manual.[23]

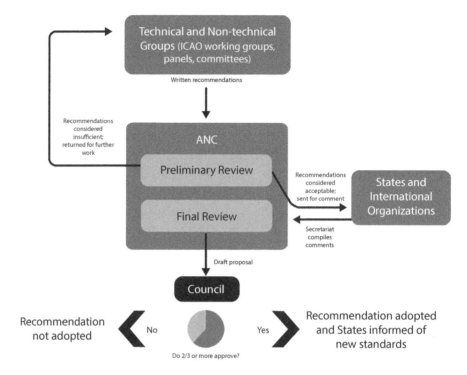

Figure 1.4 The standards-making process

Bilateral and National Regulations

So far in this chapter, we have explored aspects of international aviation regulations – particularly the multilateral Chicago Convention. However, the regulation of civil aviation has bilateral and national components in addition to the multilateral. The final section of this chapter will explore bilateral agreements – those established between two States – as well as national regulations specific to a single State.

Bilateral Regulation

Unlike domestic laws within a country, or multilateral agreements like the Chicago Convention, bilateral regulations come into being when a State proposes an agreement directly with another State. Although the Chicago Convention made several notable accomplishments, it did not specify how air traffic rights would be exchanged between nations. Therefore, the negotiation and exchange of traffic rights are accomplished through bilateral regulations called *air service agreements* (ASAs).

ASAs are legal documents created primarily to serve States' economic concerns – they give States the privileges of carrying passengers or cargo to points over, to, or from other States' territories. These traffic rights can be thought of as *market access* privileges[24] and typically have constraints related to frequency of flights, passenger and cargo capacity limits, and pricing guidelines. ASAs are developed through a series of meetings between State representatives, where negotiations take place to define how aviation trade will be coordinated between the two States.[25] Globally, the result is a highly complicated labyrinth of ASAs with more than 4 000 worldwide,[26] many of which are in the midst of negotiation, amendment, or dispute resolution at any given time. There is also the occasional termination.

Did You Know?

A *supra-State* is an organization made up of a group of States that is given the authority by its members to act as a single body. An example of a supra-State is the European Union. Agreements between a supra-State and another country are considered bilateral.

Keep in mind that it is often in a country's best economic interest to increase international air traffic, yet every nation wants to do this in a way that does not compromise the success of its own domestic airlines. Interestingly, ASAs include a *nationality rule*, which requires that airlines be owned and controlled by their home State. That is to say, an American company or citizen cannot own a majority share in Air Canada, nor could a German company or citizen own a majority share in Qatar Airways. This nationality rule ensures that airlines remain domestic entities and prevents the creation of a globalized airline.

As you can imagine, the negotiation of bilateral agreements can be a complicated and time-consuming process. The International Air Transport Association (IATA), with more than 270 member airlines around the world, plays an important role in assisting nations in the development of ASAs through cooperation with ICAO, airlines, and other international groups.

Originally, IATA had responsibility for setting fares for international routes around the world (subject to governmental approval); however, many States now implement open skies policies. *Open skies* refers to liberal bilateral agreements that can be thought of as free trade for international aviation. The concept of open skies was initiated through the United States' airline deregulation, which occurred in the late 1970s. Domestically, the United States eased restrictions on access, routes, and pricing to promote competition among US-based airlines. The concept was that free trade within aviation would increase competition among carriers, which would lead to lower fares and better service for passengers.

Following domestic deregulation within the United States, the US began negotiating open skies bilateral agreements with other countries, with 11 signed between 1978 and 1980.[27] The open skies concept has been accepted as part of modern aviation, with many liberal bilateral agreements in force around the world. Open skies bilateral agreements allow foreign airlines nearly limitless access to a country and the freedom to set their own fares.

National Regulation

In addition to multilateral and bilateral regulation, international air law is governed by national regulation. National regulation refers to a country's sovereign right to control its territory and the airspace above it, taking into account that country's international obligations under multilateral and bilateral agreements.

Although the multilateral Chicago Convention stipulates SARPs, and bilateral ASAs allow market access to other States, a large component of air law remains: the actual implementation of SARPs and the creation of domestic regulations. These tasks are carried out by a State's civil aviation authority (CAA). CAAs must carefully consider domestic laws and regulations in addition to international standards and agreements.

National regulation of international air transport involves three activities, each of which includes enforcement actions as required:

- licensing – the granting or withholding of permission to conduct air transport activities;
- legislation – law-making, policy making, and the writing of regulations; and
- ad hoc authorization – day-to-day decision-making on specific matters, such as permitting a single flight or approving a particular tariff.[28]

National regulation is structured within an organization – such as a government entity that serves as the nation's civil air transport authority – and by the legal framework of that country. Aviation professionals are required to learn the intricate details of their country's regulatory structure, and entire books are dedicated to this for each country.

Table 1.2 Civil aviation authorities of ICAO Council States

Australia	Civil Aviation Safety Authority www.casa.gov.au
Brazil	Agência Nacional de Aviação Civil www.anac.gov.br
Canada	Transport Canada www.tc.gc.ca
China	Civil Aviation Administration of China www.caac.gov.cn
France	Direction générale de l'Aviation civile www.developpement-durable.gouv.fr
Germany	Luftfahrt-Bundesamt www.lba.de
Italy	Ente Nazionale per l'Aviazione Civile www.enac.gov.it
Japan	Civil Aviation Bureau. Ministry of Land, Infrastructure, Transport and Tourism www.mlit.go.jp/en/koku
Russian Federation	Ministry of Transport www.mintrans.ru
United Kingdom	Civil Aviation Authority www.caa.co.uk
United States	Federal Aviation Administration www.faa.gov

Source: ICAOk, n.d.

International Organizations

International organizations play important roles within international air law. Organizations fall under three main categories:

1. International governmental organizations (IGOs): IGOs such as ICAO and the European Union (EU) have the power to adopt binding international law that governs international air transport, through consent of their members.

   - The Chicago Convention allows ICAO to propose and enact amendments to the Convention and manage annexes with included SARPs.

   - As a supranational organization, the EU is more agile in its law-making ability, and its actions are binding for its member States. The EU negotiates bilateral agreements, such as ASAs, on behalf of its members.

2. International non-governmental organizations (INGOs): INGOs, including IATA and Airports Council International (ACI) among others, can influence regulations through education of government officials and urging them to rethink aviation regulations.

   - IATA has straddled the fence between IGO and INGO by providing structure for pricing and routing of international air services and brokering agreements between member airlines and governments.

3. National governmental authorities and trade associations include CAAs and trade groups such as Airlines 4 America.

   - Domestically, a CAA has responsibility over the management and control of aviation activities within the sovereign airspace of its State. These authorities often manage regulations, security, and air navigation, among other issues.

   - Domestic trade associations are active within a country and represent a specific group's interests (e.g., owner and pilots' associations, air transport groups, and airport management groups, among others).

Conclusion

Every few seconds, an aircraft takes off or lands somewhere around the world. Every aspect of that flight relies upon standardization on a global scale – from aircraft manufacturing standards to airport facilities to licensing and training requirements. This standardization is accomplished through international air law, which is made up of a dense network of multilateral, bilateral, and domestic regulations.

Globally, the multilateral Chicago Convention has the most significant impact as it results in millions of professionals applying practices based on SARPs developed through ICAO.

International air law is a dynamic sphere of practice, continually evolving to meet the needs of modern aviation. Although cooperative rule-making can involve political challenges, the model established within the international aviation community is recognized as a successful example of what can be accomplished when the world comes together to support a common goal.

Key Points to Remember

1. After the Wright brothers' successful flight on December 17 1903, the perception of aviation evolved from something mystical at the turn of the century to an ordinary mode of transportation by the end of World War II.

2. *International air law* refers to the rules and regulations that impact global air transport. International air law encompasses three types of regulation: *multilateral*, *bilateral*, and *national*. Multilateral aviation agreements apply to three or more States; bilateral aviation agreements exist between two States; and national regulations apply domestically within a single State.

3. The first significant multilateral aviation regulations were developed at the Paris Conference and the Havana Conference. The 1919 Paris Conference resulted in the *Convention Relating to the Regulation of Aerial Navigation* signed by 37 States on October 13 1919. It included 43 legal articles outlining technical, operational, and organizational agreements concerning civil aviation, and created the International Commission for Air Navigation (ICAN) to manage and create air regulations. The 1928 Havana Convention resulted in international cooperation specific to the Americas and was ratified by 11 States, but global uniformity was lacking.

4. The 1944 Chicago Conference convened on November 1 1944 and involved 52 States. The result was the *Convention on International Civil Aviation*, which

   - established ICAO as the international organization responsible for civil aviation (replacing the Paris Convention's ICAN);

   - defined universal rules associated with airspace sovereignty, navigation, aircraft airworthiness and registration, global SARPs, as well as annexes and a standard form of bilateral agreement for exchange of air routes; and

   - incorporated 12 annexes (today, there are 19 annexes) on a variety of issues from personnel licensing to safety management.

5. ICAO is managed by a Secretariat and led by a Secretary General who functions as ICAO's chief executive officer. High-level policies and the ICAO work programme are developed within Assembly meetings, where all 192 contracting States convene every three years. The Council, a permanent governing body, is made up of representatives from 36 States, each of which serves a three-year term. The ANC, an advisory board with 21 members, is responsible for the ICAO work programme and the creation of *standards and recommended practices* (SARPs).

6. When an issue is identified and a response is required, the *Air Navigation Commission* (ANC) employs the standards-making process. Since the Chicago Convention, over 12 000 SARPs have been incorporated within 19 annexes and five PANS. Standards are specifications that *must* be applied consistently around the

globe. Recommended practices are specifications that *should* be applied consistently around the globe, as possible.

7. Bilateral regulations are initiated when a State proposes an agreement directly with another State. *Air service agreements* (ASAs) are common bilateral agreements in civil aviation that involve negotiation and exchange of air traffic rights.

8. National regulation, carried out by a State's *civil aviation authority* (CAA), refers to a country's sovereign control over its own territory and overlying airspace, taking into account multilateral and bilateral agreements. CAAs are involved in licensing, legislation, and ad hoc authorization.

9. International organizations, which play important roles in shaping international law, can be organized into three categories. *International governmental organizations* (IGOs), such as ICAO, have the power to adopt binding international legal rules through consent of their members; *international non-governmental organizations* (INGOs), such as IATA, can influence regulations through education and lobbying of government officials; and *national governmental authorities*, such as CAAs and trade associations, are active domestically and influence national regulations.

Table 1.3 Acronym rundown

ACI	Airports Council International
ANC	Air Navigation Commission
APAC	Asia and Pacific (air navigation region)
ASA	air service agreement
CAA	civil aviation authority
ESAF	Eastern and Southern African (air navigation region)
EU	European Union
EUR/NAT	European and North Atlantic (air navigation region)
IATA	International Air Transport Association
ICAN	International Commission for Air Navigation
ICAO	International Civil Aviation Organization
IFALPA	International Federation of Air Line Pilots' Associations
IGO	international governmental organization
INGO	international non-governmental organization

(*Continued*)

Table 1.3 (Continued)

MID	Middle East (air navigation region)
NACC	North American, Central American, and Caribbean (air navigation region)
NOTAM	notices to airmen
PANS	procedures for air navigation services
PICAO	Provisional International Civil Aviation Organization
SAM	South American (air navigation region)
SARPs	standards and recommended practices
SUPPs	regional supplementary procedures
UN	United Nations
WACAF	Western and Central African (air navigation region)

Chapter Review Questions

1.1   Explain why aviation grew so rapidly between 1903 and 1945, including the key events that you believe led to these advancements.

1.2   How is the Chicago Convention important in aviation today? List three ways in which the Convention impacts modern aviation.

1.3   What is the International Civil Aviation Organization (ICAO)? How did it come to exist? What is its role in today's aviation landscape?

1.4   How was the accomplishment of the Wright brothers revolutionary? Did they do it alone, or did they build upon the success of others? Explain your answer.

1.5   In the age of security risks in aviation, are freedoms of the air becoming more difficult for States to agree upon, or easier? Explain your answer.

1.6   What is the name of your State's civil aviation authority (CAA)? What challenges might it face in finding the balance between abiding by national laws and adhering to ICAO's standards and recommended practices (SARPs)?

1.7   Does your State have an *open skies* policy? Should it? Provide three arguments for an open skies policy, and three arguments against one.

1.8   Research an example of a bilateral aviation agreement that has benefitted both countries. Provide evidence of the benefits.

1.9   Provide an example of an international governmental organization (IGO), an international non-governmental organization (INGO), a national governmental authority, and a trade association that influence civil aviation in your State.

AN AIRCRAFT IN THE WRONG PLACE AT THE WRONG TIME[1]

On 1 September 1983, Korean Air Lines flight 007 (KAL 007) was on a leg from New York City to Seoul, with a fuel stop in Anchorage, Alaska. KAL 007 was a Boeing 747 with 246 passengers and 23 crew members on board. During the cruise portion of its final leg, the aircraft deviated more than 320 kilometres (200 miles) from its intended path and flew into what was then Soviet airspace – over the Kamchatka Peninsula, a secret military facility. KAL 007 unknowingly found itself within prohibited Soviet airspace at the height of the Cold War.

Soviets were sensitive to the activities of rival countries during the Cold War. Coincidentally, on 1 September 1983, Soviet air commanders had been tracking an American spy plane (Air Force Boeing RC-135) that had been flying wide circles just in and out of radar range east of the Kamchatka Peninsula. At one point, the American spy plane and KAL 007 flew so close together that they merged on Soviet radar. The American spy plane left Soviet radar coverage and KAL 007 flew directly towards the southern edge of the Peninsula. Soviet military personnel monitoring the radar screens assumed that KAL 007 was the American spy plane.

While being closely monitored by Soviet controllers, KAL 007 passed over the Kamchatka Peninsula, flew out over international waters, and towards Sakhalin Island (another Soviet territory just north of Japan). Sakhalin radar controllers designated KAL 007 as a military target, prompting them to use interception and engagement rules for military activity (rather than abiding by international civil aviation rules).

Two Soviet Su-15 fighters were deployed. Twenty minutes later, while flying at 30 000 feet, a fighter pilot made visual contact with KAL 007 and took up a position behind the airliner. The fighter pilot transmitted messages to ground control, stating there were four engines, but did not clearly identify the aircraft as an airliner. Hearing that the plane had four engines, ground control understood this as confirmation of Soviet military suspicions that this was the American spy plane. Soviets tried to make radio contact on an emergency frequency, but the crew on board KAL 007 was not monitoring that channel.

Unaware of the situation, the KAL 007 crew was performing routine operations, in radio contact with controllers in Japan. The fighter pilot flashed his lights and fired warning shots, trying to force the aircraft to land, but the crew of KAL 007 neither saw nor heard the warnings. Soviet controllers were anxiously communicating with their superiors about to do, as the aircraft was about to cross out of Soviet airspace back over international waters.

In an unlucky coincidence, the KAL 007 flight received instructions from Tokyo Air Traffic Control to climb to 35 000 feet. Observing this action, the Soviet fighter pilot assumed it was an evasive manoeuvre. The fighter reported the airliner's manoeuvre to Soviet commanders and was instructed to destroy the aircraft. The Soviet fighter pilot launched two missiles. One missile exploded near KAL 007, breaking a hole in the fuselage and causing the jet to lose cabin pressure. The KAL 007 crew tried to control the descent, but the aircraft stalled and fell into the Sea of Japan, killing all 269 souls on board.

It took several days for the Soviets to admit to shooting down the aircraft. At that point, the Soviets announced that their action was simply a response to a deliberate provocation by the United States of America.

CASE STUDY

There were several immediate reactions to the event:

- The International Federation of Air Line Pilots' Associations (IFALPA) called for a boycott of all flights to Moscow and several States refused landing rights to Aeroflot (Russian Airlines) for weeks after the event.

- Groups around the world denounced the Soviet Union – actions included burning Soviet flags and boycotting Soviet liquor.

- With American and Soviet tensions high:

  - the US publicly condemned the Soviets, focusing on what was done to the plane; and

  - the Soviets criticized the Americans, focusing on what the plane was doing.

- Newspaper and magazine articles and books were published with wide-ranging theories about the events that led up to the event. Did Soviets know it was a passenger-carrying civil aircraft or did they legitimately mistake it for the military aircraft? Why didn't controllers warn the KAL 007 pilots? Was the American spy plane aware of the KAL 007 aircraft – and if so, would the United States bear some responsibility for allowing the flight to continue?

As ICAO is focused primarily on the safety of aviation, it made sense for the international community to turn to them for leadership. For the first time since the intense negotiations at the Chicago Convention, ICAO found itself in the middle of heated debate between rival States.

ICAO held an extraordinary session of the Council where it was agreed that an armed attack against a civil aircraft was incompatible with international civil aviation (and humanity in general), and that the Soviets had not adequately considered the safety of passengers and crew on board the aircraft. The ANC was instructed to study ICAO documents to find ways to prevent similar disasters in the future.

These council resolutions were debated at the Twenty-fourth ICAO Assembly in 1983. States had lengthy heated debates over the issues, with the Soviets asserting that the Americans had deliberately sent KAL 007 into their airspace and, therefore, the United States ultimately held responsibility. However, most States condemned the actions of the Soviets, and Council resolutions were passed that prohibited member States from using armed force against civil aircraft.

ICAO undertook three major initiatives:

1. ICAO determined that an amendment to the Chicago Convention was required to assert that the use of weapons against civil aircraft was unacceptable. This amendment was ultimately passed by the ICAO Assembly in 1984.

2. The ANC was tasked with reviewing all technical aspects and documentation (Chicago Convention, annexes, manuals, and so on) to see if they could be improved to prevent this type of disaster. This resulted in revisions to several annexes as well as the *Manual Concerning Interception of Civil Aircraft* in 1984.

CASE STUDY

CASE STUDY

3. The Secretary General was tasked with investigating the KAL 007 accident and reporting back to Council. This last task was incredibly difficult, as ICAO had no power to enforce its decisions or subpoena information or testimony – and the key States involved were reluctant to share. The Soviets resisted the investigation and stated that they had not found the cockpit voice recorder or the flight data recorder. When the final report was submitted to Council, it was inconclusive as a number of elements were missing. The final report could not conclude how or why KAL 007 flew off course, though pilot error was suspected. The report identified where the Soviets were at fault – the fighter pilot had mistaken KAL 007 as an American spy plane and did not clarify the aircraft identification before firing the missiles. Council members condemned the Soviet Union for failing to cooperate with search and rescue activities and the ICAO investigation. Though condemned in the court of public opinion, the Soviets were not punished in any official way.

In December 1990, seven years after the accident, a Soviet foreign minister apologized to South Korea for the shooting of KAL 007, published interviews with the Soviet fighter pilot, and announced that they had looked for espionage equipment in the wreckage but found none. In 1993, the Soviets turned over to ICAO the flight recorders, which for 10 years they had denied having. The ICAO Council was asked to complete its investigation.

The resulting report determined there had been no equipment malfunction on KAL 700, the aircraft was certified and maintained, and the crew was fit to fly. Human error, on both sides, led to the event. The KAL 007 crew failed to note the autopilot was in an incorrect mode, which led to the deviation from the flight path, and they lacked the situational awareness to notice how far off course they had drifted. On the Soviet side, the fighter pilot fired before making thorough efforts to identify the aircraft. The ICAO Council declared the investigation complete.

This event had a transformative effect on ICAO – suddenly, it found itself on an international stage where disputes between member States could be brought forward. ICAO demonstrated leadership, along with technical and political expertise, while maintaining focus on its core objective of promoting civil aviation safety.

Note

1 Degani, 2001; Mackenzie, 2010; Pearson, 1987

Case Study Questions

As a student learning about ICAO and international aviation, consider the following questions:

1.10 As discussed, States have sovereign control over the airspace overlying their territories. However, how much control do you think is appropriate? In the tragic KAL 007 accident, Soviet actions were condemned, but would they have been justified if the aircraft shot down had been the military spy plane?

1.11   Before the flight recorders were analysed, some theorized that KAL 007 pilots had intentionally flew over Soviet airspace to take a more direct route to their destination in order to save fuel. If that had been the case, what rights would the Soviets have had to defend their airspace? What would have been a reasonable response to an aircraft violating a State's restricted airspace?

1.12   ICAO's response to the event reflected the relationship between Council, the Assembly, and the ANC. However, it also placed Council members in a position where they were going beyond acting as the voices of individual States to expressing a unified opinion on an event. Could this be a slippery slope? What challenges could arise from Council members moving away from speaking as individual States and towards expressing opinions as a unified organization?

References

Degani, A., 2001. *Korean Air Lines flight 007: Lessons from the past and insights for the future.* Ames Research Center: NASA.

Havel, B. F. & Sanchez, G. S., 2014. *The principles and practice of international aviation law.* New York: Cambridge University Press.

Henkin, L., 1979. *How nations behave: Law and foreign policy.* 2nd ed. New York: Columbia University Press.

ICAO, 1944. *Convention on Civil Aviation (Chicago Convention).* Chicago: International Civil Aviation Organization.

ICAO, 2004. *Manual on the regulation of international air transport, Doc 9626.* Montreal: International Civil Aviation Organization.

ICAO, 2015a. *1928: The Havana Convention.* [Online] Available at: www.icao.int/secretariat/Postal History/1928_the_havana_convention.htm

ICAO, 2015b. *1944: The Chicago Conference.* [Online] Available at: www.icao.int/secretariat/Postal History/1944_the_chicago_convention.htm

ICAO, 2015c. *Air Navigation Commission: Special 200th session commemorative review.* [Online] Available at: www.icao.int/about-icao/AirNavigationCommission/Documents/ANC-200_final_web. pdf

ICAO, 2015d. *The 1919 Paris Convention: The starting point for the regulation of air navigation.* [Online] Available at: www.icao.int/secretariat/PostalHistory/1919_the_paris_convention.htm

ICAO, 2015e. *The Paris Convention of 1910: The path to internationalism.* [Online] Available at: www. icao.int/secretariat/PostalHistory/1910_the_paris_convention.htm

ICAOa, n.d. *Air Navigation Bureau.* [Online] Available at: www.icao.int/safety/airnavigation/Pages/ default.aspx

ICAOb, n.d. *Air Navigation Commission.* [Online] Available at: www.icao.int/about-icao/AirNaviga tionCommission/Pages/default.aspx

ICAOc, n.d. *Air Transport Bureau.* [Online] Available at: www.icao.int/secretariat/air-transport/Pages/ default.aspx

ICAOd, n.d. *Bureau of Administration and Services.* [Online] Available at: www.icao.int/secretariat/ Administration/Pages/default.aspx

ICAOe, n.d. *Council States 2014–2016.* [Online] Available at: www.icao.int/about-icao/Pages/council-states-2014-2016.aspx

ICAOf, n.d. *How does ICAO develop standards and procedures for international civil aviation?.* [Online] Available at: www.icao.int/about-icao/FAQ/Pages/icao-frequently-asked-questions-faq-12.aspx

ICAOg, n.d. *Legal Affairs and External Relations Bureau.* [Online] Available at: www.icao.int/secretariat/legal/Pages/default.aspx

ICAOh, n.d. *Making an ICAO standard.* [Online] Available at: www.icao.int/safety/airnavigation/pages/standard.aspx

ICAOi, n.d. *Regional offices.* [Online] Available at: www.icao.int/Pages/Contact_us.aspx

ICAOj, n.d. *Secretary General.* [Online] Available at: www.icao.int/secretariat/Pages/default.aspx

ICAOk, n.d. *Governments and government related.* [Online] Available at: www.icao.int/Pages/Links.aspx

Mackenzie, D., 2010. *ICAO: A history of the international civil aviation organization.* Toronto: University of Toronto Press.

Pearson, D. E., 1987. *KAL 007: The cover-up.* New York: Summit Books.

Warner, S. M., 1993. Liberalize open skies: Foreign investment and cabotage restrictions keep non-citizens in second class. *The American University Law Review,* 43(277), pp. 287–323.

Notes

1 ICAO, 2015e
2 ICAO, 2015e
3 ICAO, 2015d
4 ICAO, 2015a
5 ICAO, 2015b
6 ICAO, 2015b
7 ICAO, 2015b
8 ICAO, 2015b
9 ICAO, 2015b
10 ICAO, 1944, p. 2
11 ICAOj, n.d.
12 ICAOa, n.d.
13 ICAOc, n.d.
14 ICAOg, n.d.
15 ICAOd, n.d.
16 ICAOi, n.d.
17 ICAOe, n.d.
18 ICAO, 2004, p. 3.4–1
19 ICAO, 2015c
20 ICAOb, n.d.
21 Mackenzie, 2010
22 ICAOh, n.d.
23 ICAOf, n.d.
24 Havel & Sanchez, 2014
25 Havel & Sanchez, 2014
26 Havel & Sanchez, 2014
27 Warner, 1993, p. 290
28 ICAO, 2004, p. 1.1–1

1 The term 'aircraft' describes:
   a. Aeroplanes
   b. Aeroplanes and helicopters
   c. Balloons and blimps
   d. All of the above.

2 In general, piston engines are used within smaller general aviation aircraft.
   a. True
   b. False

5 Aircraft maintenance mechanics, technicians, and engineers (AMMTEs) must be licensed before they may participate in any maintenance activities.
   a. True
   b. False

3 International regulations are in place which define standards and recommended practices for:
   a. Airworthiness
   b. Maintenance programmes
   c. Registration markings
   d. All of the above.

4 The materials used to build aircraft have natural and predictable lifespans, therefore both scheduled and unscheduled maintenance are required.
   a. True
   b. False

Answer Key: 1. d; 2. a; 3. d; 4. a; 5. b.

Learning science suggests that thinking through a few questions before you begin studying new material, even if you answer incorrectly, results in improved learning and retention.
Give it a try!

CHAPTER 2

Aircraft

CHAPTER OUTCOMES

At the end of this chapter, you will be able to . . .

- Identify and name several categories of aircraft.
- Outline the basic process that enables aircraft to fly and to be controlled while airborne.
- Name two types of engines and describe their key differences.
- Explain how international regulations impact the design and airworthiness of aircraft.
- Describe the training, roles, and work environments of aviation maintenance professionals.
- Use your understanding of aviation maintenance to discuss a case study on China Airlines flight 611, a crash that resulted from metal fatigue.

Introduction

Chapter 1 provided a brief history of aviation and a look at the international law that governs the industry. Chapters 3 to 10 in this textbook will detail a range of operations and issues within the aviation industry. However, before digging deeper into these topics – everything from navigation processes to security protocols – it seems appropriate to examine the one thing at the core of the entire industry: the aircraft.

Without machines capable of performing safe and reliable air transportation, the aviation industry would not exist. This chapter will explore types of aircraft and how they are designed, built, and maintained. The critical role of aviation maintenance professionals will also be discussed. This sector of the industry is sometimes called the *civil aerospace sector*, which includes the manufacture and maintenance of aircraft systems, components, and engines.

Categories of Aircraft

Aircraft is a general term used to describe a wide range of flying machines. These machines can be categorized based on factors including 1) how lift is achieved, 2) whether engines provide power, and 3) whether wings are fixed in place or rotate (as on helicopters). The two main categories are:

- *lighter-than-air* aircraft, which achieve lift by capturing lighter-than-air gas, such as hot air or helium; and

- *heavier-than-air* aircraft, which incorporate some form of wing that, when moved through the air, produces lift.

Within these two categories of aircraft, there is a wide range of sizes, configurations, and types. Figure 2.1 outlines how ICAO defines aircraft categories.

Aircraft in the lighter-than-air category are easy to recognize. They include airships (blimps), captive balloons (which are tethered to the ground), and free balloons (which are untethered and travel in the direction the wind pushes them). The main difference between balloons and blimps is that balloons have no directional control and travel with the wind, while blimps are powered and have *control surfaces* (discussed later in this chapter), which allow the pilot to steer the aircraft.

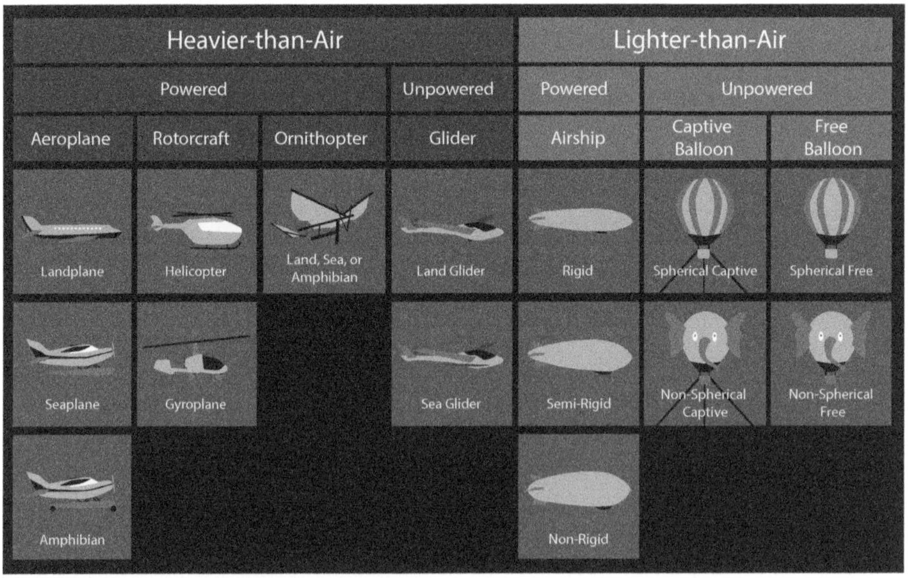

Figure 2.1 ICAO categories of aircraft
Source: Adapted from ICAO, 2012

The heavier-than-air category, however, contains a more diverse collection of aircraft. These include:

- *Aeroplanes* – engine-powered, controlled, heavier-than-air machines that derive lift from fixed wings. An aeroplane is what people most often think of when they hear *aircraft*. Aeroplanes are also called *fixed-wing aircraft*, distinguishing them from helicopters with rotating wings. Aeroplanes can be configured with wheels to land on the ground (landplanes), with floats to land on water (seaplanes,) or with a combination of floats and wheels that allows them to operate in both environments (amphibians).

- *Rotorcraft* – aircraft that have rotating (rotor) blades. Rotor blades can be thought of as spinning wings, with a teardrop cross-sectional shape. When they rotate, they produce a low-pressure area above the rotors resulting in lift. Helicopters incorporate an engine that powers the rotor blades, while gyroplanes have an unpowered rotor blade along with a powered propeller.

- *Ornithopters* – aircraft that flap their wings like a bird to achieve lift. These aircraft are mostly experimental and do not play a significant role in civil aviation.

- *Gliders* – unpowered aircraft that require assistance (such as from a tow aeroplane) to become airborne. Once airborne, gliders can take advantage of air currents to maintain and even gain altitude.

2.1 Language of Aircraft

Airworthy refers to an aircraft that is safe to fly.

A **fuselage** is the body of an aircraft – the main section that holds pilots, passengers, and luggage – not including the wings or tail section. **Empennage** refers to the tail section of an aeroplane.

Fleet refers to all the aircraft operated by a particular company.

In aircraft, a **stall** is an aerodynamic condition (not directly related to engine function) that occurs when the aircraft wing reaches a critical angle to the airflow, causing a sudden loss of lift.

The term **avionics** refers to electronic systems on board an aircraft. Generally, the term is associated with the instrumentation in the cockpit.

The **cockpit** (also called the **flight deck**) is where the pilot's chair and flight controls are located, from which the pilot flies the aircraft.

The line (or **the flight line**) refers to aircraft and personnel engaged in typical operations within a company. When maintenance work is completed **on the line**, it means the repairs are completed during the course of typical operations without having to remove the aircraft from service. Repairs done on the line can be compared to replacing your car's windshield wiper in your driveway – it's not something that requires a trip to the mechanic's shop.

Overhaul of equipment refers to an intensive repair than cannot be done on the line; overhaul involves disassembly, cleaning, and inspection.

The Basics of Flight

Professionals entering the aviation industry, even those who may not be involved in operating aircraft, should have a basic understanding of how aircraft fly including the structures that allow them to be controlled while airborne and their means of power. The following section introduces a simple wing design and an aeroplane's control surfaces; however, many variations exist with more complicated wing designs and methods of controlling flight (such as for rotary-wing aircraft).

How Do Aeroplanes Fly? Lift, Weight, Thrust, and Drag

For those first learning about how aeroplanes fly, it's helpful to begin with the impact of the four forces on an aircraft: lift, weight, thrust, and drag.

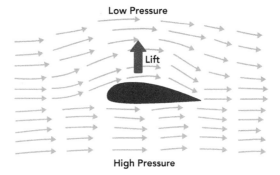

Figure 2.2 Aircraft wing

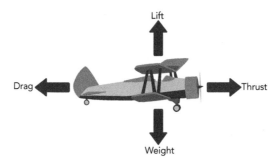

Figure 2.3 Four forces

The wings of an aircraft are designed with the front edge tipped up slightly (or the upper half of the wing is built with a slight curvature) so that air flows more quickly over the top surface of the wing and more slowly over the bottom when an aircraft moves through the air. The faster flowing air produces low pressure, while the slower moving air below the wing causes higher pressure. The difference between the high- and low-pressure areas results in an upward force of *lift* on the wings. For an aircraft to fly, the force of the *lift* needs to be greater than the aircraft's *weight*. Weight is the force resulting from gravity on the aircraft.

Drag is the force that works against an aircraft moving through the air. Drag occurs when the surfaces of an aircraft come into contact with the air, resulting in friction. Smooth aircraft surfaces produce less friction and therefore result in less drag. *Thrust* is the force that causes an aircraft to move through the air. Thrust is produced by an aircraft's engine(s) and when thrust is stronger than drag, the aircraft accelerates in the direction of the net force.

How Are Aeroplanes Controlled in Flight?

When lift and thrust overcome weight and drag, an aircraft becomes airborne – flight is achieved! Although this is an amazing feat in itself, a major challenge remains: the aircraft

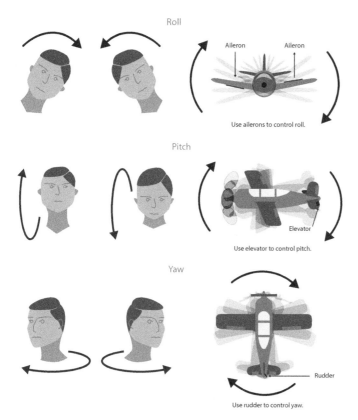

Figure 2.4 Flight control surfaces and the three axes of rotation

must be controlled while in flight. The three-dimensional movement of an airborne aircraft is easiest to understand by considering the three axes of rotation: *yaw*, *pitch*, and *roll*.

Aircraft have moveable *flight control surfaces* that cause rotation around one of these three axes of rotation. When a pilot moves the yoke (control wheel) or presses on a foot pedal, it causes the connected control surface on the body of the aeroplane to move. In traditional fixed-wing aeroplanes, there are three flight control surfaces:

- *Ailerons* control roll. Ailerons are the control surfaces on the outer edge of each wing that move in opposite directions. When a pilot moves the yoke left or right (much like turning a car's steering wheel), one aileron moves up and the other down. This increases the lift on one wing and decreases it on the other, resulting in a roll.

- The *elevator* controls pitch. The elevator is a control surface located on the horizontal part of the tail fin (also called the *horizontal stabilizer*). When a pilot pushes forward on the yoke or pulls it towards their chest, the elevator moves up or down, which decreases or increases lift on the aircraft's tail section. This results in the nose of the aircraft pitching up or down.

- The *rudder* controls yaw. The rudder is located on the aircraft's vertical tail fin (also called the *vertical stabilizer*). The pilot pushes foot pedals to move the rudder from side to side, exerting a yaw force on the aircraft.

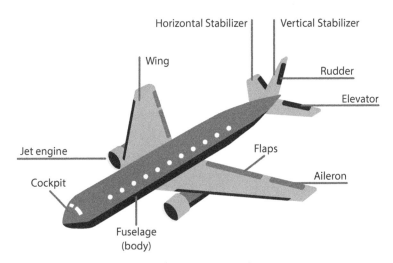

Figure 2.5 Aeroplane structure and flight control surfaces

Engines

Piston Engines

Dirigible airships, the first controllable flying machines, used steam engines for thrust. However, these coal-powered steam engines were so heavy that they were impractical for use in heavier-than-air aircraft. In the late 1800s, automotive engineers patented the first internal combustion engines. Americans Samuel Langley and the Wright brothers adapted this technology for their aircraft designs. From the Wright brothers' first successful flight on 17 December 1903 until the 1930s, the gas-powered internal combustion engine was the only means of generating thrust on aeroplanes.

Internal combustion engines are also called *piston engines* within the aviation industry. These engines, based on the same technology that powers most automobiles, are still used to power most small aircraft today.

Internal combustion engines are made up of several cylinders, with each cylinder containing pistons connected to a crankshaft. The basic workings of a four-stroke piston engine comprise the following four stages:

1. Intake stroke – the piston moves downward creating low pressure within the cylinder. Fuel and air are drawn into the cylinder through the intake valve.

2. Compression stroke – the intake valve closes, and the piston pushes upward compressing the fuel and air mixture within the contained cylinder.

3. Power stroke – a spark ignites the fuel and air mixture releasing heat and energy, pushing the piston down and exerting force on the crankshaft. This energy powers the aircraft's propeller and the compression strokes of other cylinders.

4. Exhaust stroke – the exhaust valve opens and the piston moves upward, pushing the exhaust out of the engine.

These piston engines are primarily used to power general aviation (GA) aircraft – those outside of the airlines or military – and are manufactured by companies such as Lycoming, Rotax, and Continental Motors.

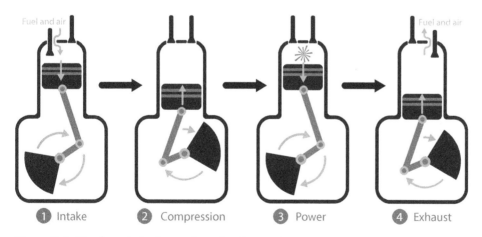

1 Intake   2 Compression   3 Power   4 Exhaust

Figure 2.6 The four-stroke internal combustion engine

Turbine Engines

In the early 1700s, Sir Isaac Newton theorized that a rearward-directed explosion could push a machine forward. This theory, based on his third law of motion – *for every action there is an equal and opposite reaction* – was applied to the invention of the first jet engine by Frank Whittle, a British Royal Air Force officer who patented the technology in 1930. Subsequently, jet engines were introduced into aviation. Early pioneers in this area include German Hans von Ohain, who was responsible for the Heinkel He 178 aircraft (1939) and General Electric, which produced the American jet plane XP-59A (1942).

Jet engines entered service in civil aviation in the 1950s and quickly demonstrated their advantages over piston engines. Jets were significantly more powerful and durable than their piston engine counterparts. More powerful engines led the industry into the 'Jet Age', allowing aircraft manufacturers to build aircraft that were bigger, faster, and capable of travelling farther.

Jet engines are also called *gas turbine engines* or simply *turbines*. These engines are made up of a fan, compressor, combustor, turbine, and nozzle. The engine takes in air with a large spinning fan. The air then flows through a compressor (a series of spinning blades that rotate around a shaft), which squeezes the air and increases air pressure. The air then flows into a combustor where fuel is sprayed into the high-pressure air and a spark ignites the mixture. The burning gases release energy, passing through spinning turbine blades (which power the compressor), and push out the back of the engine through the nozzle. This causes thrust, which pushes the aircraft forward.[1]

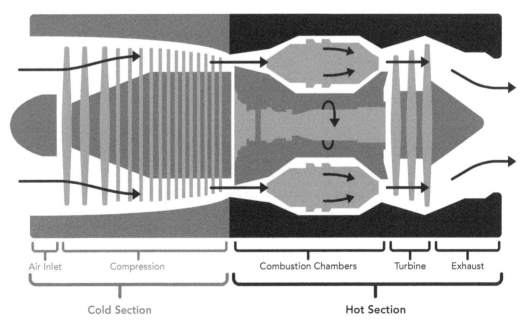

Figure 2.7 Gas turbine engine

There are various types of jet engines:[2]

- *Turbojet* – the simplest jet engine that follows the process described above. Air flows through a compressor, is ignited in the combustor, and is pushed out through the back of the engine, producing thrust to push the aircraft forward.

- *Turbofan* – a modified turbojet with a large fan at the front of the engine, which improves thrust at lower speeds and makes the engine quieter. Much of the air coming into the engine flows on the outside of the compressor and combustion chamber and exits the engine cold (or is mixed with the hot air exhaust), thereby increasing thrust without using more fuel.

- *Turboprop* – a jet engine attached to a propeller. After combustion, the air turns the turbine, which is connected by a shaft to the propeller. Some regional airlines use turboprop aircraft.

- *Turboshaft* – an engine that operates like a turboprop, but which powers the rotor blades of a helicopter instead of a propeller.

- *Ramjet* – a jet engine that contains no moving parts (rotating fan, compressor, or turbine). Airflow results from the air being 'rammed' into the engine as it moves forward at a high speed. This engine requires some sort of assisted take-off to get it up to speed before it is functional. This type of engine is used for space vehicles while in Earth's atmosphere.

A few of the companies that manufacture aircraft jet engines are Pratt & Whitney, GE Aviation, and Rolls-Royce.

Did You Know?

The structure of jet engines makes them vacuum cleaners in the sky. As they suck in air, they can also suck in airborne objects such as birds. This is a problem at low altitudes during take-offs and landings.

Quieter engines and larger intake fans on popular turbofan engines increase the likelihood of bird strikes. Engine manufacturers test their products to ensure they can handle bird strikes, and attempt to ensure that if the engines fail, the failures are 'contained' (i.e., if the rotating compressor and turbine blades break off the engine, the broken pieces are not projected into the airframe or into the cabin).

The danger of bird strikes became well understood by the public in 2009 after US Airways flight 1549 successfully ditched (made a water landing) on New York's Hudson River following multiple bird strikes that caused loss of thrust from both engines.

International Regulations

With a basic understanding of how aircraft fly, we turn to considering how aircraft around the world are built, maintained, and operated safely throughout international airspace. An aeroplane may be designed in one country, assembled in another, registered to fly and entered into passenger service in a third country, and eventually sold to an operator in a fourth! Therefore, it is essential that a uniform set of international standards exists to support safety and airworthiness on a global scale.

To accomplish this objective, standards and recommended practices (SARPs) are issued by ICAO to control critical aspects of aircraft design, maintenance, and operation. ICAO's SARPs apply to States based on each one's relationship to the aircraft:

- *State of Design* – the country with jurisdiction over the organization responsible for the design of the aircraft type (i.e., the country where the manufacturer who drafted the aircraft design plans is located).

- *State of Manufacture* – the country with jurisdiction over the organization responsible for the final assembly of the aircraft, engine, or propeller.

- *State of Registry* – the country where the aircraft is registered.

- *State of Operator* – the country where the operator (i.e., airline) has its principal place of business.

Referring to various countries as 'State of . . .' is common in international aviation, but it does take some getting used to. On an international scale, the use of this terminology is important as it identifies which States have the various responsibilities for the safety and airworthiness of aircraft.

Did You Know?

ICAO's SARPs apply to international aviation only. In situations where an aircraft is operated exclusively within its home country, the civil aviation authority (CAA) of that country has regulatory authority. Consider a home-built aircraft flown recreationally over short distances. It would be impractical for strict international regulations to be applied to these aircraft and, therefore, domestic regulations are applied instead.

Airworthiness: Annex 8

Airworthy is a term used in the aviation industry to describe an aircraft and its components that are in a safe condition to fly. ICAO publishes SARPs associated with airworthiness in Annex 8.[3] The application of these international standards begins when a manufacturer develops plans to build a new aircraft, and these standards continue to be applied throughout the life of the aircraft. It is crucial that aircraft are designed, developed, built, and operated in a manner that meets international standards for safety and reliability.

Type Certificate

Before a new type of aircraft can legally fly or be sold, it must receive a *type certificate*. Type certificates are issued by the civil aviation authority (CAA) in the State of Design. The aircraft manufacturer submits drawings, reports, and evidence that the aircraft meets airworthiness requirements. The aircraft is then subjected to a series of inspections as well as ground and flight tests. When enough evidence has been analysed, and the aircraft is deemed compliant with airworthiness requirements, a type certificate for the aircraft model is issued. Once a type certificate is issued, the aircraft design cannot be changed.

Did You Know?

Evaluating airworthiness, for the purpose of issuing a type certificate, involves the assessment of a wide variety of factors, including flying qualities, structural design, engine design and installation, crew operating environment, operating limits, and crashworthiness of aircraft (to improve accident survivability). To meet the requirements of airworthiness, the manufacturer's plans must also respect security in design (including the identification of a *least-risk bomb location*, where an explosive could detonate with the lowest risk to the aircraft).

Certificate of Airworthiness

Once an aircraft enters into service, periodic inspections and maintenance must be completed to maintain an aircraft's airworthiness. Unlike a type certificate, which applies to all aircraft of the same model, a *certificate of airworthiness* (CofA) is granted on an aircraft-by-aircraft basis.

The State of Registry is responsible for assessing airworthiness and issuing certificates of airworthiness. An aircraft would lose its CofA if its operator failed to abide by the scheduled aircraft maintenance programme or if the aircraft incurred severe damage.

If a problem with a specific aircraft type is discovered after it has entered service, the CAA will issue an *airworthiness directive* (AD). ADs typically describe weaknesses in the aircraft design that impact safety and prescribe maintenance solutions to resolve the concerns. Aircraft operators must abide by the ADs to ensure that their aircraft maintain a valid CofA.

Operation of Aircraft: Annex 6

Annex 6 outlines the SARPs associated with the operation of aircraft.[4] It is the aircraft operator's responsibility to ensure that every aircraft in its fleet is maintained as airworthy, has the operational and emergency equipment required, and has a valid CofA.

The operator must produce and manage several pieces of documentation and make them available to the CAA in the State of Operator for inspection. These include:

- *Maintenance control manual* – a reference document for the CAA, which outlines the maintenance activities the operator will conduct. Its main purpose is to describe the operator's procedures to ensure all scheduled and unscheduled maintenance is performed on time and satisfactorily.

- *Maintenance programme* – a detailed listing of scheduled maintenance tasks to be performed on each aircraft, for use by maintenance and operational personnel, which must be approved by the State of Registry.

- *Maintenance records* – detailed records on maintenance-related activities for each aircraft that includes time in service (hours, days, and/or take-off and landing cycles), compliance with airworthiness information, repairs, time since last overhaul, compliance with maintenance programme, and signings of maintenance releases.

Aircraft Marks and Type Designators: Annex 7

Every aircraft has identifying marks painted in a prominent location on its body. These marks are mandatory and are based on ICAO SARPs from Annex 7: Aircraft Nationality and Registration Marks.[5]

Aircraft markings begin with a *nationality common mark* that identifies all aircraft from a specific State (see Table 2.1 for examples) and are followed by a series of *registration marks*

Table 2.1 Nationality common marks of ICAO Council States

Australia	VH
Brazil	PP, PR, PT, PU
Canada	C, CF
China	B
France	F
Germany	D
Italy	I
Japan	JA
Russian Federation	RA
United Kingdom	G
United States	N

Source: ICAO, 2015

(letters or numbers) unique to each specific aircraft. Registration marks, including the nationality common mark, are distributed by the State of Registry. A fireproof identification plate containing the marks must also be secured inside every aircraft.

When an aircraft is registered with a CAA, the operator will receive a *certificate of registration* that contains the common and registration marks as well as the aircraft serial number and owner's details.

Aircraft Type Designators

Aircraft are given names by their manufacturers, such as *Boeing 737* or *Airbus 320*, but these names can cause confusion because they don't obviously indicate whether the craft is large or small, powered or unpowered, and so on. To allow for quick understanding of aircraft type, in both written flight plans and in verbal radio communications, ICAO's Doc 8643 has specified aircraft type designators for all aircraft.[6]

These standardized short forms (not more than four characters) are used in aircraft flight plans, in air traffic control communication, and by aircraft operators. The International Air Transport Association (IATA) publishes a separate system of codes used for the general public in airline timetables.

Table 2.2 Aircraft type designators

	Wake Turbulence Category (WTC)	
H (heavy)	aircraft with take-off mass of 136 000 kilograms (300 000 pounds) or more	
M (medium)	aircraft with take-off mass less than 136 000 kilograms (300 000 pounds) and more than 7000 kilograms (15 500 pounds)	
L (light)	aircraft with take-off mass less than 7000 kilograms (15 500 pounds)	
	Aircraft Type	
First Character	Second Character	Third Character
L (landplane) S (seaplane) A (amphibian) H (helicopter) G (gyrocopter) T (tilt-wing)	# (number of engines)	P (piston engine) T (turboprop/turboshaft engine) J (jet engine) E (electric engine)
	Examples (manufacturer/model, designator, WTC, description)	

Boeing 737-300, B733, M, L2J – medium WTC landplane with two jet engines
Airbus A-380-800, A388, H, L4J – heavy WTC landplane with four jet engines
Bell 206 A Jetranger, B06, L, H1T – low WTC helicopter with one turboprop (or turboshaft) engine
Cirrus SR-20, SR20, L, L1P – low WTC landplane with one piston engine

Source: ICAO, 2009

2.2 How Does It Work? Weight and Balance

Weight is an important consideration in the design and certification of aircraft. Manufacturers build aircraft with a specified maximum take-off weight (MTOW) associated with the structural strength and lifting properties of the wings. Another key consideration in the design process, closely related to weight, is the aircraft's centre of gravity (CG).

To understand CG, imagine an aircraft balanced in mid-air suspended by a rope. The point where the rope would be attached to the aircraft is the CG. Now consider what would happen if heavy bags were loaded into the tail section of the aircraft. To keep the aircraft balanced, the rope would have to be reattached in a position farther back as the CG would have moved backwards. An aircraft's CG continually changes depending upon the distribution of weight. The challenge is that if the CG is too far forward or aft, it may become very difficult (or impossible) for a pilot to control the flight.

The ideal CG, which gives the pilot the greatest control over the aircraft, is carefully calculated. However, the take-off weight of every flight varies based on fuel, passengers, catering, and cargo loads, among other factors. To accommodate this variability, manufacturers specify a safe range for the CG. Aircraft operators, specifically the pilot-in-command, are responsible for calculating the weight and balance of each flight before take-off to ensure the CG falls within the safe range.

Case Study: Air Midwest Flight 5481 – Weight and Balance

On January 8 2003, Air Midwest (operating as US Airways Express) flight 5481 was a Raytheon (Beechcraft) 1900D aircraft scheduled to travel from Charlotte, North Carolina to Greer, South Carolina in the United States. There were two crew members and 19 passengers on board the aircraft.

As is required, the crew calculated the weight and balance of the aircraft (including the passengers, baggage, and fuel). Following their company's standard operating procedure (SOP), they used standard weights – about 80 kilograms (175 pounds) per passenger and about 9 kilograms (20 pounds) per bag – to determine that the CG was in the manufacturer's safe range. The ramp agent told the pilots that a few bags were particularly heavy, between 31 and 37 kilograms (70 and 80 pounds), and that the cargo hold was 98 per cent full by volume. The pilot said it wasn't a problem as there was a child aboard, which would allow for extra baggage weight.

Just after take-off from Charlotte, the nose pitched up dangerously high causing the airflow over the wing to be disrupted and the aircraft to stall. This resulted in a loss of lift and the plane dove towards the ground. The aircraft crashed into a US Airways hangar and the impact ignited the fuel on board the aircraft. All 19 passengers and two crew members died in the accident.

In the subsequent investigation, the National Transportation Safety Board (NTSB) identified two factors that contributed to the accident.

1. One of the cables that connected the pilot's controls to the elevator had been improperly adjusted by maintenance, limiting the elevator's range of motion. The technician performing the maintenance was a subcontractor (receiving training on the 1900D aircraft) who had skipped nine steps from the maintenance manual, including the step that required the range of motion to be checked.

2. Witnesses implied the aeroplane was heavily loaded, but the cockpit voice recorder (CVR) revealed that the pilots calculated and approved the aircraft as within the safe CG range. Pilots used the standard weights of about 80 kilograms (175 pounds) per passenger and about 9 kilograms (20 pounds) per bag. Investigators, wanting to double-check the calculations, weighed burnt luggage retrieved from the wreckage and contacted the next of kin to get actual weight of passengers. From these calculations (based on their best estimates), they determined the aircraft was about 150 kilograms (over 300 pounds) heavier than the standard-weights estimate. The aircraft was tail-heavy and outside of its safe CG range.

After take-off, when the crew retracted the landing gear, the CG moved even further aft, causing the nose to pitch up. With the reduced elevator control resulting from the maintenance error, the crew was unable to lower the nose and the aircraft stalled. The NTSB suggested that the Federal Aviation Administration (FAA) revise average passenger weights as they had not done so since the 1930s even though the average weight of Americans has increased over that time. In a subsequent survey, operators determined that a more accurate average weight for travellers was 88 kilograms (195 pounds) not 80 kilograms (175 pounds) and that baggage was being underestimated by about 2.3 kilograms (5 pounds) per bag. It has yet to be determined if future technology can more accurately weigh travellers and their baggage to inform more accurate weight and balance calculations.

Original Equipment Manufacturers

The term *original equipment manufacturer* (OEM) refers to an organization that builds products. In aviation, OEMs manufacture aircraft and engines. There are many OEMs within aviation, some with a history stretching back to the earliest days of the industry. OEMs employ a variety of aviation professionals including aerospace engineers, management teams, manufacturing tradespeople, sales and marketing personnel, and test pilots among others.

As an introduction, this section will review Boeing and Airbus who are leaders in the major airline market, Bombardier and Embraer whose aircraft support the regional airline and corporate aviation markets, and a variety of other OEMs whose aircraft are primarily used within the general aviation (GA) sector. Many of these manufacturers also develop military aircraft. However, as this book is an introduction to civil aviation, military aviation is beyond the scope of this discussion.

Major Airline Manufacturers

Boeing

In 1916, William E. Boeing incorporated his aeroplane company, which became known as Boeing. The first aircraft the company produced was the Model C, which they tested in Seattle and began shipping to the United States Navy in 1917. Today, the Boeing company is a leading manufacturer of aircraft for major airlines.[7]

The name Boeing is synonymous with its 7-series commercial aircraft (such as the 747 or 737) used by airlines around the globe. However, the company is also the largest manufacturer of military aircraft, provides maintenance and support services around the

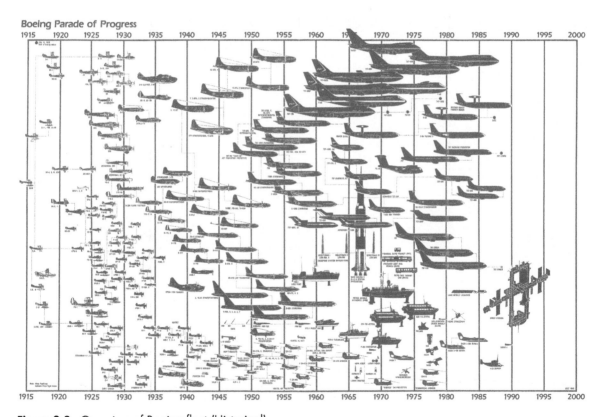

Figure 2.8 Overview of Boeing fleet (Historical)

Source: Copyright Boeing

world, and develops satellites and equipment for space operations, as well as networking technologies.[8]

In 2015, Boeing's revenue was US$96 billion. Boeing Commercial Airplanes made up US$66 billion of the company's 2015 profits and employs about 83 000 people. The company is headquartered in the Puget Sound region of Washington in the northwestern United States. Boeing aeroplanes make up about half of the world's airline fleet, with more than 10 000 in service.[9]

Airbus

In 1967, federal ministers from France, Britain, and Germany agreed that their nations would collaborate in the production of an airbus to increase European presence in the OEM industry. In 1969, an agreement was signed to launch the A300 – '300' was chosen because the original aircraft blueprint included three hundred seats. The A300 was to be the world's first twin-engine wide-bodied passenger jet (wide-body describes an aircraft with two aisles in the cabin as opposed to narrow-body aircraft with only one aisle). This marked the launch of Airbus,[10] now a direct competitor to Boeing in the manufacture of aircraft for major airlines.

Airbus aircraft can be recognized by their names that begin with 'A3' – A320, A330, A350, and A380, among others. Today, an A320 takes off or lands every two seconds somewhere around the world.

To compete with Boeing's established products, Airbus incorporated new technology wherever it would have economic, safety, or operational benefits.[11] Over the years, these innovations have included technology that changed how cockpits were managed, such as:

- the elimination of the third pilot (or *flight engineer*) position, which resulted in the modern two-pilot forward-facing cockpit;

- *glass cockpits* where traditional analogue steam-gauge instruments were replaced with digital screens that displayed flight and navigation information;[12] and

- *fly-by-wire* (FBW) controls where the mechanical linkage between the pilot's controls and flight surfaces (such as ailerons on the wings) were eliminated and replaced by a computer that calculates the desired control surface movements.

Did You Know?

Fly-by-wire technology, now standard throughout the industry, offers several advantages. Eliminating rods and cables reduces aircraft weight, which lowers fuel burn. Also, because a computer 'drives' the control surfaces, pilots can fly various types of aircraft in the same way regardless of size or weight, which significantly lowers training time and costs.

In addition to commercial aircraft, Airbus is also the leading manufacturer of helicopters and an expert in defence and space technologies.

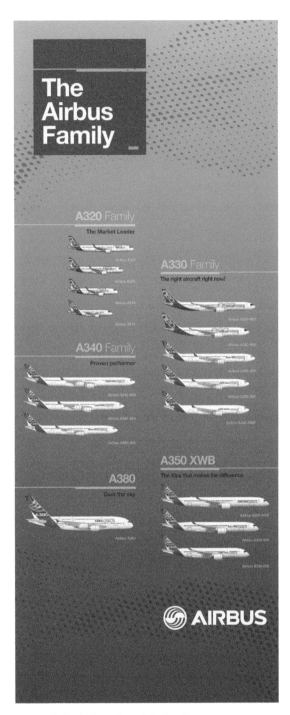

Figure 2.9 Airbus fleet (Modern)

Source: Copyright Airbus

In 2015, Airbus employed more than 135 000 people and had a revenue of €45.8 billion (not including helicopter or space revenues).[13] The company is headquartered in Toulouse, France. In 2016, Airbus hit the milestone of delivering its 10 000th aircraft, which went to Singapore Airlines.

Regional Airline Manufacturers

Bombardier

Joseph-Armand Bombardier built and designed his first 'snow vehicle' in 1922 in Quebec, Canada and grew his invention into a snowmobile company. The Bombardier organization expanded into the railway business in the 1970s and into aircraft manufacturing in the 1980s.[14]

In 1989, Bombardier launched the 50-seat Canadair Regional Jet (CRJ) programme, targeting the regional airline market. The CRJ entered the market in 1992 and grew into a family of regional aircraft. Bombardier also manufactures Global, Learjet, and Challenger business jets.[15] In 2015, Bombardier's aerospace sector had revenues of US$11.2 billion and employed 31 200 people.[16]

Embraer

In 1969, Embraer (originally Empresa Brasileira de Aeronáutica S.A) was founded with support from the Brazilian government. The company's first effort was a turboprop aircraft for the military and civilian market called the Bandeirante (a 15- to 21-seat aircraft). The company went on to develop the EMB 326 Xavante jet trainer and ground attack aircraft for the Brazilian government, the EMB 400 Urupema glider, and the EMB 200 Ipanema agricultural crop duster.[17] Today, Embraer is known for its regional jets based on the ERJ 145 platform and executive jets including the Phenom, Legacy, and Lineage families.[18]

In 2014, Embraer earned over US$6 billion in revenue and employed more than 19 000 people.[19]

General Aviation Manufacturers

The General Aviation Manufacturers Association (GAMA) is an industry trade group representing manufacturers of general aviation aircraft. (Remember that *general aviation* refers to operations outside of the military or airlines.) Figure 2.11 summarizes the annual data associated with shipment of aircraft in the GA sector, as reported by GAMA.

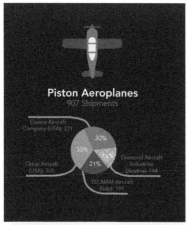
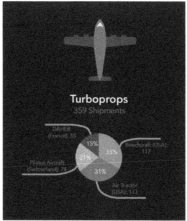
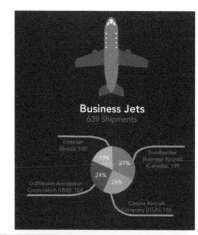
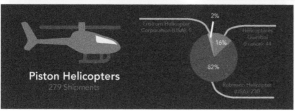
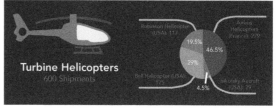

Figure 2.10 General aviation aircraft shipments, 2015

Source: Adapted from GAMA, 2016

2.3 International Coordinating Council of Aerospace Industries Associations (ICCAIA)

In 1972, ICCAIA was formed to represent, with a single voice, six regional aerospace manufacturer associations from Europe, the United States, Brazil, Canada, Japan, and Russia. ICCAIA's member companies include aircraft and engine manufacturers, ground and space system manufacturers, avionics and parts manufacturers, and component suppliers.

ICCAIA's mission is to support the development of standards, grow capacity, harmonize regulations, and offer technical expertise.

2.4 Commercial Space

Commercial space is a relatively new sector in aviation that merges commercial air transport with space operations. The mission of commercial space is to develop vehicles and spaceports to accommodate personal space flights (essentially *space tourism*). Since the space age began, only 557 human beings have travelled above Earth's atmosphere.[1] The commercial space sector seeks to offer this opportunity to more people.

But where does air travel end and space travel begin? Although there is no exact boundary where Earth's atmosphere abruptly ends, there are three ways in which *space* has been defined:

1. NASA defines space as 81 kilometres (50 miles) above Earth;

2. the Kármán line, a more internationally recognized boundary, defines space at 100 kilometres (62 miles) above Earth;

3. space can also be defined as the lowest Earth orbit altitude that can support satellites, which is about 160 kilometres (100 miles) above Earth.

The Commercial Spaceflight Federation (CSF) is an industry association that promotes human space flight. The member companies of CSF collaborate to promote safety and to share best practices and expertise. Members include commercial spaceflight and spaceport developers and their suppliers. CSF offers an internship programme for students interested in space business or policy.[2]

Although there are many companies active within this sector, a few commercial space companies worth noting include SpaceX, Virgin Galactic, Blue Origin, Bigelow Aerospace, and Space Adventures.

Technology is rapidly evolving, but currently space tourist fares are extremely expensive. As of 2016, Virgin Galactic requires deposits beginning at US$250 000, and complete trips cost upward of US$60 million per passenger so it is unlikely that this opportunity will be accessible to the average person in the short term. But who knows what the future holds?

Notes

1 Virgin Galactic, 2016   2 CSF, 2015

Aircraft Maintenance Mechanics, Technicians, and Engineers

Maintenance engineers can be considered the first aviation professionals as industry pioneers had to first design and build an aircraft before anyone had the opportunity to become a pilot! Historical aircraft required many hours of maintenance work for every hour of flight time, as nearly everything on the aircraft needed to be overhauled on a regular basis. As technology has advanced and aircraft systems have grown in complexity, many hours of maintenance are still required to maintain airworthiness. This task has evolved to require a team of maintenance professionals to keep a large aircraft in service. Maintenance is a crucially important – as well as time-consuming and expensive – element of aviation operations.

The structure of an aircraft, along with all the systems and components it contains, is made of materials with a natural and predictable lifespan. Eventually, materials fatigue and need to be repaired or replaced. Although manufacturers choose the most durable materials feasible, the lifespan of these materials may be shortened by stresses from the physical world. These stresses include environmental factors (such as heavy turbulence, weather, and pressure forces on the airframe), design limitations (structural weak points), deterioration of materials (corrosion), or human error (such as a hard landing by a pilot).

2.5 Metal Fatigue

The best way to understand metal fatigue is to think about bending a paperclip back and forth until it eventually breaks. Metal fatigue is the result of hundreds or thousands of load cycles that repeatedly put pressure on metal until it eventually weakens and cracks. When aircraft components break down as a result of metal fatigue, they leave distinctive smooth edges as opposed to metal being ripped apart by aerodynamic loads in an accident, which leave rough, jagged metal edges.

The task of maintaining airworthiness falls on aircraft maintenance mechanics, technicians, and engineers (AMMTEs). AMMTEs work within every sector of the industry from the initial production of an aircraft by an OEM, to airlines and corporate flight departments, and throughout the wide range of general aviation operations.

Did You Know?

The terminology used to describe maintenance professionals varies around the world. In the United States, they are called *aircraft maintenance technicians* (AMTs) while in Europe and other areas they are referred to as *aircraft maintenance engineers* (AMEs) or simply *engineers*.

AMMTEs maintain airworthiness by regularly servicing, maintaining, and overhauling airframes, engines, avionics, and components. Maintenance work includes both *preventative scheduled maintenance* at regular recurring intervals (based on hours of flight time, take-off/

landing cycles, or days) and *unscheduled maintenance* in response to a pilot-identified defect, airworthiness directive (AD), or another unexpected snag.[20]

In 1948, ICAO published the first licensing standards for maintenance professionals in Annex 1. However, a licence is not an absolute requirement for AMMTEs as they typically work in teams with licensed professionals supervising unlicensed mechanics.

Maintenance teams generally include *licensed certifying AMMTEs* who have the authority to sign a maintenance release (approving an aircraft's return to service), *licensed AMMTEs*, and *unlicensed apprentices* or *mechanics*. Although a licence is not required to perform aviation maintenance, unlicensed professionals must be supervised by licensed AMMTEs, and they do not have certifying authority to return an aircraft to service after maintenance work is complete. In some parts of the world, the term *licensed aircraft maintenance engineer* (LAME) is used to distinguish licensed from unlicensed AMMTEs.

Aviation maintenance training follows an apprenticeship model, whereby experienced licensed AMMTEs work alongside trainees to pass on their knowledge and skills. The apprenticeship is supplemented with a variety of classroom-based courses that cover topics such as air law and airworthiness, aircraft general knowledge, engineering, maintenance, and human performance. To earn a licence, a trainee must complete practical training and pass written examinations. This highly specialized trade requires a combination of advanced systems knowledge along with hands-on technical skills. It typically takes between two and four years of experience inspecting, servicing, and maintaining aircraft or its components to be eligible for a licence.[21]

AMMTEs can only exercise the privileges of their licence if they have the proper endorsement for the type of work they are doing; they can receive an endorsement, or *rating*, for a specific aircraft type, avionic system, airframe, or component[22] or for a level of work complexity.[23] Throughout their careers, AMMTEs may seek additional qualifications to expand the privileges of their licence.

Figure 2.11 AMMTE (photo)

The licensed certifying AMMTE has the most authority on the maintenance team – he or she is responsible for signing the *maintenance release*. A maintenance release must be completed and signed to certify that work was performed satisfactorily and in accordance with the maintenance organization's procedures manual.[24] On the maintenance release, AMMTEs must sign with their own identity, which is an indication of the importance of this credential. The maintenance release certifies that the aircraft is safe to fly and to return to active service. As these machines are often valued at hundreds of millions of dollars and carry hundreds of passengers, certifying a maintenance release is an enormous responsibility. To hold certifying privileges, licensed engineers must have years of experience and complete training courses. In some countries, it takes up to seven years for an AMMTE to earn the right to certify.

Did You Know?

AMMTEs track their experience using logbooks. They carefully document the work they have completed, including the aircraft type and registration, job number, and hours spent on the task, then have the logbook signed by the certifying AMMTE on the job. These logbooks are important records required for AMMTE career progression.

Maintenance Work Environments

To organize maintenance work and to issue appropriate licences and certifications, most States categorize maintenance by *technology group* (specific aircraft systems, technologies, engines, avionics, hydraulics, and so on) and by *maintenance work environment.* Maintenance work environments include *line, base,* and *shop.*[25]

- *Line maintenance* refers to work conducted on the aircraft on the flight line, while the aircraft is in active service. This work includes daily checks, turnaround maintenance, and short interval checks.

- *Base maintenance* refers to work carried out after an aircraft is brought into a maintenance hangar (i.e., it is no longer *on the line*). This work involves corrosion control, painting, and engine changes, among other modifications.

- *Shop maintenance* describes maintenance tasks performed on aircraft parts that have been removed from the aircraft. Shop maintenance can take place in electronics or avionics shops, engine shops, and component shops, among other places. These shops may be physically separate places or combined, depending on the operation.

> ### Did You Know?
>
> Airlines operate *maintenance control centres*, where AMMTEs are available to support pilots and help troubleshoot issues that might occur while the aircraft is in flight.

Within these groups, subcategories exist for licensing purposes by CAAs. Depending on the type and complexity of the work, an AMMTE will need a specialty-rating endorsement added to his or her licence. Some privileges are company specific and not transferrable between employers.

For example, many countries align their AMMTE licensing structure with the European Aviation Safety Agency's (EASA's) category system:[26]

- Cat-A licences allow engineers to certify simple tasks and cabin defects;

- Cat-B1 licences allow engineers to certify airframes, engines, and electrical work;

- Cat-B2 licences allow engineers to certify avionics (cockpit instruments and other non-mechanical systems);

- Cat-B3 licences allow engineers to certify piston aircraft up to 2000 kilograms (4400 pounds); and

- Cat-C licences allow engineers to certify base maintenance (requires several years' experience at the B-level).

Approved Maintenance Organizations

Approved maintenance organizations (AMOs) are facilities that offer maintenance services to air operators. Annex 6 specifies international requirements for AMOs. AMOs must be approved by a CAA and that approval is typically limited to a particular type of maintenance (aircraft structure, avionics, engine, propeller, and so on) and must indicate the certifying maintenance professional who has the authority to sign the maintenance release. Maintenance organizations are also commonly referred to as *maintenance repair and overhaul* (MRO) facilities. Therefore, it is common to see them expressed as AMO/MRO.

Managing Maintenance Disruptions

Aviation companies try to optimize the use of each of their aircraft – they don't maintain several back-up aircraft that sit idly by in case one aircraft in the fleet requires extra time at the AMO/MRO. Therefore, when unscheduled maintenance is required, the removal of an aircraft from service can cause wide-ranging scheduling problems that can impact an entire organization. Aviation companies try to mitigate these disruptions by employing:

1 A *minimum equipment list* (MEL) – the MEL identifies the essential equipment that must be functional for the aircraft to remain in service. If a component is inoperative,

but is not crucial for the flight and therefore not on the MEL, the aircraft can safely remain in service and the repair will be deferred to a later time. A secondary back-up instrument in the cockpit is an example of a repair that can be deferred because the aircraft can safely fly without it. By contrast, a functional fuel gauge would be on the MEL – an aircraft would have to be removed from service immediately so that it could be repaired.

2 *Line replaceable units* (LRU) – a piece of equipment that has been designed to be removed (and replaced quickly and easily) in case of failure to minimize delay. The transponder in the cockpit is an example of an LRU – it can be replaced quickly using standard connection points and attachments without having to send the aircraft to the hanger for maintenance.

3 *Equipment redundancy* – some aircraft are built with redundant back-up units, so that if one fails, the other will take over.[27] For example, most aircraft have both a primary and a back-up radio in the cockpit.

Maintenance Documentation

As modern aircraft are incredibly complex machines, manufacturers create systems and resources to support the work of AMMTEs. These include check sheets and certification systems, electronic systems that monitor aircraft functions and present alerts when malfunctions are detected, and databases for logging defects (in order to identify systemic weaknesses), among many others.[28]

A unifying feature of all maintenance work is the reference documentation used when inspecting, testing, removing, replacing, and certifying aircraft components. AMMTEs have a massive amount of documentation that they must reference continually. The documentation includes information from the aircraft and engine manufacturers, airlines, and their CAA. (See Table 2.3)

Table 2.3 Maintenance documentation

Manufacturer Maintenance documentation		
Maintenance planning data/document	MPD	listing of scheduled maintenance and intervals (daily, flight hours, and take-off/landing cycles)
Component maintenance manual	CMM	comprehensive maintenance details for components (parts of aircraft that are removed for repair in the shop) built by the aircraft manufacturer
Vendor manual	VM	comprehensive maintenance details for components (parts of aircraft that are removed for repair in the shop) built by external vendors

(Continued)

Table 2.3 (Continued)

Manufacturer Maintenance documentation

Master minimum equipment list	MMEL	listing of the minimum equipment that must be operational for an aircraft to remain in service.
Fault isolation manual	FIM	decision flow charts that guide maintenance professionals through troubleshooting to identify the source of a problem
Fault reporting manual	FRM	questions/diagrams used by flight crew (pilots) to analyse and describe a problem to maintenance professionals
Illustrated parts catalogue	IPC	listing of all parts, and their location, on an aircraft
Schematic diagram manual	SDM	detailed schematic diagrams of hydraulic, electronic, and electrical systems
Wiring diagram manual	WDM	detailed diagrams of aircraft wiring and electrical structures
Task card	TC	checklist used by engineers to complete specific tasks (without having to haul the entire AMM around)
Service bulletin	SB	notices issued to airlines by manufacturers to identify a modification that should be implemented by airline maintenance to improve safety, efficiency, and/or performance
Service letter	SL	

Airline Maintenance documentation

Operations specifications	Ops Specs	high-level overview of maintenance that organizes and references other company documentation
Aeroplane maintenance manual	AMM	comprehensive listing of basic maintenance for aircraft and on-board equipment
Technical policies and procedures manual	TPPM	the primary source for an airline's maintenance activities, which defines how all maintenance work takes place (often includes the inspection manual)
Minimum equipment list	MEL	an airline-customized list of the minimum equipment that must be operational for an aircraft to remain in service (based on the manufacturer-issued MMEL, which may include options that don't apply to all airline configurations)

Domestic Regulatory documentation (from civil aviation authority)

Civil aviation authority regulations	*Term varies by State*	listing of domestic laws associated with airworthiness and aircraft operation
Airworthiness directive	AD	specifications for legally required repairs or modifications to aircraft (issued when faults are identified, sometimes after an incident or accident)

Source: Adapted from Kinnison, 2004

Did You Know?

Aircraft maintenance requires a massive amount of documentation. Some say that proper maintenance documentation weighs more than the aircraft itself! For example, the Boeing 767 aeroplane maintenance manual (AMM) is 20 000 pages, which would weigh about 125 kilograms (275 pounds).

Conclusion

There is a wide variety of modern aircraft around the world, ranging from airliners that carry hundreds of passengers to single-pilot gliders. Regardless of size or structure, all aircraft that cross State borders must abide by international regulations established by ICAO. Standards and recommended practices are developed to ensure uniform safety and airworthiness of aircraft around the world. In addition to adhering to ICAO regulations, each aircraft operator must also abide by the domestic regulations of its civil aviation authority.

Modern aircraft are designed and built for safety and airworthiness; future maintenance is a consideration from the beginning of the design process. All materials have a natural life and will eventually need repair or replacement to maintain airworthiness. Aircraft manufacturers therefore create scheduled maintenance programmes for each aircraft to describe how it must be maintained over the course of its life. This initial maintenance programme is supplemented with additional unscheduled maintenance tasks that may result from the discovery of an issue that must be remedied throughout an entire aircraft fleet. (This is the aviation equivalent of a recall on a model of car.[29])

The work of conducting scheduled and unscheduled maintenance is the responsibility of an approved maintenance organization's (AMO's) aircraft maintenance mechanics, technicians, and engineers (AMMTEs). Without qualified and competent AMMTEs, aviation growth would not be possible. With the predicted growth in air traffic, many new AMMTEs will be required. Boeing estimates that between 2016 and 2035, the industry will require 679 000 new AMMTEs worldwide.[30]

Key Points to Remember

1. *Aircraft* is a general term that includes both lighter-than-air vehicles (balloons and blimps) and heavier-than-air craft (aeroplanes, helicopters, gliders, and so on).

2. Aeroplane wings create lift as they are pushed through the air by thrust produced by the engines. When lift and thrust overcome the weight of the aircraft and its drag resistance, the aircraft becomes airborne. Aeroplanes have moveable control surfaces that allow them to be controlled while airborne: ailerons control roll, the elevator controls pitch, and the rudder controls yaw.

3. Broadly, aircraft engines can be divided into two categories: pistons and turbines. Piston engines are used primarily in smaller GA aircraft, while turbines power larger commercial and military aircraft.

4. International regulations related to aircraft design and operation are found in three annexes of the Chicago Convention.

    - Annex 8 specifies the requirements for:

        - *type certificates*, which must be issued before an aircraft can legally fly or be sold; and

        - *certificates of airworthiness*, which must be maintained throughout the life of an aircraft with regularly scheduled maintenance.

    - Annex 6 specifies how operators must manage and document their maintenance programmes.

    - Annex 7 specifies regulations related to registration markings required on all aircraft.

5. Boeing and Airbus are the main manufacturers of aircraft for large commercial airlines, while Bombardier and Embraer manufacture regional and business aircraft. Many smaller manufacturers are involved in developing GA aircraft.

6. Weight and balance calculations are an important consideration in the design and operation of aircraft.

7. Scheduled and unscheduled maintenance is required on all aircraft because the materials used to build them have a natural and predictable lifespan and are subject to environmental stressors. Aircraft maintenance mechanics, technicians, and engineers (AMMTEs) are the aviation professionals responsible for this maintenance.

8. AMMTEs may be licensed certifying personnel, licensed, or unlicensed. If licensed, they require endorsements for the specific type of work they are doing (associated with an aircraft type, avionic system, airframe, or component). Only licensed certifying AMMTEs have the authority to sign a maintenance release to return an aircraft to service after maintenance work is complete.

9. Maintenance work is categorized based on technology group (aircraft systems, technologies, engines, avionics, and so on) and by work environment (line, base, and shop).

10. As aircraft systems are highly complex, maintenance work requires a massive amount of reference documentation and careful record-keeping.

Table 2.4 Acronym rundown

AD	airworthiness directive
AME	aircraft maintenance engineer
AMM	aeroplane maintenance manual
AMO	approved maintenance organization
AMT	aircraft maintenance technician
AMMTEs	aircraft maintenance mechanics, technicians, and engineers
CAA	civil aviation authority
CG	centre of gravity
CMM	component maintenance manual
CofA	certificate of airworthiness
CSF	Commercial Spaceflight Federation
CVR	cockpit voice recorder
EASA	European Aviation Safety Agency
FAA	Federal Aviation Administration
FBW	fly-by-wire
FIM	fault isolation manual
FRM	fault reporting manual
GA	general aviation
GAMA	General Aviation Manufacturers Association
IATA	International Air Transport Association
ICAO	International Civil Aviation Organization
ICCAIA	International Coordinating Council of Aerospace Industries Associations
IPC	illustrated parts catalogue
LAME	licensed aircraft maintenance engineer
LRU	line replaceable units
MEL	minimum equipment list

(*Continued*)

Table 2.4 (Continued)

MMEL	master minimum equipment list
MPD	maintenance planning data/document
MRO	maintenance repair and overhaul
MTOW	maximum take-off weight
NASA	National Aeronautics and Space Administration
NTSB	National Transportation Safety Board
OEM	original equipment manufacturer
Ops Specs	operations specifications
SARPs	standards and recommended practices
SB	service bulletin
SDM	schematic diagram manual
SL	service letter
SOP	standard operating procedure
SRM	structural repair manual
TC	task card
TPPM	technical policies and procedures manual
VM	vendor manual
WDM	wiring diagram manual
WTC	wake turbulence category

Chapter Review Questions

2.1 What is an airworthiness certificate? What elements are evaluated by a civil aviation authority (CAA) before one is issued? What is an airworthiness directive? In your own words, describe how both contribute to international aviation safety.

2.2 Explain how birds can be a risk to aviation. Why is the issue particularly dangerous for turbofan engines?

2.3 Describe why the role of the AMMTE is crucial in aviation. Why does the industry require several levels of AMMTE: some unlicensed, others licensed, and the most senior licensed with the authority to sign maintenance releases?

2.4 Compare an aircraft's control surfaces that cause roll, pitch, and yaw with the controls used to manoeuvre an automobile. What makes aircraft more challenging to control?

2.5 Aircraft require several systems acting simultaneously to maintain and control flight. Refer to Figures 2.5, 2.6, and 2.7, which illustrate the components of an aeroplane and the workings of aircraft engines, to explain how Air Midwest flight 5481 lost control (as discussed in Case Study: Air Midwest Flight 5481).

2.6   What are some of the pros and cons of aircraft design and maintenance being regulated by each State rather than through the central authority of ICAO? Justify your answers.

2.7   How does the concept of airworthiness support international aviation safety? Do you think adhering to airworthiness regulations can be challenging for operators? Explain.

2.8   Why are there only a few major aviation OEMs? What challenges might new companies face when trying establish themselves as a major aircraft manufacturer?

2.9   Do you believe turbine engines are superior to piston engines for aeroplane flight? Why are most airline aircraft turbine-powered while GA aircraft are piston-powered?

2.10   Independently research an aviation accident that was associated with a loss of control and explain what happened to the aircraft, making reference to the four forces of flight: weight, thrust, lift, and drag.

2.11   Why is the concept of 'State of . . .' important in international aviation? Provide an example of a case in which your State (or a nearby State) would be considered each of the following, and explain why:

   a.   State of Design – a manufacturer whose aircraft design plans were drafted in your State.

   b.   State of Manufacture – an aircraft manufacturer within your State.

   c.   State of Registry – an aircraft that is registered within your State (consider both general aviation and airline aircraft).

   d.   State of Operator – an air operator (such as an airline or general aviation company) with a base of operations in your State.

CASE STUDY

CHINA AIRLINES FLIGHT 611 – THE DANGERS OF METAL FATIGUE[1]

On 25 May 2002, China Airlines flight 611 (CI 611) left Chiang Kai-shek International Airport in Taipei headed to Chek Lap Kok Airport in Hong Kong. The aircraft was a Boeing 747–200 carrying 206 passengers and 19 crew members. The weather that day was sunny with light winds. About 30 minutes into the flight, as the climbing aircraft reached an altitude of 34 900 feet, air traffic controllers monitoring the flight saw the aircraft's radar return split into four pieces (indicating that the aircraft itself had broken into pieces – the radar was sensing them as separate objects). The radar return then disappeared completely. No distress calls were transmitted by the pilots, which is odd because, at that altitude, pilots typically have enough time to troubleshoot and transmit a distress call if there is a problem with the plane.

Search and rescue teams were deployed and within a few hours they located human remains and aircraft wreckage floating in the sea, approximately 45 kilometres (27 miles) northeast of the Penghu Islands. All passengers and crew members on board perished in the accident. Salvage operations were conducted. The flight recorders were recovered, but they had both mysteriously stopped recording when the aircraft seemed to break apart in mid-air.

Investigators initially suspected explosives or a fuel tank explosion, but both possibilities were ruled out as no soot or explosive residue were found. Similarly, analysis of the wreckage helped them rule out a cargo door opening, over-pressurization of the cabin, and hazardous cargo as contributing factors.

Investigators then began to focus on structural failure as the cause of the accident. They completed a ballistic analysis, based on the location of the wreckage found on the ground, and determined that the tail section had been the first part of the airframe to separate. This focused their efforts on the aft portion of the airframe.

On one large piece of wreckage, investigators found cracks caused by metal fatigue, which led them to question the repair history of the 747. After reviewing the history of the aircraft, they learned that China Airlines had accepted delivery of the aircraft in 1979 and that the next year, the aircraft had suffered a *tail strike* (which is when an aircraft lands with its nose too high, causing the tail section to make contact with the runway). The tail strike left a section of damaged, scratched skin along the bottom of the aft portion of the aircraft. Mechanics repaired this section by sanding the area and installing a doubler. Much like a patch on a ripped pair of jeans, a *doubler* is a piece of aircraft skin installed over a damaged or cracked portion to reinforce the airframe. However, the repair was not carried out as specified in Boeing's structural repair manual (SRM) – the damaged skin was not removed and the doubler was much too small.

Multiple fatigue cracks developed from the scratches over time. However, the doubler remained in place until the accident, concealing the cracks that had grown unnoticed during thousands of take-off and landing cycles over 22 years. Accident investigators found one fatigue crack that was 64.5 centimetres (25.4 inches) long. Ultimately, when a crack becomes long enough that the remaining cross section of metal can't support the load, the entire aircraft structure fails. Investigators estimated that a continuous crack of at least 180 centimetres (71 inches) would have been required to cause CI 611 to break apart.

Investigators concluded that on the day of the CI 611 accident, when the aircraft reached 35 000 feet, the differential pressure between the cabin and the outside environment caused the pre-existing cracks to grow long enough to cause unstable separation and a loss of cabin pressure. This severed the flight recorder wiring before any anomalies could be recorded. The fuselage began separating and shedding debris until the structural integrity could no longer support the loads and the entire tail section separated from the aircraft. Unstable forces led to separation of all four aircraft engines while the remaining portions of the aircraft, including the wings attached to the forward section of the fuselage, crashed into the water.

Following the accident, China Airlines' other four Boeing 747–200 aircraft were grounded for safety checks. The NTSB, which had assisted with the investigation, released a safety recommendation in 2003 that indicated how improper repairs to an aircraft may hide damage and allow metal fatigue damage and fracturing to grow unnoticed over time. The FAA subsequently issued an AD that all Boeing 747 aircraft in the United States should be visually inspected for cracks. Aircraft with doublers installed were to be carefully inspected for scratches (either visually or using ultrasound technology to 'see' beneath the doublers).[2]

Notes

1 Adapted from ASC, 2005   2 FAA, 2003

Case Study Questions

Drawing on the information you learned in this chapter, consider the following questions:

2.12   Who holds ultimate responsibility in a situation like that of the CI 611 crash? Consider the various responsibilities of ICAO, the CAA, the aircraft manufacturer, the airline, and the individual maintenance engineer who completed the faulty repair.

2.13   How can an airline maintain best practices in maintenance without sacrificing the economics of their business? What is the proper balance between safety and profitability?

2.14   How is it possible that a physical defect like the one described above can exist for so long without anyone noticing it? Do you believe that human beings naturally become complacent over time? How can an organization promote critical thinking among maintenance professionals rather than maintaining the status quo?

2.15   It took investigators a great deal of time to search through more than 20 years' worth of maintenance records before they identified the critical tail strike. With modern technology, can you think of a more efficient and effective method for storing and searching maintenance records?

References

Airbus, 2015. *What the group achieved in 2015.* [Online] Available at: www.airbusgroup.com/int/en/investors-shareholders.html

Airbus, 2016. *The success story of Airbus.* [Online] Available at: www.airbus.com/company/history/

ASC, 2005. *In-flight breakup over the Taiwan Strait northeast of Makung, Penghu Island, China Airlines flight CI 611, Boeing 747-200, B-18255, May 25, 2002.* Taiwan: Aviation Safety Council.

Boeing, 2016a. *100 Years of Boeing.* Seattle: Boeing.

Boeing, 2016b. *Long-term market: Current market outlook 2016–2035.* [Online]

Available at: www.boeing.com/commercial/market/long-term-market/pilot-and-technician-outlook/

Bombardier, 2016. *History*. [Online] Available at: www.bombardier.com/en/about-us/history.html

CSF, 2015. *Annual report 2015*, Washington, DC: Commercial Spaceflight Federation.

Embraer, 2011. *Who we are*. [Online] Available at: www.embraer.com/en-US/ConhecaEmbraer/
TradicaoHistoria/Pages/default.aspx

Embraer, 2015. *Embraer in numbers*. [Online] Available at: www.embraer.com/en-US/Conheca
Embraer/EmbraerNumeros/Pages/Home.aspx

FAA, 2003. *Airworthiness directives; Boeing model 747 series airplanes*. Washington, DC: Federal Aviation Administration.

General Aviation Manufacturers Association, 2016. *Shipment data for year 2015*. [Online] Available at: www.gama.aero/media-center/industry-facts-and-statistics/shipment-database?page=show_
year&tab=year&type1=all&year=2015&quarter=1&type=189&comp_id=&submit=Go

ICAO, 2006. *Procedures for air navigation services: Training, Doc 9868*, Montreal: International Civil Aviation Organization.

ICAO, 2009. *Aircraft type designators, Doc 8643/37*. Montreal: International Civil Aviation Organization.

ICAO, 2010. *Annex 8 to the Convention on International Civil Aviation: Airworthiness of aircraft, 11th ed.* Montreal: International Civil Aviation Organization.

ICAO, 2011. *Annex 1 to the Convention on International Civil Aviation: Personnel licensing, 11th ed.* Montreal: International Civil Aviation Organization.

ICAO, 2012. *Annex 7 to the Convention on International Civil Aviation: Aircraft nationality and registration marks, 6th ed.* Montreal: International Civil Aviation Organization.

ICAO, 2015. *Aircraft nationality marks, national emblems and common marks*. Montreal: International Civil Aviation Organization.

ICAO, 2016. *Annex 6 to the Convention on International Civil Aviation: Operation of aircraft, 10th ed.* Montreal: International Civil Aviation Organization.

Khee, L. Y., 2009. Evolution of aircraft maintenance training. *Journal of Aviation Management*, pp. 9–16.

Kinnison, H. A., 2004. *Aviation maintenance management*. New York: McGraw-Hill.

NASA, 2014. *How does a jet engine work?*. [Online] Available at: www.grc.nasa.gov/www/k-12/UEET/StudentSite/engines.html

NTSB, 2004. *Loss of pitch control during takeoff Air Midwest Flight 5481 Raytheon (Beechcraft) 1900D, N233YV, Charlotte, North Carolina, January 8, 2003*. Washington, DC: National Transportation Safety Board.

The Atlantic, 2016. *A century in the sky*. [Online] Available at: www.theatlantic.com/sponsored/boeing-2015/a-century-in-the-sky/652/

Virgin Galactic, 2016. *Human spaceflight*. [Online] Available at: www.virgingalactic.com/human-spaceflight/

Notes

1 NASA, 2014
2 NASA, 2014
3 ICAO, 2010
4 ICAO, 2016
5 ICAO, 2012
6 ICAO, 2009
7 The Atlantic, 2016
8 Boeing, 2016a
9 Boeing, 2016a
10 Airbus, 2016

11 Airbus, 2016
12 Airbus, 2016
13 Airbus, 2015
14 Bombardier, 2016
15 Bombardier, 2016
16 Bombardier, 2016
17 Embraer, 2011
18 Embraer, 2011
19 Embraer, 2015
20 Kinnison, 2004

21 ICAO, 2011, ICAO, 2016, p. 9–5
22 ICAO, 2011
23 ICAO, 2006, p. 4–3
24 ICAO, 2016, p. 8–1
25 ICAO, 2006
26 Khee, 2009
27 Kinnison, 2004, pp. 11–13
28 Khee, 2009
29 Kinnison, 2004
30 Boeing, 2016b

1. The term 'civil aviation' describes:
   a. Airlines and space-operations
   b. Airlines and general aviation (GA)
   c. Airlines, GA, and military aviation
   d. Airlines, GA, military, and space-operations.

2. International regulations outline requirements for 1) training and licensing of aviation professionals, and 2) issuing operating certificates to air operators.
   a. True
   b. False

5. In general, dangerous goods (those which are flammable, corrosive, or toxic) are prohibited from travelling by air.
   a. True
   b. False

3. There are more aircraft and pilots active within the airline sector than in the general aviation sector.
   a. True
   b. False

4. The following aviation professionals require licenses:
   a. Pilots
   b. Pilots and air traffic controller officers (ATCOs)
   c. Pilots, ATCOs, and dispatchers
   d. Pilots, ATCOs, dispatchers, and flight attendants.

Answer Key: 1. b; 2. a; 3. b; 4. c; 5. b.

Learning science suggests that thinking through a few questions before you begin studying new material, even if you answer incorrectly, results in improved learning and retention.
Give it a try!

CHAPTER 3

Operations

CHAPTER OUTCOMES

At the end of this chapter, you will be able to . . .

- Describe the international regulations impacting aviation operations.

- Summarize how aviation professionals are trained and licensed.

- Explain several categories of aviation operations within the general aviation sector and the airline sector.

- Discuss the safety considerations related to transporting dangerous goods by air.

- Apply what you have learned to analyse the Colgan Air flight 3407 accident, which had a direct impact on aviation regulations in the United States.

Introduction

The primary purpose of aviation operations is the transport of people and goods by air. The direct economic contribution of civil aviation worldwide is estimated at US$2.7 trillion, and the industry provides 10 million jobs around the world: 220 000 for air navigation service providers, 450 000 in airport operations, 1.1 million in civil aerospace, 2.7 million in airlines, and 5.5 million in airport services.[1]

For students of aviation, the scope and complexity of operations can be daunting. In discussing millions of jobs, one can be overwhelmed trying to understand how the various groups and roles exist and interact. To simplify matters, we tend to distinguish between the *civil aerospace sector* (or aircraft manufacturing) discussed in the previous chapter, and *civil aviation operations*, which includes airlines, general aviation, air navigation service providers, and aviation cargo.[2]

In this chapter, we will discuss the licensing of aviation professionals as well as several aviation operations within general aviation and airlines. While air navigation service falls under the umbrella of civil aviation operations, we will discuss navigation (and the role of air traffic controllers) in Chapter 4 of this book.

International Regulations

International regulations impact every aspect of civil aviation. In this chapter, we will look specifically at Annex 1: Personnel Licensing and Annex 6: Operation of Aircraft, two annexes of the Chicago Convention that relate to civil aviation operations.

Annex 1 specifies the requirements for training and licensing pilots, air traffic controllers, maintenance engineers, and dispatchers. Training and licensing are vital for ensuring safety throughout the entire aviation network. Licensing criteria are detailed later in this chapter.

Annex 6 outlines safe operating practices for aircraft involved in international air transport. One of the key standards specified in Annex 6 is that an *air operator certificate* (AOC) is required for organizations that wish to operate a commercial air operation. An AOC is a formal document that can be thought of as an organization's *licence to do business* – without one, it cannot legally be involved in aviation operations.

To get an AOC, an air operator needs to demonstrate to the issuing authority of the State of the Operator (the country where the company's primary place of business is located) that its organization, training, flight operations, ground handling, and maintenance are adequate for safe operations. If, after an evaluation, the company is found to be capable of conducting safe operations, an AOC is given out by the issuing authority within that State, typically the civil aviation authority (CAA).

3.1 The Language of Operations

Operators are companies that offer some type of air service (i.e., that operate aircraft).

Ab initio is a Latin term meaning 'from the start', which describes the new cadet approach to training pilots. Ab initio programmes accept and train people with no previous piloting experience.

When pilots work professionally, they are said to be **flying the line**, which means that they are actively involved in bidding routes each month and flying trips for an operator. Sometimes pilots are 'pulled from the line' and then 'returned to the line', perhaps to complete annual recurrent training for a few weeks before returning to active flying duties.

A **check airman** (or **check pilot**) is an operator's senior training pilot who is qualified to observe other pilots' performance, in the cockpit or in a simulator, to ensure competence. A **check ride** refers to the act of a check pilot riding along on a flight and observing pilot performance. Check airmen observe pilot performance in the cockpit while sitting in the **jump seat**, also called an **auxiliary crew station**. In the cockpit, the jump seat is typically located between and behind the Captain and first officer (FO) seats, and can be folded down as needed. Flight attendants also use jump seats in the cabin during take-offs and landings.

Operators occasionally conduct **ferry flights**, which are delivery (non-revenue) flights to move aircraft from one location to another. If a pilot rides along on a ferry flight, to get to a location where he or she will be piloting a revenue flight, it is called **deadheading** or **POSTECH** (positioning (technical) pilot crew).

Licensing Aviation Professionals

To promote standardization and efficiency around the globe, ICAO publishes standards and recommended practices (SARPs) for the licensing of aviation professionals in Annex 1: Personnel Licensing. Licences are required for pilots, air traffic controllers, maintenance engineers, and dispatchers. Note that flight attendants do not require a licence.

Licensing is mandated to ensure that flight and ground crew members possess the competency to perform their professional duties. Licensing authorities within contracting States are responsible for issuing licences, but the minimum qualification standards for licences are specified by SARPs in Annex 1. Licensing authorities are responsible for assessing qualifications, issuing licences and ratings, designating approved examiners, approving training courses and simulators used for training, and validating licences issued by other States.[3]

Did You Know?

ICAO SARPs in Annex 1 describe the *minimum* standards for licences. Many countries choose to enforce stricter criteria, and some use different terms to describe their licences. However, at their core, all licences are based on the same standards published by ICAO.

As some States choose to enforce licensing requirements beyond the minimums, holding a licence in one country does not grant you the same professional privileges globally. In general, a licence is valid for private flying in foreign countries, but the validity of a licence must be confirmed before it can be used for commercial operations.

The licensing process can be complicated, but in general several key criteria must be met:

- *Prerequisites* – to be licensed, an individual must meet requirements of minimum age, experience (typically measured in flight hours for flight crew and years of duty for ground personnel), and medical fitness.

- *Training* – operational, classroom, and often simulator training is used to develop professional competence as a person works towards a desired licence.

- *Demonstration of competency* – with prerequisites met and training complete, a person must demonstrate professional competence before a licence can be issued. Competence is generally evaluated through a written exam and an operational assessment (such as an in-aircraft flight test).

- *Currency* – once a professional licence is earned, it must be kept current. As knowledge and skill fade over time if they are not practised, ICAO requires that professionals 'exercise the privileges' of their licence on an ongoing basis to maintain the validity of their credentials. This means that professionals must continually practise and complete assessments of their knowledge and skill throughout their careers.

Table 3.1 Licensing of aviation professionals

Pilot licences	Privileges	Experience required	Additional requirements
Private pilot licence (PPL)	To act unpaid as pilot-in-command (PIC) of an aircraft on non-revenue flights during daylight hours in good (visual) weather conditions	*General aviation aircraft* – 40 flight hours with 10 hours of solo flight time including 5 hours of solo cross-country *Airship* – 25 flight hours	Minimum age: 17 English language proficiency Class 2 medical
Commercial pilot licence (CPL)	To exercise all privileges of a PPL and to be paid to function as PIC of commercial flights certified for single-pilot operations (or co-pilot of a two-pilot aircraft) Notes: The CPL can be considered the *junior* professional licence for pilots. Night rating is required for night flying.	*Aeroplane* – 200 flight hours including 100 hours as PIC, 20 hours of cross-country, and 10 hours of instrument flight *Helicopter* – 150 flight hours including 35 hours as PIC, 10 hours of cross-country, and 10 hours of instrument flight *Airship* – 200 flight hours, including 50 hours as an airship pilot, 30 hours as PIC with 10 hours of cross-country, 10 hours of night flight, and 40 hours of instrument flight	Minimum age: 18 English language proficiency Class 1 medical

Pilot licences	Privileges	Experience required	Additional requirements
Airline transport pilot licence (ATPL)	To exercise all privileges of a CPL and to act as PIC of aircraft that require more than one pilot Notes: The ATPL can be considered the *senior* professional licence for pilots. Instrument rating privileges are included with the ATPL–aeroplane.	*Aeroplane* – 1500 flight hours, including 500 hours as PIC, 200 hours of cross-country, 100 hours of night flight, and 75 hours of instrument flight *Helicopter* – 1000 flight hours, including 250 hours as PIC, 200 hours of cross-country, 50 hours of night flight, and 30 hours ofinstrument flight	Minimum age: 21 English language proficiency Class 1 medical
Multi-crew pilot licence (MPL)	To exercise all privileges of a PPL and to act as co-pilot of an aircraft that requires a co-pilot Notes: Additional experience is required to exercise privileges of a CPL. Equivalent to an ATPL but restricted to multi-crew operations. Instrument rating privileges are included for multi-crew operations only.	240 hours as pilot flying and pilot-not-flying (actual and simulated flight) Must achieve advanced level of competency	Minimum age: 18 English language proficiency Class 1 medical

Other professional licences	Privileges	Experience required	Additional requirements
Aircraft maintenance mechanics, technicians, and engineers (AMMTEs)	To exercise the privileges of the licence as prescribed by the State Note: As discussed in Chapter 2, licensed AMMTEs require additional endorsements to complete certain types of tasks, and to have certifying privileges.	4 years' experience in inspection, servicing, and maintaining aircraft or components 2 years' experience for a restricted licence	Minimum age: 18
Dispatcher	To control and supervise flights To provide briefings and assist the pilot-in-command in identifying a safe and expeditious route for the flight	2 years' service as flight crew member, meteorologist in organization dispatching aircraft, or air traffic controller OR 1 year as an assistant dispatcher OR completion of a training course	Minimum age: 21

(Continued)

Table 3.1 (Continued)

Other professional licenses	Privileges	Experience required	Additional requirements
Air traffic control officer (ATCO) Licence Ratings • Aerodrome control • Approach control procedural • Approach control surveillance • Area control procedural • Area control surveillance • Approach precision radar control	To provide or supervise control service for the airport or unit for which they are rated Notes: As will be discussed in Chapter 4, ATC licences themselves carry no privileges – they must be added with ratings. Ratings become invalid when controller has ceased to exercise privileges for a period that shall not exceed 6 months.	Completion of training course and at least 3 months of service in actual air traffic under supervision of rated ATC Completion of training course AND Experience working under supervision of rated air traffic controller (within 6 months preceding application) as follows: *Aerodrome control rating*: 90 hours or 1 month within an aerodrome control service at the unit for which the rating is sought *Approach control procedural, approach control surveillance, area control procedural, and area control surveillance ratings*: at least 180 hours or 3 months of control service for which the rating is sought *Approach precision radar control rating*: not fewer than 200 precision approaches (at least 50 carried out at the unit and on the equipment for which the rating is sought)	Minimum age: 21 English language proficiency Class 3 medical

Source: ICAO, 2011a

Pilot Licensing

As training and licensing of maintenance professionals and air traffic controllers are covered in Chapters 2 and 4, respectively, the remainder of this discussion will focus primarily on pilot licensing.

Licences are endorsed with ratings; a rating adds a certain privilege to a licence. Consider that the holder of a commercial pilot licence (CPL) could act as the pilot-in-command (PIC) of a small Cessna 172 or the first officer (FO) of a large Boeing 737 – although the licence is the same, piloting different types of aircraft that range widely in complexity requires special training and additional class and/or type ratings.

Ratings and Other Requirements

Pilots can't act as a PIC of an aircraft unless they have the appropriate *class rating* and *type rating* for that specific aircraft. Aircraft are grouped into *classes* (single-engine or multi-engine; land or sea). A pilot's licence indicates a class, and that pilot can operate all small aircraft within that class (e.g., all small single-engine land aircraft). Additional qualifications – such as a multi-engine rating or seaplane rating – are required for pilots to operate other classes of aircraft.

However, for more complicated types of aircraft, it is neither safe nor feasible for pilots to receive a blanket rating that allows them to operate several types within a class. Therefore, pilots flying complex aircraft need an individual *type rating* for each aircraft type, in addition to the *class rating*. Examples of complex aircraft, that require individual type ratings, include 1) aircraft weighing more than 5700 kilograms (about 12 500 pounds), 2) aircraft requiring a minimum of two pilots, 3) helicopters, and 4) powered-lifts (aircraft capable of vertical take-offs and landings). Type ratings require a pilot to demonstrate competent performance within normal and emergency procedures.

Pilot licences can be endorsed with other ratings that grant additional privileges to the pilot. These include

- *instructor rating* – required for pilots to act as a flight instructor;

- *night rating* – required for pilots to operate between sunset and sunrise; and

- *instrument rating* – required for flights conducted without visual reference to the ground.

Did You Know?

When conducting an *instrument flight* – generally because poor weather conditions prevent them from seeing the ground – pilots navigate by referencing their on-board instruments and rely on an air traffic controller to keep them a safe distance from other flights and obstacles. These flights are operated under *instrument flight rules*, which will be discussed in more detail in the next chapter. Instrument ratings can be added to PPL, CPL, or ATPL–helicopter licences. The ATPL–aeroplane licence automatically includes an instrument rating.

Other considerations in pilot licensing include

- *Age* – pilots are not permitted to fly a single-pilot international commercial route if they have reached their 60th birthday, or a two-pilot international commercial route if they have reached their 65th birthday.[4]

- *Hours* – a crucial aspect of demonstrating professional experience is the documentation of a pilot's flight hours. Pilots diligently track each hour they fly in a logbook, including a description of the aircraft and type of flying conducted. PIC hours are particularly valuable, referring to time when the pilot is legally responsible for the safe operation of the flight. PIC time is earned when a pilot is flying solo (as the only pilot) or serving as Captain in an aircraft that requires more than one pilot.

- *English language proficiency* – English is designated as the international language of aviation. All pilots on international routes (as well as air traffic controllers who handle international traffic) must be operationally capable of communicating in the English language, including both aviation phraseology and plain conversational dialogue.

3.2 Simulators

Simulators are an increasingly popular tool for pilot training, as they offer significant financial and safety advantages over in-aircraft training. To be used for training, a simulator must be approved by the CAA's licensing authority.

The term *fidelity* describes how accurately a simulator represents the real-world experience. High-fidelity simulators use very high-resolution displays and are set on motion platforms that replicate aircraft movement. Some even incorporate enhanced features like smoke machines to practise managing on-board fires. For most airlines, initial type training is conducted exclusively in high-fidelity simulators, meaning that the first time an airline pilot flies the actual aircraft type, he or she will have a load of paying passengers in the cabin.

Medical Assessments

For a pilot (or air traffic controller) licence to be valid, it must be accompanied by a current medical assessment. Licences and medical assessments involve separate evaluations – licences are issued based on professional competence while medical assessments are issued to indicate a person is healthy enough to do his or her job. Medical assessments are conducted by doctors with specialized training in aviation medicine, who have been made a designated medical examiner (DME) by their State.

Medical assessments evaluate physical and mental health, visual and colour perception, and hearing. Licence holders are responsible for informing a DME if their medical status changes (such as after a recent surgery or when beginning a new medication).

ICAO designates three levels of medical assessments, with Class 1 assessments (for professional pilots) having the most rigorous criteria. Individuals are reassessed at predetermined intervals throughout their professional career, with older adults requiring more frequent medicals (as we are more likely to experience health problems with age).

Table 3.2 Classes of medical assessments as specified by ICAO.

Medical assessment class	Licence type	Period of validity
Class 1	Commercial pilot licence, multi-crew pilot licence, and airline transport pilot licence	12 months until 40th birthday 6 months after 40th birthday
Class 2	Flight navigator licence, flight engineer licence, private pilot licence, glider pilot licence, free balloon pilot licence	60 months until 40th birthday 24 months after 40th birthday 12 months after 50th birthday
Class 3	Air traffic controller licence	48 months until 40th birthday 24 months after 40th birthday 12 months after 50th birthday

Source: ICAO, 2011a, pp. 6-1-6-17
Note that the terminology used to describe medical assessment classes, and their validity periods, vary slightly between States, but they will be based upon the minimum standards established by ICAO.

Psychoactive Substances

Licence holders are not permitted to use psychoactive substances as these can impair a person's mental processes and create a safety risk. Psychoactive substances include alcohol, opioids, cannabinoids, sedatives and hypnotics, cocaine, other psychostimulants, hallucinogens, and volatile solvents. Coffee and tobacco are excluded from this category as their use is generally considered acceptable.[5]

Licence holders can't exercise the privileges of their licence, either temporarily or on an ongoing basis, while under the influence of any substance. For example, a pilot who consumed alcohol is *temporarily* unsafe to fly, while a habitual user of drugs or alcohol may be *permanently* unsafe until he or she has completed treatment.

3.3 Royal Aeronautical Society

The Royal Aeronautical Society (RAeS), founded in 1866, is an international professional organization dedicated to supporting the aerospace community. The RAeS has over 22 500 members around the world. Through the Society's Young Persons' Network, a variety of resources are available including educational tools, job boards, bursaries, and resources for career support and development. See www.aerosociety.com for more details about the RAeS.

Aviation Operations

Worldwide, there are three main categories of aviation operations: general aviation, airlines, and military aviation.

General aviation (GA) refers to all operations that fall outside of commercial airlines and military aviation. This sector comprises a wide variety of operations, including flight instruction, corporate flight departments, medical transport, and personal (recreational) flying. A subset of GA is *aerial work* – commercial and private operations of which the primary mission is not to carry passengers between two points. Aerial work includes surveying, photography, agriculture, and search and rescue activities.[6]

Airlines are organizations that provide commercial air transport of passengers, cargo, or both. Most airlines are scheduled operations (as opposed to *charter operations*, which operate on-demand flights). In this definition of airlines:

- *commercial* means the organization charges fares and operates for profit;

- *scheduled* means that the times and dates of flights are determined in advance; and

- *air transport* means that people and goods are moved by aircraft.

Military aviation is the use of aircraft to support military activities, and may include combat, reconnaissance, airlifts, humanitarian aid, and logistical support. As previously noted, the focus of this text is on international *civil* aviation, and therefore a detailed discussion of military aviation is beyond the scope of this book.

Did You Know?

Aviation is sometimes described as a *paramilitary* industry, as aspects of military culture are woven into the fabric of civil aviation (for example, rank structure between Captain and first officer, uniforms for airline pilots, and so on). Although not part of a State's armed forces, professionals in the civil aviation industry share a deep respect for the contributions and service of their military counterparts.

General Aviation

General aviation refers to professional and private aviation activities that are not part of the airline or military sectors. This includes flight instruction, corporate flying, aerial work, small commuter operations, most helicopter operations, and pleasure flying. In total, approximately 350 000 aircraft and 700 000 pilots are involved in GA activities around the world (as compared to 60 000 aircraft and 400 000 pilots employed by airlines).[7]

When most people think of general aviation, the image of a small piston-engine aircraft operating from a rural airport on a personal sightseeing flight comes to mind. Although pleasure flying is an important component, it accounts for only about one-quarter of GA activities; the majority of activities within this sector are related to professional services.

The rights and interests of aircraft owners and personal flyers are represented by groups called *aircraft owners and pilots associations* (AOPAs). There are 77 AOPAs around the globe, with many countries having a national association. These associations do important work to preserve airspace for private use, promote safety practices, distribute educational materials, maintain the economic viability of smaller regional airports, and advocate for the interests of aircraft owners. The International Council of Aircraft Owner and Pilot Associations represents these national organizations at ICAO.

In the United States, and in some other areas, the term *fixed-based operator* (FBO) is used to describe a company providing general aviation services at an airport. FBOs typically offer flight instruction along with aircraft tie-downs, refuelling, aircraft rentals, and maintenance. Interestingly, the term *fixed-based operator* stretches back to the earliest days of aviation at the end of World War I. As pilots and aircraft returned from the war, some dishonest aviators earned a reputation for taking money from civilians for flying lessons and then flying away at night, leaving with the money without having provided any services. This led reputable GA businesses to adopt the term *fixed-based operator* to assure the public that the company was fixed to a particular airport, and not a *fly-by-night operation*.

A crucial part of GA is flight instruction, as almost all pilots start their training within the GA sector. In addition, FBOs and personal flyers are often active in outreach programmes to spark an interest in aviation in the next generation. There are a variety of outreach programmes wherein small aircraft owners or flight schools offer free flights to children and their local communities. GA plays an important role in the development of the next generation of aviation professionals, inspiring many young people to pursue careers in aviation.

Business Aviation

Business aviation is an important aspect of general aviation. Corporate flight departments allow for the rapid on-demand transport of staff with private aircraft (piston-engines, turbo-props, jets, and helicopters). For companies with high-priced executives or more than one major centre of operation, this mode of transportation offers several advantages. In general, trips are more direct, flights can use smaller airports that are closer to company offices, airport line-ups and security congestion are avoided, and flights can depart on demand to one or multiple destinations.

Globally, business aviation travel is on the rise with a 1.4 per cent annual growth rate.[8] Although there tends to be a higher accident rate within the GA sector, business aviation practices are highly professional and thus business aviation has an excellent safety record comparable to airlines.[9]

3.4 National Business Aviation Association

The National Business Aviation Association (NBAA) serves as a voice for business aviation. A United States-based organization, the NBAA was founded in 1947 and works to make business aviation efficient, productive, and successful. The NBAA offers a variety of student and professional scholarships. See www.nbaa.org for details about the organization.

There are a few options for organizations interested in business aviation:

- *Full ownership* – some corporations will purchase aircraft and manage their own aviation operations. These corporate flight departments employ pilots, as well as maintenance and dispatch professionals. Other organizations enter into joint-ownership relationships (partnerships, timeshares, or interchange agreements) with another company to share costs.

- *Fractional ownership* – for companies that want access to a business aircraft but don't need a full corporate flight department, fractional ownership is an attractive option. With this option, an aircraft management company oversees aircraft services (maintenance, pilot training and scheduling, and administration). However, the aircraft is shared by several corporations that each own a percentage of the aircraft (as little as a one-sixteenth share of an aircraft). This allows the cost of the aircraft to be shared among several companies, with each getting a representative number of hours of use of the aircraft.

- *Charter* – a third option for companies with only occasional need for a business aircraft is charter. Charter operations allow companies to purchase on-demand flights, allowing instant access to an aircraft without having to maintain aviation operations themselves. Some companies that frequently charter flights will purchase a block of time at a discounted rate, which is called *block charter*.

To better understand how a company would choose between these three options, consider a Cessna Citation X (a twin-engine jet that holds nine passengers and two pilots), which costs US$21.6 million new and between US$7.7 and 15.5 million used.[10] Table 3.3 below shows an example of the financial considerations under each option.

Looking at the above scenario, if a company's aircraft utilization is more than 240 flight hours, the lowest operation cost is associated with full ownership. For companies that require less flight time, fractional ownership and charter options make more financial sense.

Table 3.3 Business aviation ownership options

	Full ownership	Fractional ownership	Charter
Total cost	$876 608/year	$537 540/year	$460 000/year
Availability	unlimited	about 100 hours/year (assuming 5-year contract with a 1/8 share)	100 hours/year (used in 25-hour increments)
Cost per mile	$18.61	$11.41	$9.77
Paid returns	No mandatory paid returns	Mandatory paid returns	Possible mandatory paid returns
Tax implications	Depreciation available	Depreciation available	Depreciation not possible

Source: Cox, 2010

Airlines

In 2016, the world's airlines safely moved 3.3 billion people on 32.8 million flights through a network of 52 000 routes.[11] The airline industry is in the business of transportation.

Did You Know?

Every day, airlines carry 9.8 million passengers on 104 000 flights, and move US$17.5 billion worth of cargo.[1]

Note

1 ATAG, 2016, p. 5

Airlines can be grouped broadly into major airlines and regional airlines. Major airlines have international (sometimes global) route structures and operate a fleet of large jets. Major airlines can provide passenger service, cargo service, or both, and generally operate under either a *traditional* or *low-cost carrier* (LCC) business model.

- Traditional carriers (also called *legacy* or *full-service* carriers) offer a range of amenities, such as meal service, in-flight entertainment, checked baggage allowance, and full customer service departments. These carriers typically use a hub-and-spoke model (discussed later in this chapter) to organize their networks.

- Low-cost carriers (or *no-frills* carriers) take advantage of strategic business decisions to offer lower fares to passengers (e.g., using only one type of aircraft, which reduces pilot training and aircraft maintenance costs). These carriers also tend to offer fewer amenities and generally use a point-to-point (P2P) model to organize their network.

Regional airlines, on the other hand, operate smaller aircraft (usually with fewer than 120 seats). A regional airline may be an independent company or a subsidiary of a major airline. Regional airlines operate networks over shorter distances than major airlines.

Airline Pilots

In the 1960s, a typical airline flight had five flight crew: two pilots (Captain and first officer), a flight engineer, a flight navigator, and a radio operator. New technologies available to modern airlines have reduced the pilots' workload to allow for safe operations with only two pilots – the Captain who is pilot-in-command (PIC) and sits in the cockpit's left seat, and the first officer who is second-in-command and sits in the cockpit's right seat.

Technologies simplify many activities and often allow crew members to function as systems managers and decision makers rather than control operators. Annex 1 specifies licensing requirements for flight engineers, navigators, and radio operators; however, the reality is that these roles are rarely used in modern aviation and are becoming obsolete.

Did You Know?

For every aircraft in its fleet, an airline will require between 10 and 30 pilots, depending on the type of flying conducted by the airline.

For pilots who aspire to fly for airlines, the traditional career path requires successful completion of training, an *hours-building* period in GA to gain experience, and a few years flying with a regional carrier before being eligible to interview with a major airline. However, with the growing demand for pilots in recent years, some airlines have launched *ab initio* cadet programmes. *Ab initio* is a Latin term meaning 'from the start' and refers to a process whereby airlines hire people with no aviation background and pay for their training to become pilots for their airline. These cadet pilot programmes are increasingly popular in areas experiencing tremendous growth in aviation, including some States in Asia. The traditional and ab initio processes are detailed in Figure 3.1.

Although traditional and cadet programmes represent two distinct pilot pathways, the projected shortage of pilots has resulted in some hybrid models. Some airlines are establishing cadet programmes that see pilots join the companies directly after completing training, skipping the hours-building phase entirely.

3.5 Women in Aviation International

Women represent only about six per cent of airline pilots around the world. Women in Aviation International (WAI) is a non-profit organization that supports female aviators and offers millions in scholarships to fund training costs. See www.wai.org for information about the organization.

Once pilots begin working for an airline (rather than in general aviation), they can expect to belong to a union and to abide by a seniority system. Airline pilots are usually supported by powerful unions that look out for their best interests. Although there are domestic unions, and some airlines have their own company-specific unions, about 100 000 pilots around the world are members of the International Federation of Air Line Pilots' Associations (IFALPA). The role of IFALPA is to be a global advocate of the piloting profession.[12]

As well as being unionized, almost all airlines use a seniority system for pilots. When a pilot is hired at an airline and completes airline-specific training, he or she is given a seniority number and begins working in the most junior position (generally, flying the smallest aircraft as an FO). As more senior pilots retire or move on, and new recruits are hired, the pilot moves up on the seniority list and has more flexibility over his or her schedule (i.e., can avoid working weekends and holidays). Keep in mind that most pilots must bid for their trip schedules each month, and those with highest seniority get first choice. With higher seniority, pilots have a few options:

- Upgrade to Captain on current aircraft type – earn Captain's pay but have lower seniority than other Captains, which results in less control over their schedules.

- Upgrade to FO on a larger aircraft – earn more pay associated with flying a larger aircraft, but have little control over their schedules (as the FO with the lowest seniority flying that aircraft type).

- Remain FO on current aircraft type – gain relative seniority (i.e., seniority compared to other FOs who fly that aircraft); although they won't be paid as much as Captains or FOs of larger aircraft, they will have more control over their schedules.

Unfortunately, seniority is usually not transferrable to a new airline – a pilot who flew for 20 years with an airline that went bankrupt would begin at the bottom of the seniority list when hired by a different airline.

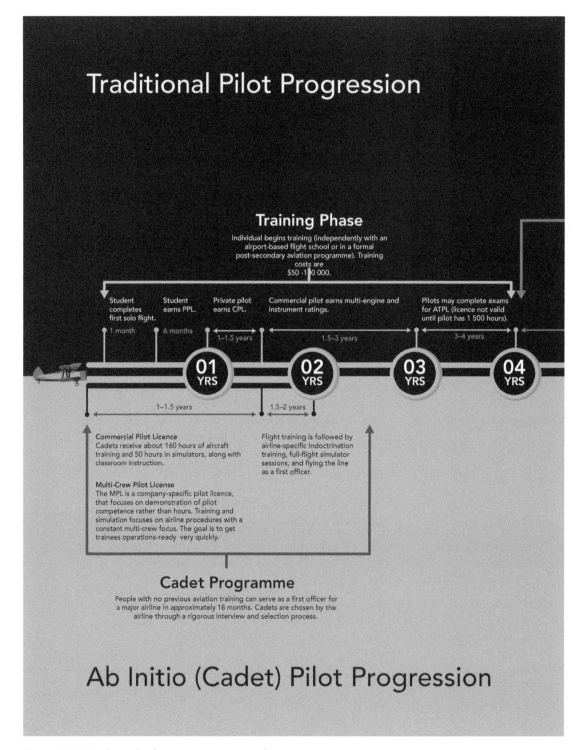

Figure 3.1 Traditional pilot progression vs cadet programme

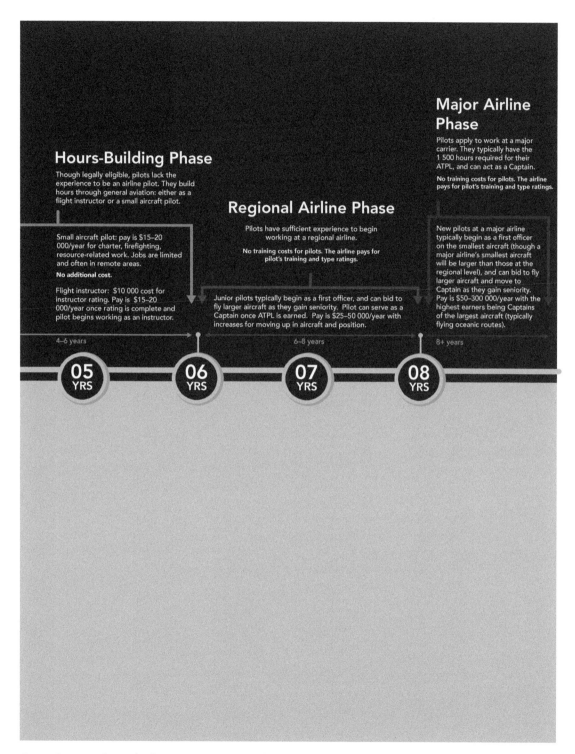

Hours-Building Phase

Though legally eligible, pilots lack the experience to be an airline pilot. They build hours through general aviation: either as a flight instructor or a small aircraft pilot.

Small aircraft pilot: pay is $15–20 000/year for charter, firefighting, resource-related work. Jobs are limited and often in remote areas.

No additional cost.

Flight instructor: $10 000 cost for instructor rating. Pay is $15–20 000/year once rating is complete and pilot begins working as an instructor.

4–6 years

Regional Airline Phase

Pilots have sufficient experience to begin working at a regional airline.

No training costs for pilots. The airline pays for pilot's training and type ratings.

Junior pilots typically begin as a first officer, and can bid to fly larger aircraft as they gain seniority. Pilot can serve as a Captain once ATPL is earned. Pay is $25–50 000/year with increases for moving up in aircraft and position.

6–8 years

Major Airline Phase

Pilots apply to work at a major carrier. They typically have the 1 500 hours required for their ATPL, and can act as a Captain.

No training costs for pilots. The airline pays for pilot's training and type ratings.

New pilots at a major airline typically begin as a first officer on the smallest aircraft (though a major airline's smallest aircraft will be larger than those at the regional level), and can bid to fly larger aircraft and move to Captain as they gain seniority. Pay is $50–300 000/year with the highest earners being Captains of the largest aircraft (typically flying oceanic routes).

8+ years

05 YRS   **06 YRS**   **07 YRS**   **08 YRS**

Figure 3.1 Traditional pilot progression vs cadet programme (Continued)

Case Study: Eastern Air Lines Flight 212 – The Sterile Cockpit Rule

On 11 September 1974, Eastern Air Lines flight 212 was on an instrument approach to Charlotte, North Carolina in dense ground fog. Unfortunately, the aircraft crashed just 5.3 kilometres (3.3 miles) short of the runway. Of the 82 people on board the aircraft, only 11 passengers and two crew members survived the crash.[1]

The National Transportation Safety Board (NTSB) investigators determined that the accident was not related to malfunctions of the aircraft or ground facilities, which led them to look more closely at the human factors that may have caused the accident.

The cockpit voice recorder (CVR) revealed that up until two minutes and 30 seconds before impact, the pilots had been passionately discussing matters not related to the flight (with topics ranging from used cars to politics to attempts to spot a local amusement park). None of the required altitude call-outs were completed by the Captain.

The investigators determined that non-pertinent conversations caused the crew to become distracted and reflected poor cockpit management and a casual attitude towards the flight. The probable cause of the accident was the crew's lack of awareness at critical points in the approach, resulting from the non-pertinent conversations.[2]

In 1981, the Federal Aviation Administration (FAA) enacted the *sterile cockpit rule* prohibiting non-flight-related conversation during critical phases of flight. Europe followed suit with similar regulations and ICAO supports the practice internationally through Doc 9870, *Manual on the Prevention of Runway Incursions*. The common application of the sterile cockpit rule is the prohibition of non-flight-related conversation below 10 000 feet, with exceptions for flights that cruise below this altitude. Standard operating procedures (SOPs) also limit other distractions during critical flight phases, such as restricting cabin crew entry onto the flight deck and non-pertinent calls from dispatch.

Sterile cockpit procedures have evolved to include pilots' use of personal electronic devices, which are considered a distraction and therefore prohibited during critical phases of flight.

Notes

1 NTSB, 1975   2 NTSB, 1975

Airline Professionals

Airlines are staffed by a variety of professionals beyond pilots, including flight attendants, dispatchers, maintenance engineers, customer service personnel, and a management team, among others.

Flight Attendants

In the 1920s, when the first airlines began to offer commercial air transport, the passenger experience was unpleasant. Air travel was loud, cold, and often smelled of fuel. Moreover, rough turbulence was common as they didn't have the modern guidance around storms that we have today. Flight attendants (also called *cabin crew*) were introduced to make the travel experience more pleasant. The first cabin crew were teenage boys, and these were followed by adult men in the late 1920s. Soon the required qualifications shifted to accept females who were trained nurses. It was only when nurses were required to support World War II[13] that regulations changed once again to accept women without nursing qualifications.

Cabin crew often act in a hybrid role between flight crew and airline marketing – they must balance the safety of a flight while serving as the face of the airline's customer service and marketing initiatives.

Training programmes for cabin crew typically last between five and 12 weeks, and the training costs are covered by the airline. Training covers a variety of areas including aircraft and cabin familiarization, aviation medicine, safety skills for emergency evacuations (cabin depressurization, safety briefings, and firefighting), as well as customer service standards for economy and business classes.[14]

Simulators play an important role in cabin crew training. While pilots train in simulators that replicate the flight deck, flight attendants train inside a section of an aircraft cabin and galley. The cabin crew training simulator can move in a way that mimics turbulence and can even fill with smoke for training in how to manage on-board fires.

Annex 1 of the Chicago Convention does not require cabin crew to be licensed, unlike other aviation professionals. In Europe, the European Aviation Safety Agency's (EASA) regulations incorporate an *attestation* for cabin crew, which serves as a licence. The attestation must be held by all cabin crew active in the European Union, which allows experienced cabin crew to move between European airlines without having to restart their training. (They do still need to complete the airline-specific aspects of training.)

Most States do not have any form of cabin crew licence as it is regarded as unnecessary and an administrative burden. However, the International Transport Workers' Federation (ITF), the union representing cabin crew, is fighting for a licence with the expectation that it will lead to improved pay and recognition.[15] Flight attendant salary currently ranges from about US$15 000 per year for junior flight crew for a regional airline to US$60 000 per year for senior flight crew for a major airline.

Airline Dispatchers

The role of an airline dispatcher is in some ways similar to that of an air traffic controller (discussed in the next chapter) – they are both ground-based positions that involve inter-action with and provision of services to in-flight aircraft. However, dispatchers work for an airline and their role is to help pilots identify the safest and most expeditious route for a trip. Flight dispatchers require meteorological training to identify hazardous weather patterns and chart safe courses to avoid such hazards. Airline dispatchers require a licence, as speci-fied in Annex 1 of the Chicago Convention.

Unlike GA flying where the pilot is responsible for flight planning, within an airline it is the flight dispatcher who completes the flight plan and delivers it to the Captain for approval. The flight plan is prepared with consideration for safe routing, fuel require-ments, maintenance limitations, take-off and landing weights, weather, and NOTAMs (*notices to airmen* of flight hazards such as clear air turbulence). The dispatcher is respon-sible for signing the dispatch release, without which the pilot does not have authority to depart.

If anything changes during the course of a flight, such as an unanticipated weather event, the dispatcher contacts the Captain in flight to amend the flight plan. The dispatcher is also responsible for informing the company and the public of any resulting schedule changes. Dispatchers are well-paid professionals who earn between US$30 000 and US$150 000, with variations based on seniority and between airlines.

Airline Management

Worldwide, the airline industry has a reputation for being exciting, glamorous, and tech-nologically advanced. Perhaps less glamorous is the reality of a history of slim profit margins for airlines and cyclical periods of profits and subsequent losses. It is the role of the aviation management team to set the strategic direction of the airline towards profitability.

However, the financial success of an airline is often outside the control of management, as it is linked to the likelihood that the travelling public will choose to fly. These decisions are influenced by factors such as the economy, global health scares (such as SARS: severe acute respiratory syndrome), and public fears following terrorist attacks.

As illustrated in Figure 3.2, profitability in the airline industry is cyclical, with a few years of profits followed by a period of losses. Both the 9/11 terrorist attacks (and subse-quent escalation in fuel costs) and the 2008 financial crisis resulted in periods of significant losses for airlines, which were followed by rebounds. Because of the cyclical nature of the industry, it is particularly important for airlines to track and strategically manage performance indicators.

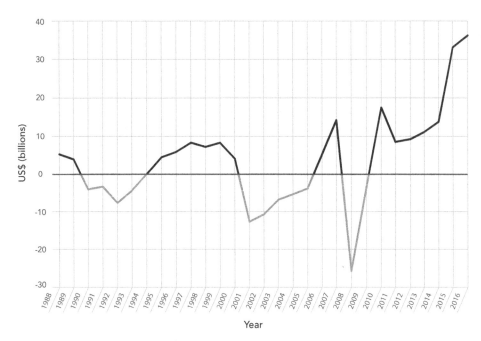

Figure 3.2 Global airline profits and losses

Source: Adapted from Doganis, 2006, p. 5; ICAO, 2006, p. 15; IATA, 2016a, p. 1

Performance Indicators

Passenger ticket sales represent the primary source of revenue for airlines. However, to measure the success of an airline, various performance indicators must be considered. These indicators are set out in Table 3.4.

Table 3.4 Performance indicators for measuring airline success

Performance indicator	Significance	Method of calculation
available seat kilometres (ASK)	the passenger capacity of a flight	multiply the number of seats available to be sold on an aircraft by the distance flown on a flight leg
revenue passenger kilometres (RPK)	the passenger traffic on a flight	multiply the number of fare-paying passengers on a flight leg by the distance flown on that leg
passenger load factor[1]	percentage of seats sold on a flight	divide the RPK by the ASK
unit costs	total cost of each flight leg.	add all fixed and variable costs associated with a particular flight leg
yield (average unit revenue)	the profitability of a flight leg (expressed in cents per kilometre)	total revenue collected on a flight leg divided by the RPK

[1] Airlines can calculate *planned load factors* (predicted for a flight leg), *actual load factors* (based on real data for a flight leg), and *break-even load factors* (the point at which the costs of a flight are balanced with the revenue). When the actual exceeds the break-even load factor, the airline makes a profit on the flight; when the actual is below the break-even load factor, the airline loses money on the flight.

Airlines must balance their capacity (ASK) with actual sales (RPK) to stay in business. Achieving the ideal balance between load factors and yield can be tricky. For example, a low-cost carrier that sold tickets for $1 would likely sell out. The flight in question would have a passenger load factor of 100 per cent; however, the yield would be very low as the airline collected very little revenue. On the other hand, if an airline sold out its first-class cabin at very high fares but flew with an empty economy cabin, the load factor would be low but the yield could be high. To attain profitability, airlines must find a strategic balance that considers both revenue and costs.

> **Did You Know?**
>
> Cargo airlines calculate similar metrics to passenger airlines. They use *revenue tonnes per kilometre* (RTK) for the tonnes of cargo sold per flight leg and *freight tonne kilometres* (FTK) as a measure of cargo traffic. *Weight load factors* measure the ratio of RTK to the available cargo capacity, and *freight yields* are calculated by dividing total revenue for a flight by the FTK.

Costs

There are three general categories of operational costs for airlines. *Flight operation costs* refer to the cost of operating an aircraft; *ground operation costs* include the cost of ground crew, maintenance, and airport facility fees; and *system operation costs* comprise costs for the variety of activities that support the success of the airline (sales, promotions, and administration). Note that flight attendants' salaries are considered system operation costs, while pilots' salaries are included in flight operation costs.

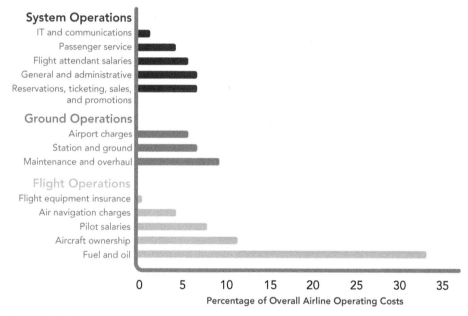

Figure 3.3 Airline costs

Source: Adapted from IATA, 2013

As shown in Figure 3.3, there are many costs that passenger fares must cover for an airline to be profitable. The fact that fuel and oil costs – an airline's greatest expense – are extremely variable contributes to the cyclical profit and loss periods.

A variety of other operational and industry-wide considerations impact aviation operations and performance, several of which are outlined in Figure 3.4 below.

Trends and Practices in Operations

Aviation growth in Asia

Populations in China and South Asia have grown faster than in other areas of the world, resulting in dynamic growth of aviation in these areas. China is expected to become the largest market for aviation by 2029 and India, the third largest (with the United States coming in second).

Consolidations, alliances, and code-sharing

Large profitable airlines tend to acquire failing competitors, leading to the expansion of large airlines, as well as increased multinational alliances (such as Star, **oneworld**, and SkyTeam) (Doganis, 2006, p. 9). Increasingly, airlines within an alliance will code-share flights. Code-sharing is an agreement between two airlines whereby they both market the same flight under their own airline names, as part of their published schedule.

Hub-and-spoke (H&S) models vs. point-to-point (P2P) models

Traditional hub-and-spoke models are based around an airline's designated hub airport (which serves as a home base for operations). Passengers flow from feeder cities to the hub airport, and then out to destination cities. Delays or bad weather at a hub can impact an airline's entire operation.

Low-cost carriers tend to use a point-to-point (P2P) model that connects cities directly, rather than transferring through hub airports. This results in fewer connections and less travel time for passengers; however, without a central hub, these airlines may not offer certain trips that passengers seek.

P2P
A → B
C → D
E → F

H&S
A   B
C ↔ Hub ↔ D
E   F

Liberalization

The increase in bilateral open skies agreements, discussed in Chapter 1, have led to the creation of new domestic and international airlines, with fewer route restrictions and more freedom to set pricing.

Low-cost carriers

LCCs represent a threat to traditional carriers. These airlines use a P2P model and innovative business practices to keep costs low and morale high. The LCC model began with the United States' Southwest Airlines and grew to companies around the world including Europe's Ryanair and easyJet, Malaysia's AirAsia, and Brazil's GOL, among others. Though not all LCCs are successful, their low fares allow them to capture market share from legacy carriers.

Figure 3.4 Trends and practices in operations

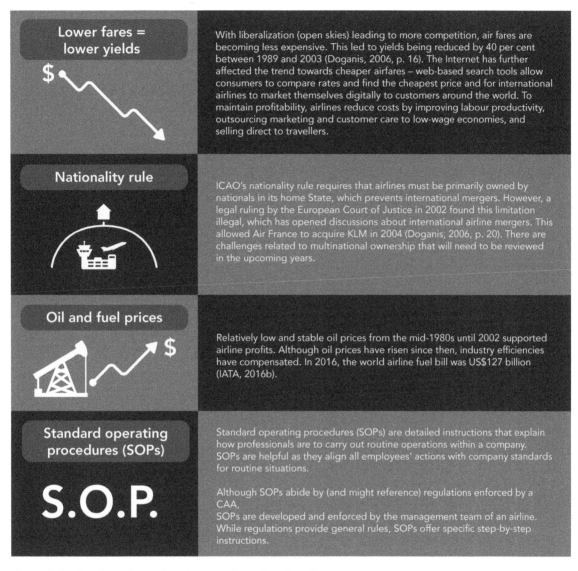

Lower fares = lower yields

With liberalization (open skies) leading to more competition, air fares are becoming less expensive. This led to yields being reduced by 40 per cent between 1989 and 2003 (Doganis, 2006, p. 16). The Internet has further affected the trend towards cheaper airfares – web-based search tools allow consumers to compare rates and find the cheapest price and for international airlines to market themselves digitally to customers around the world. To maintain profitability, airlines reduce costs by improving labour productivity, outsourcing marketing and customer care to low-wage economies, and selling direct to travellers.

Nationality rule

ICAO's nationality rule requires that airlines must be primarily owned by nationals in its home State, which prevents international mergers. However, a legal ruling by the European Court of Justice in 2002 found this limitation illegal, which has opened discussions about international airline mergers. This allowed Air France to acquire KLM in 2004 (Doganis, 2006, p. 20). There are challenges related to multinational ownership that will need to be reviewed in the upcoming years.

Oil and fuel prices

Relatively low and stable oil prices from the mid-1980s until 2002 supported airline profits. Although oil prices have risen since then, industry efficiencies have compensated. In 2016, the world airline fuel bill was US$127 billion (IATA, 2016b).

Standard operating procedures (SOPs)

Standard operating procedures (SOPs) are detailed instructions that explain how professionals are to carry out routine operations within a company. SOPs are helpful as they align all employees' actions with company standards for routine situations.

Although SOPs abide by (and might reference) regulations enforced by a CAA,
SOPs are developed and enforced by the management team of an airline. While regulations provide general rules, SOPs offer specific step-by-step instructions.

Figure 3.4 Trends and practices in operations (Continued)

OPERATIONS

Table 3.5 Airlines from ICAO Council States. Note that the fleet numbers provided are approximate and include mainline fleet only, not the fleet of subsidiaries.

State/Airline	Operations	Alliance	Fleet
Australia/ Qantas[1]	Founded in 1920, Qantas is an acronym for Queensland and Northern Territory Aerial Services (the airline's original name). The company operates subsidiary brands (Qantas Freight, QantasLink and Jetstar) allowing for regional, domestic, and international passenger and cargo operations.	**one**world Alliance	120 aircraft (Airbus A380, A330; Boeing 737, 747, 717; Bombardier Q400)
Brazil/LATAM[2]	LAN (Linea Aérea Nacional) began operations in Chile in 1929 while Brazil's TAM (Transportes Aéreos Regionais) began flying in 1976. In 2012, LAN and TAM formed a joint operation known as LATAM Airlines Group (which includes LATAM Cargo).	**one**world Alliance	160 aircraft (Airbus A319, A320, A321, A350; Boeing 767, 777[3])
Canada/ Air Canada[4]	Air Canada's predecessor (Trans-Canada Air Lines) began operations in 1937, becoming Canada's national airline, and was renamed Air Canada in 1964. Divisions include Air Canada (mainline), Air Canada Express, Air Canada Rouge, and Air Canada Cargo.	Star Alliance	170 aircraft (Boeing 777, 787, 767; Airbus A330, A321, A320, A319; Embraer E190)
China/China Southern Airlines[5]	China Southern Airlines was founded in 1988 after the restructuring of the Civil Aviation Administration of China. It acquired and merged with other airlines to become the carrier with the largest fleet in The People's Republic of China. China Southern Airlines is headquartered in Guangzhou. The airline's parent company is China Southern Air Holding Company, a state-owned enterprise. China Southern Cargo is the airline's cargo subsidiary.	SkyTeam Alliance	700 aircraft (Airbus A380, A330, A321, A320, A319; Boeing 787, 777, 747, 757, 737)

State/Airline	Operations	Alliance	Fleet
France/Air France[6]	In 2004, Air France and the Netherlands-based KLM merged (but both brands were preserved) and are known as the Air France–KLM group. Air France was created in 1933 when five French airlines merged (Air Union, Air Orient, Société Générale de Transport Aérien, CIDNA and Aéropostale). KLM began operations in the Netherlands in 1919 under the name Koninklijke Luchtvaartmaatschappij. The Air France–KLM group is one of the largest airlines in the world. It has eight subsidiary brands including regional airlines (Hop!, Martinair, and Transavia), maintenance, and cargo forwarding.	SkyTeam Alliance	225 aircraft* (Airbus A380, A340, A330, A321, A320, A319, A318; Boeing 777, 747)
Germany/Lufthansa[7]	The first Lufthansa was founded in 1926, but was dissolved after World War II by the Allies. In 1953, the Aktiengesellschaft für Luftverkehrsbedarf (Luftag) airline was founded and bought the name and trademark of the original Lufthansa. The Lufthansa Group's passenger airlines include Germanwings, Eurowings, SWISS, and Austrian Airlines with equity interests in Brussels Airlines and SunExpress.	Star Alliance	275 aircraft (Airbus A319, A320, A321, A330, A340, A380; Boeing 737, 747)
Italy/Alitalia[8]	Alitalia–Aereolinee Italiane Internazionali began operations in 1947. The modern Alitalia is a blend of the Italian words ali (wings) and Italia (Italy). Facing bankruptcy in 2008, the airline was acquired from Compagnia Aerea Italiana (CAI). In 2014, the United Arab Emirates' Etihad Airways acquired 49 per cent ownership in Alitalia from CAI. Alitalia's subsidiaries include Alitalia Cargo and the regional airline Alitalia CityLiner.	SkyTeam Alliance	100 aircraft (Airbus A319, A320, A321, A330; Boeing 777)

(Continued)

Table 3.5 (Continued)

State/Airline	Operations	Alliance	Fleet
Japan/All Nippon Airways[9]	In 1952, Japan Helicopter & Aeroplane Transports Co. was established to restore air transportation services disrupted by World War II. In 1957, the company changed its name to All Nippon Airways (ANA), which has grown to be the largest airline in Japan, surpassing its competitor Japan Airlines. Restructuring under ANA Holdings, Inc., the group has a number of subsidiaries including the low-cost carrier Vanilla Air, ANA Wings, Air Japan, airport ground support, maintenance, training and ANA Cargo.	Star Alliance	210 aircraft (Airbus A320; Boeing 737, 767, 777, 787)
Russian Federation/Aeroflot[10]	Aeroflot, which means 'air fleet' in English, originated in 1923. Following the dissolution of the Soviet Union, the airline went through several changes. Much of its Soviet-made fleet was replaced with Boeing, Airbus, and modern Russian aircraft. Although it now has some private ownership, the airline is 51 per cent owned by the Russian government. Aeroflot has several subsidiary airlines including Pobeda Airlines, Rossiya Airlines, Aurora Airlines, and Sherotel.	SkyTeam Alliance	190 aircraft (Airbus A320, A321, A330; Boeing 737, 777; Sukhoi Superjet 100)
United Kingdom/British Airways[11]	In 1919, Aircraft Transport and Travel Limited, the predecessor to British Airways (BA), began the world's first daily international air service. BA was famous for its supersonic Concorde flights, which could travel from London to New York in less than 3.5 hours (instead of the typical 8 hours). However, the Concorde was retired in 2003. BA merged with Iberia (an air carrier from Spain) in 2011 to form the International Airlines Group (IAG). BA has several subsidiaries including BA CityFlyer, OpenSkies, and British Airways World Cargo.	**one**world Alliance	270 aircraft (Airbus A318, A319, A320, A321, A380; Boeing 747, 767, 777, 787)

State/Airline	Operations	Alliance	Fleet
United States/ American Airlines[12]	In the late 1920s and early 1930s, more than 80 small airlines were acquired and merged to form American Airways. The company was renamed American Airlines in 1934. American Airlines' parent company (AMR Corporation) filed for bankruptcy protection in 2011. This led to a major restructuring, which included a merger with US Airways in 2013. The merger led to the creation of a new holding company called American Airlines Group, Inc., and resulted in the airline becoming the largest in the world. American Airlines' regional partner is American Eagle.	**one**world Alliance	930 aircraft (Airbus A319, A320, A321, A330; Boeing 737, 757, 767, 777 787; Embraer ERJ-190; McDonnell Douglas MD-80)

*Note that this number includes mainline and cargo Air France fleets, but not regional or KLM fleets.

1 Qantas, n.d.
2 LATAM, 2016
3 Planespotters.net, 2016
4 Air Canada, 2016
5 China Southern Airlines, 2016
6 Air France, 2015

7 Lufthansa Group, n.d.
8 Alitalia, n.d.
9 ANA Group, n.d.
10 Aeroflot, n.d.
11 British Airways, n.d.
12 American Airlines, n.d.

3.6   International Air Transport Association

The International Air Transport Association (IATA) is a trade association that represents airlines around the world to support a safe, secure, and profitable air transport industry. Founded in Havana, Cuba, in April 1945, IATA now has 54 offices in 53 countries with its head office located in Montreal, Canada.

IATA activities include aviation lobbying, identifying key industry priorities, reducing costs, launching communication campaigns, and distributing training and services. IATA's membership includes 265 airlines from 117 countries, representing 83 per cent of global air traffic.[1]

Note

1 IATA, 2016c

Dangerous Goods

More than half of all cargo moved through the global transportation network is dangerous – toxic, flammable, corrosive, explosive, or even radioactive! These goods serve important needs in society: consider the explosive charges used in mining operations, compressed gases or infectious substances used in medical research, or batteries that power mobile devices and computers. Air transportation is often chosen to transport these goods – rather than road, rail, or sea – as it is typically the most expeditious mode of transport. The safe transport of dangerous goods is a crucial consideration in aviation cargo operations.

Did You Know?

In 2016, there were several instances of mobile phones catching fire on board aircraft due to faulty batteries. In response, the FAA formally advised passengers to keep their devices turned off and not to stow them in checked baggage. Several Australian airlines banned these mobile devices from their aircraft, while others deployed flame-containment bags throughout their fleets to contain defective devices.

To ensure the safety of civil aviation, materials capable of posing a risk to safety, health, property, or the environment are classified as *dangerous goods* and subject to mandatory safe handling requirements.

Air travel (as compared to other modes of transportation) complicates the transport of dangerous goods as they are subject to changes in atmospheric pressure between ground and flight altitudes, vibrations, turbulence, and other conditions naturally associated with air travel. These environmental stresses can lead to reactions within dangerous materials. To reduce the risk to civil aviation, ICAO publishes SARPs for the transportation of dangerous goods in Annex 18,[16] along with detailed technical instructions in a separate document.

Annex 18 contains a list of goods that are always prohibited from air travel, as well as a short list of dangerous goods that may not be transported by air without an exemption (such as infected live animals). Annex 18 also outlines how these materials must be packed, labelled, documented, and periodically inspected. The pilot-in-command of an aircraft transporting dangerous goods must be informed, in writing, of any dangerous materials on board.[17]

For transportation purposes, dangerous goods are categorized into nine hazard classes by the United Nations Committee of Experts on the Transport of Dangerous Goods. Each of the nine classes, listed below, has different rules for transport:

1. explosives
2. gases
3. flammable liquids
4. flammable solids (including those that can spontaneously combust and those which emit flammable gases when in contact with water)
5. oxidizing materials and organic peroxides
6. infectious substances
7. radioactive materials
8. corrosive substances
9. miscellaneous dangerous goods including those that are environmentally hazardous

Case Study: Asiana Airlines Flight 991 – A Dangerous Goods Accident[1]

On 28 July 2011, a Boeing 747–400F aircraft operated by Asiana Airlines left Incheon, Republic of Korea, for Shanghai, China. As this was a scheduled cargo flight, the two pilots were the only souls on board the aircraft. The Captain and FO were both properly trained and licensed pilots.

The aircraft had been loaded with 58 265.8 kilograms (128 454 pounds) of cargo, loaded in 30 pallets and five containers. The cargo was distributed throughout the aircraft among 11 cargo positions on the lower deck and 24 on the main deck. Two of the main deck pallets contained dangerous goods including flammable and corrosive liquids (paints and resins) and lithium-ion batteries.

About an hour into the flight, the pilots reported smoke in the cockpit to air traffic control, declared an emergency, and requested a descent. Controllers cleared the aircraft to descend and the FO requested a diversion to Jeju Airport, stating that there was a fire on board the aircraft. Controllers approved the diversion. Several minutes later, the Captain reported 'Rudder control . . . flight control, all are not working . . .' and the FO transmitted, 'We have heavy vibration on the airplane, may need to make an emergency landing, emergency ditching . . .' Subsequent attempts to contact the pilots on the radio were unsuccessful. Eighteen minutes after the pilots' initial report of smoke in the cockpit, the aircraft crashed into the sea. Both pilots were fatally injured and the cargo shipments on board the aircraft were destroyed.

The aircraft wreckage was distributed within an area 3 kilometres by 4 kilometres (1.86 by 2.49 miles) at a depth of about 85 metres (279 feet) underwater. The location was about 130 kilometres (81 miles) west of Jeju Airport. No signal from the underwater locator beacon was detected. Search teams struggled with terrible weather (including seven typhoons) as they combed the area for weeks after the accident.[2] They undertook a complicated effort that used salvage ships, remotely operated vehicles, submarine rescue ships, trawling boats, and divers. These efforts recovered about 40 per cent of the aircraft and 15 per cent of the cargo, but failed to recover either the flight data recorder or the cockpit voice recorder.

Investigation revealed that the aircraft's maintenance history included no faults or corrective actions related to this accident and that the centre of gravity was in accordance with the flight manual (i.e., cargo was properly loaded and distributed within the aircraft). It was concluded that the aircraft was destroyed by in-flight fire and subsequent impact forces resulting from crashing into the sea, yet the source of the fire was still unknown.

After examination, investigators and Boeing experts determined that a fire developed on or near the pallets containing dangerous goods, as these areas had the most severe fire damage. The fire quickly escalated, producing so much energy that some dangerous goods were found on the top surface of the right wing, 30 meters (98 feet) away from where they were stored. The on-board fire caused some pieces of the aircraft fuselage to bend outwards while others separated from the aircraft mid-air. This accident illustrated how the safety risk posed by dangerous goods must be carefully managed by industry professionals.

According to the investigators, there was no action the pilots could have taken to prevent the crash and Asiana Airlines had accepted, stored, and loaded the dangerous goods in accordance with regulations and procedures.

Subsequent safety recommendations issued following this accident included ensuring that dangerous goods (flammable liquids and lithium-ion batteries) are segregated and loaded in separate unit load devices (e.g., pallets), and that they are equipped with a fire extinguishing system. Lithium-ion batteries were identified as a source of risk as they can act as an ignition source or fuel an existing fire, and may be subject to overheating while in transit.[3]

Notes

1 ARAIB, 2015
2 Kaminski-Morrow, 2012

3 NTSB, 2016, p. 2

Conclusion

Civil aviation is an industry with a global economic impact of US$2.7 trillion.[18] Civil aviation operations vary widely, from a small aircraft owner conducting a sightseeing flight to a major airline operating a fleet of hundreds of large aircraft and employing thousands of pilots. This sector supports the livelihoods of millions of dedicated and hard-working professionals who collectively accomplish more than 100 000 flights every day.

On a global scale, the number of flights per day is predicted to double to about 200 000 by the year 2030.[19] To successfully manage this tremendous growth, aviation professionals will be required to continually innovate in their practices as well as invest in the training and retention of the next generation of aviation professionals.

Key Points to Remember

1. Civil aviation includes both the aviation sector (airlines, general aviation, air navigation service providers, and aviation cargo) and the civil aerospace sector (aircraft, systems, and engine manufacture and maintenance).

2. International regulations outline the training and licensing requirements of aviation professionals (Annex 1) and the issuance of air operator certificates (Annex 6).

3. Pilots, air traffic controllers, maintenance engineers, and dispatchers all require licences. Pilot licences include private pilot licences, commercial pilot licences, multi-crew pilot licences, and airline transport pilot licences. Ratings can be added to licences to grant additional privileges.

4. Licensing requires that several criteria be met including *prerequisites* (age, experience, medical fitness), *training* (operational, classroom, and simulator training), *demonstration of competency* (passing an exam or test), and *currency* (licence holders must exercise the privileges of their licence regularly). English language proficiency is also required for international operations. Medical assessments must be completed on a regular basis for a licence to remain valid.

5. Aviation operations include both *general aviation* and *airlines*. (Military aviation is also included but is beyond the scope of this text.) GA refers to all professional and private aviation activities that are not part of the airline or military sectors, including flight instruction, business aviation, most helicopter operations, and pleasure flying. There are more planes and pilots involved in GA than with airlines, and GA makes a variety of important contributions to aviation and to society, including the creation of many job opportunities.

6. Airlines are air transport companies in the business of moving people and cargo between two points. Major airlines are international operations with fleets of large jets. Regional airlines operate smaller jets (usually fewer than 120 seats) over a smaller network of routes.

7. Airline flights are usually operated by two pilots: a Captain and a first officer. Pilots usually require six to eight years' experience flying before they are eligible to work at a major airline. However, new cadet training programmes allow pilots to fly for major airlines in less than two years.

8. Flight attendants (or *cabin crew*) provide customer service to airline passengers, and in the rare case of an emergency, assist passengers with safety issues and evacuations.

9. Airline dispatchers help pilots identify the safest and most expeditious route for their trip. Dispatchers are ground-based personnel employed by airlines.

10. Airline management has operational oversight and makes strategic decisions to ensure profitability. Management considers key performance indicators including *passenger load factor*, which is the percentage of seats sold on a flight leg and *yield*, which refers to the profitability of a leg. For an airline to turn a profit, revenue must offset the costs of operations, which include flight, ground, and system costs. The single greatest cost to an airline is fuel and oil, making up over 30 per cent of total costs.

11. There are several current trends impacting airline operations, including the liberalization of international routes, the growth of aviation in Asia, airline consolidations and alliances, and the emergence of low-cost carriers (which generally use point-to-point models rather than hub-and-spoke models).

12. Dangerous goods are those that have the potential to pose a flight safety risk. In Annex 18, ICAO specifies SARPs for the transport of dangerous goods by air.

Table 3.6 Acronym rundown

AMMTE	aircraft maintenance mechanics, technicians, and engineers
AOC	air operator certificate
AOPA	aircraft owners' and pilots' associations
ASK	available seat kilometres
ATC	air traffic control
ATPL	airline transport pilot licence
CAA	civil aviation authority
CPL	commercial pilot licence
CVR	cockpit voice recorder
DME	designated medical examiner
EASA	European Aviation Safety Agency
FAA	Federal Aviation Administration
FBO	fixed-base operator
FO	first officer
FTK	freight tonne kilometre
GA	general aviation
H&S	hub-and-spoke
IAOPA	International Aircraft Owners and Pilots Association

IATA	International Air Transport Association
ICAO	International Civil Aviation Organization
IFALPA	International Federation of Air Line Pilots' Associations
ITF	International Transport Workers' Federation
LCC	low-cost carrier
MPL	multi-crew pilot licence
NBAA	National Business Aviation Association
NOTAM	notice to airmen
NTSB	National Transportation Safety Board
P2P	point-to-point
PIC	pilot-in-command
PPL	private pilot licence
RAeS	Royal Aeronautical Society
RPK	revenue passenger kilometres
RTK	revenue tonnes per kilometre
SARPs	standards and recommended practices
SOP	standard operating procedure
WAI	Women in Aviation International

Chapter Review Questions

3.1 What is the International Air Transport Association (IATA)? How did it come to exist? What is its role in today's aviation landscape?

3.2 What is an air operator certificate (AOC)? What must be evaluated by a civil aviation authority (CAA) before one is issued? In your own words, describe how an AOC is associated with international aviation safety.

3.3 Why do you think personnel licensing was such an important issue that it became the first annex to the Chicago Convention? Explain.

3.4 Which aviation sector do you think is most important in the world today (general aviation, airlines, or military)? Justify your response.

3.5 Do you think that pilots who are deemed fit to fly by a medical screening, should be allowed to fly passengers while taking prescription medication? Why or why not?

3.6 What type of flying do you think would be most desirable for a pilot? Explain your choice.

- a flight instructor teaching student pilots at a small flight school

- a corporate pilot for a large organization with its own fleet of private aircraft

- an airline pilot for a traditional carrier

- an airline pilot for a low-cost carrier.

3.7 In airline operations, do you believe that more pilots on the flight deck result in improved safety? Why has the industry shifted from crews of four or five pilots to only two pilots (Captain and First Officer)? Is it possible that technology may evolve such that only a single pilot is required for airline operations? Explain your thoughts.

3.8 Do you agree with the concept of seniority within airlines? Is this a fair system? How might your answer, as a student, differ from the answer of an airline pilot with 20 years' experience with a company? Explain.

3.9 Should pilots be concerned with the financial performance of the airline industry as a whole (Figure 3.2)? Should they be concerned with the financial performance of the airline they work for? Why or why not?

3.10 Looking at the variety of costs faced by airlines (Figure 3.3), which are the easiest to control? Which are the most difficult? Why?

3.11 Name two factors that are outside the control of airlines but have the potential to impact airline profits in the coming year. Explain.

3.12 Would you support airlines in your State recruiting pilots through a traditional model, a cadet programme, or a hybrid model? Can you identify operators that use each of these approaches? What are the strengths and weaknesses of each?

CASE STUDY: COLGAN AIR FLIGHT 3407 – THE 1500-HOUR RULE[1]

On February 12 2009, a Bombardier DHC-8–400 aircraft was being operated by Colgan Air (a regional airline) as Continental Connection flight 3407. The flight boarded in Newark, New Jersey around 7:30 p.m., but was delayed on the ground for about two hours before departing for Buffalo Niagara International Airport in New York. The flight carried two pilots, two flight attendants, and 45 passengers.

Pilot Background

The Captain was 47 years old, held an airline transport pilot licence (ATPL) and a first-class medical, and had accumulated 3379 hours of flight experience. He earned a salary of US$55 000 per year. His training record showed a history of failed check rides. He lived in Florida and commuted to Newark for work – often requiring him to stay overnight with a friend or nap in the crew room before duty. Before the flight on February 12, the Captain had just completed a two-day trip and had spent the night sleeping in the crew room at Newark.

The First Officer (FO) was 24 years old, held a commercial pilot licence (CPL) and a first-class medical, and had accumulated 2244 hours of flight experience. She was earning US$16 000 per year. She lived in Seattle, Washington and commuted across the United States to Newark. Before the flight on 12 February, she commuted as a passenger on a flight from Seattle that had departed at about 8:00 p.m. Pacific Standard Time. Another passenger reported that she slept for about 90 minutes during the flight. After catching another flight, and reporting two more hours of sleep, she arrived at Newark the day of the Colgan Air flight at about 6:00 a.m. Before her duty time, she told another pilot that a couch in the crew room 'had her name on it' and later reported a six-hour nap to a friend via text message.

Both pilots began Colgan Air flight 3407 tired. In addition, the FO was not feeling well, but reassured the Captain 'I'm pretty tough' and continued the flight. Her sniffles were recorded throughout the flight on the CVR.

Flight, Accident, and NTSB Investigation

Although sterile cockpit procedures prohibit conversations about anything not pertinent to the flight below 10 000 feet, the pilots engaged in casual conversation throughout the flight – including throughout the final minutes of the descent – which distracted from their flight duties.

On approach to land at Buffalo, the aircraft speed reduced as the crew set up the aircraft for landing. The pilots should have recognized the reduction in airspeed, but there was a breakdown in their monitoring and workload management. A few seconds later, the stick-shaker activated, providing a warning to pilots that the aircraft was approaching stall speed, and automatically disconnecting the autopilot. Neither pilot called 'stall', which was standard operating procedure (SOP) to initiate a response to stall conditions.

Instead, the Captain did exactly the opposite of what he should have done – he raised the nose abruptly and increased thrust power. (*Note: the proper response to a stall warning is to lower the nose of the aircraft and then apply power*). The airspeed slowed further, resulting in a stall and a left-wing-down roll. The Captain continued to make inappropriate control inputs – the aircraft automatically activated a 'stick-pusher' three times in an attempt to automatically lower the nose and recover from the stall condition but the Captain fought against the input by aggressively pulling back on the controls. The FO's actions suggested a lack of understanding about the situation – while the Captain was fighting with the stick pusher she retracted the flaps without being told to do so. In general, rather than an automatic response to the incident based on training and experience, the flight crew responded with 'startle and confusion' – in the words of the National Transportation Safety Board (NTSB) – which exacerbated the problem.

The aircraft stalled and crashed into a home in Clarence Center, New York, about nine kilometres (five nautical miles) northeast of the airport. All people on board the aircraft and one person on the ground were killed.

The NTSB determined that the probable cause of the accident was the Captain's inappropriate response to the stick-shaker, leading to an aerodynamic stall. Several contributing factors were identified, including the pilots' failure to monitor the airspeed and manage the flight and to follow sterile cockpit procedures. However, the accident shed light on pilot practices that were not previously understood by the flying public – specifically the low wages of regional airline pilots and the negative impacts of fatigue.

CASE STUDY

Following the accident, families of victims lobbied the United States Congress to enforce new, stricter regulations on regional airlines, in an effort to improve safety and pilot working conditions. This led to the Airline Safety and Federal Aviation Administration Extension Act of 2010. A key component of this Act was the 1500-hour rule, which would ensure that pilots have an ATPL (which requires a minimum of 1500 hours of flight time) before they can serve as an FO. Based on ICAO regulations, the previous requirement for an FO was a CPL and an aircraft type rating (which requires a minimum of about 250 hours).

However, there were dramatic and unanticipated challenges associated with this regulation. Industry spokespeople called this an example of smoke-and-flames rule-making (i.e., that it was a quick but not well-considered regulatory response to an accident).[2] Although the Act was intended to solve a problem, the fact is that both pilots in the Colgan Air accident had more than 1500 hours' flight time so the Act would not have prevented that accident.

The impact of this Act on the American aviation industry has been far-reaching. The regional airline industry traditionally paid pilots only slightly more than minimum wage. The challenge is that regional airlines must balance crew pay with revenue from passengers' fares – and they are often in heated competition to keep ticket prices low.[3] As pilot salaries increase, it can be difficult for airlines to keep ticket costs low enough to be competitive.

The 1500-hour rule is striking because airlines in other parts of the world are taking an entirely opposite approach. ICAO has moved towards establishing competency frameworks for aviation professions. These frameworks focus on the knowledge, skill, and attitude required for pilot competence, rather than hours. The multi-crew pilot licence (MPL), which allows airlines to recruit people with no previous aviation experience and train them in 18 months to begin acting as an FO, is based on competency rather than hours.

Notes

1 NTSB, 2010
2 Collins, 2014, p. para. 1

3 Collins, 2014, p. para. 2

Case Study Questions

Making reference to this case study, and applying what you have learned in this chapter, provide informed responses to the following:

3.13 As most hours-building time takes place in small GA aircraft, does the 1500-hour rule prepare pilots to be better airline pilots? Is it possible that bad habits could be learned during that time?

3.14 ICAO annexes suggest principles for managing the risks of fatigue (such as limiting duty time and mandating rest periods along with data-driven approaches that identify and eliminate fatigue risk areas). The pilots of Colgan Air flight 3407 started their trip tired because of previous trips and commuting schedules. Is it possible to ensure that crew members are using their rest time wisely, not staying out late socializing or commuting? Should there be rules limiting commuting for crew members? Should pilots be penalized for cancelling a trip because of fatigue resulting from a bad night's sleep? What steps could or should an airline take to manage the risks of fatigue?

3.15 The sterile cockpit rule prohibits non-critical conversations when the aircraft is below 10 000 feet. The Colgan Air flight 3407 pilots did not follow this procedure and chatted casually throughout the landing. Investigation revealed that the FO sent a text message while the aircraft was on the ground awaiting take-off clearance. If you were an FO, and a Captain struck up a social conversation during landing, how would you respond? How can airlines monitor and enforce sterile cockpit procedures? Which is more important – following procedures or following the lead of the Captain?

3.16 In the United States, the Airline Safety and Federal Aviation Administration Extension Act of 2010 (called the '1500 hour rule') required FOs to have 1500 hours of flight time before flying for an airline. As the world faces a pilot shortage, what are the broader implications of this Act? Do more hours necessarily mean better skills? How can regional airlines attract experienced pilots while maintaining competitive ticket prices?

3.17 As other areas of the world are turning towards competency-based training methods (rather than the traditional hours-based approach to training), such as the MPL used in some pilot cadet programmes, does the United States' 1500-hour requirement put its regional airline industry at a competitive disadvantage? Can you think of any solutions, other than the 1500-hour rule, that might better address the risks that led to this accident?

References

Aeroflot, n.d. *Subsidiaries*. [Online] Available at: www.aeroflot.com/ru-en/about/subsidiaries

Air Canada, 2016. *Corporate profile*. [Online] Available at: www.aircanada.com/en/about/acfamily/index.html

Air France, 2015. *The company*. [Online] Available at: http://corporate.airfrance.com/en/the-company/key-figures/fleet/

Alitalia, n.d. *History*. [Online] Available at: http://corporate.alitalia.it/en/history/index.html

American Airlines, n.d. *History of American Airlines*. [Online] Available at: www.aa.com/i18n/customer-service/about-us/history-of-american-airlines.jsp

ANA Group, n.d. *ANA's history*. [Online] Available at: www.ana.co.jp/eng/aboutana/corporate/history/main2.html

ARAIB, 2015. *Crash into the sea after an in-flight fire, Asiana Airlines, Boeing 747-400F, HL7604, international waters 130 km west of Jeju International Airport, 28 July 2011*. Sejong Special Self-governing City, Republic of Korea: Aviation and Railway Accident Investigation Board.

ATAG, 2005. *The economic and social benefits of air transport*. Geneva: Air Transport Action Group.

ATAG, 2016. *Aviation benefits beyond borders*. Geneva, Switzerland: Air Transport Action Group.

Barry, K., 2007. *Working the skies: The fast-paced, disorienting world of the flight attendant.* New York: NYU Press.

British Airways, n.d. *Stronger together – British Airways and Iberia.* [Online] Available at: www.british airways.com/en-ca/information/about-ba/iag

China Southern Airlines, 2016. *About China Southern Airlines: Company profile.* [Online] Available at: http://global.csair.com/US/GB/GYNH/COMP

Collins, R., 2014. *A double tragedy: Colgan Air Flight 3407.* [Online] Available at: http://airfactsjournal. com/2014/03/double-tragedy-colgan-air-flight-3407/

Cox, J. R. C., 2010. *What your own business jet really costs – The formula explained.* [Online] Available at: www.forbes.com/sites/wheelsup/2010/06/21/what-your-own-business-jet-really-costs-the-for mula-explained/#469e3b9e3765

Doganis, R., 2006. *The airline business.* 2nd ed. London: Routledge: Taylor & Francis Group.

IAOPA, n.d. *What Is General Aviation.* [Online] Available at: www.iaopa.eu/what-is-general-aviation

IATA, 2013. *Airline cost management group (ACMG) report.* Montreal: International Air Transport Association.

IATA, 2016a. *Economic performance of the airline industry.* Montreal: International Air Transport Association.

IATA, 2016b. *Fact sheet – Fuel.* Montreal: International Air Transport Association.

IATA, 2016c. *Fact sheet –IATA.* Montreal: International Air Transport Association.

ICAO, 2006. Annual review of civil aviation 2005. *ICAO Journal,* 61(5).

ICAO, 2010. *Annex 6 to the Convention on International Civil Aviation: Operation of aircraft, 9th ed.* Montreal: International Civil Aviation Organization.

ICAO, 2011a. *Annex 1 to the Convention on International Civil Aviation: Personnel licensing, 11th ed.* Montreal: International Civil Aviation Organization.

ICAO, 2011b. *Annex 18 to the Convention on International Civil Aviation: The safe transport of dangerous goods by air, 4th ed.* Montreal: International Civil Aviation Organization.

ICAO, 2012. *Manual of civil aviation medicine, Doc 8984, 3rd ed.* Montreal: International Civil Aviation Organization.

IFALPA, 2016. *Mission statement.* [Online] Available at: www.ifalpa.org/about-us/mission-statement. html

Kaminski-Morrow, D., 2012. *Fire brought down Asiana 747F in just 18 min.* [Online] Available at: www.flightglobal.com/news/articles/fire-brought-down-asiana-747f-in-just-18min-376916/

Kearns, S. K., Mavin, T. J. & Hodge, S., 2016. *Competency-based education in aviation: Exploring alternate training pathways.* Burlington, VT: Ashgate.

LATAM, 2016. *About us.* [Online] Available at: www.latam.com/en_ca/about-us/

Lufthansa Group, n.d. *As time flies by.* [Online] Available at: www.lufthansagroup.com/en/company/ history.html

NBAA, 2015. *Business aviation fact book.* Washington, DC: National Business Aviation Association.

NTSB, 1975. *Aircraft accident report: Eastern Air Lines, Inc. Douglas DC-9-31, N8984E, Charlotte, North Carolina, September 11, 1974 (NTSB/AAR-75-9).* Washington, DC: National Transportation Safety Board.

NTSB, 2010. *Loss of control on approach, Colgan Air, Inc., operating as Continental Connection flight 3407, Bombardier DHC-8-400, N200WQ, Clarence Center, New York, February 12, 2009. (NTSB/ AAR-10/01).* Washington, DC: National Transportation Safety Board.

NTSB, 2016. *Safety recommendation (A-16-001 and -002).* Washington, DC: National Transportation Safety Board.

Planespotters.net, 2016. *LATAM Airlines Brasil fleet details and history.* [Online] Available at: www. planespotters.net/airline/LATAM-Airlines-Brasil

Qantas, n.d. *Our company.* [Online] Available at: www.qantas.com/travel/airlines/fleet-developments/global/en

Notes

1 ATAG, 2016, p. 4
2 ATAG, 2005, p. 4
3 ICAO, 2011a
4 ICAO, 2011a
5 ICAO, 2012
6 ICAO, 2010
7 IAOPA, n.d.
8 NBAA, 2015, p. 25
9 NBAA, 2015, p. 2
10 Cox, 2010, p. para 8
11 ATAG, 2016, p. 5
12 IFALPA, 2016
13 Barry, 2007
14 Kearns, et al., 2016
15 Kearns, et al., 2016
16 ICAO, 2011b
17 ICAO, 2011b
18 ATAG, 2016, p. 4
19 ATAG, 2016, p. 5

1 All air traffic control officers (ATCOs) rely heavily on their visual reference of the outside world, frequently using binoculars to spot traffic.
   a. True
   b. False

2 ATCO's communicate with aircraft exclusively by speaking over the radio.
   a. True
   b. False

5 Within which category of flight rules would a pilot require support from an ATCO?
   a. Visual Flight Rules (VFR)
   b. Instrument Flight Rules (IFR)

3 Flight Information Regions (FIRs), which are large sections of airspace, are established by the Civil Aviation Authorities (CAAs) within each State.
   a. True
   b. False

4 Within FIRs, the world's airspace is further broken down into categories, with ICAO designating seven classes of airspace (A through G).
   a. True
   b. False

Learning science suggests that thinking through a few questions before you begin studying new material, even if you answer incorrectly, results in improved learning and retention.
Give it a try!

CHAPTER 4

Navigation

CHAPTER OUTCOMES

At the end of this chapter, you will be able to . . .

- Explain why air traffic management is essential in the aviation industry.

- Differentiate between the various roles in which air traffic control officers work and the ways in which they ensure safe separation of aircraft.

- Discuss three key considerations in air navigation – communication, surveillance, and navigation, including the technologies, standards, and expectations involved in accomplishing each.

- Outline some of the ways in which the aviation industry continues to innovate with respect to the future of air navigation, making specific reference to NextGen and SESAR.

- Use your understanding of air navigation to discuss a case study on the disappearance of Malaysia Airlines flight MH 370.

Introduction

The safe navigation of an aircraft from a departure airport, across a massive stretch of ocean or land to its final destination is not accomplished by pilots in isolation. Global air traffic management (ATM) requires an international network of ground- and satellite-based navigation aids, regulations dictating the rules of the sky, and human expertise. This chapter will introduce you to the organizations, people, technologies, systems, regulations, and procedures that support global ATM.

After a dramatic increase in air traffic following World War I, the need for air traffic control became apparent. In 1920, the Aeronautics Branch in the United States issued the first form of regulations for air traffic control called Uniform Field Rules.[1] Around the same time, controllers at Croydon Airport began supporting flights in the United Kingdom, communicating with pilots using red and green take-off lights or flags, and acknowledging radio position reports.[2]

The need for air traffic control was emphasized after two passenger-carrying biplanes collided in mid-air on April 7 1922. This mid-air collision over Picardy, France[3] was a signal to the aviation industry that as more aircraft took to the skies, it would become increasingly difficult for pilots to safely separate their aircraft from obstructions and from other traffic. The burgeoning role of the air traffic control officer (ATCO) became more critical and more widely accepted.

In 1946, after an inspection of air traffic control practices in the United States, the predecessor to ICAO aligned global standards with the American rules. With this action, the foundation was set for a globally unified air navigation system.

Air Traffic Control Officers

Second only to flight crew on an aircraft, air traffic controllers are probably the most well-known aviation professionals. It is generally understood that controllers are responsible for the separation of aircraft (i.e., maintaining a safe distance between aircraft). The details of this responsibility, however, may not be understood by the general public. The job of an ATCO is embedded within a complex system, and requires quick decision-making, situational awareness, constant interaction with technology, and high attention to detail. This chapter aims to offer insight into the exciting and challenging environment of ATCO professionals.

Did You Know?

In 1923, as voice communication was becoming more popular than Morse code, Frederick Stanley Mockford, a senior radio officer at Croydon Airport in the UK, was asked to come up with a verbal equivalent to the Morse emergency code, SOS. It was important that the word not be used in common language, so *help* was not appropriate. As most of the traffic he managed was between the UK and France, he came up with the term *Mayday* by adapting the French term *m'aider*, which means 'help me'. In 1927, *Mayday* became the official voice distress call used to communicate life-threatening emergencies – it is still in use today.[1]

Note

1 Boulton, 2013

How Do They Do Their Jobs?

The primary responsibility of air traffic controllers is to maintain safe separation between aircraft and

- other aircraft;

- ground-based obstructions (e.g., buildings, towers, hills, and mountains); and

- airspace boundaries (invisible three-dimensional sections of the sky where different rules and restrictions apply).

Controllers separate aircraft in three ways, as detailed in Figure 4.1.

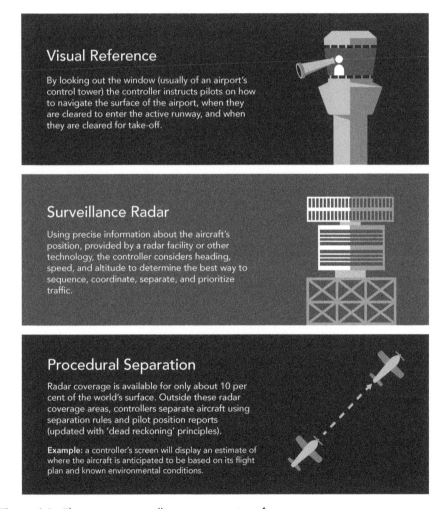

Visual Reference

By looking out the window (usually of an airport's control tower) the controller instructs pilots on how to navigate the surface of the airport, when they are cleared to enter the active runway, and when they are cleared for take-off.

Surveillance Radar

Using precise information about the aircraft's position, provided by a radar facility or other technology, the controller considers heading, speed, and altitude to determine the best way to sequence, coordinate, separate, and prioritize traffic.

Procedural Separation

Radar coverage is available for only about 10 per cent of the world's surface. Outside these radar coverage areas, controllers separate aircraft using separation rules and pilot position reports (updated with 'dead reckoning' principles).

Example: a controller's screen will display an estimate of where the aircraft is anticipated to be based on its flight plan and known environmental conditions.

Figure 4.1 Three ways controllers separate aircraft

Where Do They Work?

The three methods outlined above are applied by controllers working in four distinct roles: 1) aerodrome control, 2) approach control, 3) area control, and 4) oceanic control.

Unlike a pilot's workflow, which can be considered linear (i.e., taxi, take-off, cruise, approach, landing, and taxi), an ATCO's workflow often involves simultaneous management of several aircraft at a variety of flight stages (e.g., an approach controller works with arrivals, departures, and overflying aircraft all at the same time). The process of managing multiple flights has been described as similar to 'playing a game of ping-pong with 10 people at once'.

Trying to understand the complex world of an ATCO can be overwhelming to people who are new to aviation. As most people are familiar with the linear phases of a flight, it can be helpful to think about the ATCO's work in terms of how each one interacts with a pilot during the different phases of a flight: flight planning, taxi, take-off, climb, cruise, descent, and landing.

Table 4.1 ATCO roles

	Aerodrome control	Approach control	Area control	Oceanic control
Responsibility	Ground position: safe and efficient movement of aircraft and vehicles on airport apron and taxiways Tower position: safe and efficient landing and take-off of aircraft	Safe and efficient sequencing of arriving and departing aircraft to and from an airport terminal area	Safe and efficient separation of aircraft while they are en route (cruising)	Safe and efficient separation of aircraft navigating across the ocean
Location	In the control tower at an airport	At a unit typically located at or near an airport but may also be co-located with an area control centre	In a centrally located centre within the airspace sector	In a centre typically located near a coastline
How do they separate aircraft?	Visual reference, (often using binoculars to look out tower windows); surveillance radar; radio communications with aircraft	Surveillance radar and radio communications	Surveillance radar and radio communications	Procedural separation (aircraft location is estimated because most oceanic airspace is beyond the range of ground-based radar); radio communications

Flight Planning

Before a pilot heads out to an aircraft, a *flight plan* is generally filed. A flight plan can be filed electronically (using an air navigation service provider website) or verbally over the phone. Flight plans are usually mandatory for instrument flight rules (IFR) flights and recommended for visual flight rules (VFR) flights.

4.1 IFR and VFR

Aircraft follow different sets of rules depending on whether they are flying *visually* or via *instruments*. In visual flight conditions, navigation and separation are accomplished primarily by the pilot looking out the cockpit window. In instrument flight conditions, the aircraft may pass through clouds or other visual obstructions (such as fog or smoke). As the pilot is not able to see ground-based obstructions or other aircraft, the assistance of air traffic control is required to ensure safe separation, approach, and landing.

The contents of a flight plan are specified by ICAO, but it generally includes key details about the flight including the date, aircraft identification, flight rules, aircraft type, on-board equipment, and the desired routing of the flight. Once the flight plan has been filed with air traffic control (ATC), the details are saved in the controller's system and a *flight progress strip* (FPS) is generated. The FPS serves as a visual reference for controllers, who have one for each of the aircraft they are separating. With new technology, paper strips are increasingly being replaced by electronic flight progress strips (EFPS).

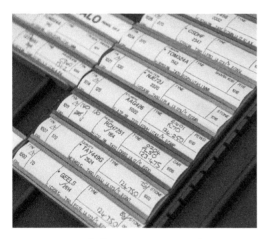

Figure 4.2a Flight progress strip (photo)

Source: "Photograph kindly provided with permission for use by NATS"

Figure 4.2b Electronic flight progress strip (photo)

Source: "Photograph kindly provided with permission for use by NATS"

Taxi

When a pilot reaches the aircraft, and is ready to begin taxiing, he or she will contact the *ground controller* on a specific radio frequency. The ground controller is located in the control tower and uses visual reference (and sometimes ground radar) to track surface movements at the airport. The ground controller will activate the flight plan, issue a clearance, and guide the aircraft to the active runway while ensuring separation from other aircraft and airport vehicles. Once the aircraft is in position – just short of the active runway – the ground controller will tell the pilot to contact *tower* on a different frequency.

Take-off

It is common for ground and tower controllers to be seated side-by-side, as they both work in the airport tower. The tower controller has authority to give the pilot permission to enter the active runway and give clearance for take-off. Shortly after take-off, tower will tell the pilot to contact *departure* on a new frequency.

Figure 4.3 Airport tower (photo)

Source: "Photograph kindly provided with permission for use by NATS"

Climb

Unlike the ground and tower controllers located in an airport tower, the departure controller is based in an approach control unit (ACU). ACUs are usually centrally located at or near the airport(s) they serve. Some ACUs are co-located with a larger area control centre (ACC).

While the tower controller uses visual reference, departure controllers rely on radar information to separate aircraft as it is not possible to see aircraft from their location – most ACUs are large rooms without any windows. The departure controller will direct the flight through the airport terminal area towards the cruise portion of the flight, at which point the

Figure 4.4a Visual controller (photo)

Source: "Photograph kindly provided with permission for use by NATS"

Figure 4.4b Area/Approach controller (photo)

Source: "Photograph kindly provided with permission for use by NATS"

pilot will be told to contact an *area controller,* who will guide the aircraft to its cruising flight level and through the en route portion of the flight.

Cruise

Area controllers work in large ACCs typically located centrally within the airspace they are responsible for. The area controller will guide the aircraft through his or her designated airspace. If the flight is travelling a long distance (i.e., through multiple flight information regions [FIRs], discussed later in the chapter), the area controller will hand the flight off to another area controller when the aircraft is approaching an airspace boundary that separates FIRs.

When an aircraft approaches its destination, the communication chain is reversed: area, approach, tower, and then ground. Note that for interoceanic flights, an oceanic controller is also included in the chain. Oceanic controllers have slightly different training, necessary to safely separate aircraft over the ocean where there is no radar coverage.

Licence

Based on Annex 1 of the Chicago Convention, ATCOs are required to earn and maintain a licence to do their jobs. In addition, they must earn ratings to move to a different role and a validation for each section of airspace or aerodrome they work in.

For example, a fully qualified controller will hold an ATC licence from a civil aviation authority (CAA) and a rating associated with a specific role (tower, approach, area, or oceanic). A qualified controller who wishes to change roles must retrain and earn a new rating – for example, to move from a tower position to an area control position. In addition, the controller must have a validation to work at a specific airport or geographic area. If the controller moves to a different geographic area, a new validation is required to demonstrate competence in the new space.

4.2   The Language of Air Navigation

An aircraft's **call sign** is its 'name' while communicating with an ATCO. An ATCO's *station position* (for example Tower or Ground) is its 'name' while communicating with an aircraft.

ATCOs grant **clearance**, which is permission for an aircraft to enter a specific area (e.g., to enter the active runway, to take off, or to land). Aircraft require a clearance to enter *controlled airspace*.

When controllers are managing aircraft positions, they often instruct pilots to *hold*. This term can be used in ground-based instructions (**hold short** means to stop just before entering a runway, and **hold position** means to stop immediately wherever an aircraft is taxiing) and air-based instructions in which aircraft are asked to enter a **holding pattern** (a pre-determined oval shaped flight path) until given clearance to proceed.

Radiotelephony (RTF) describes aviation communications over the radio, includes standard terms and phrases to reduce the confusability of instructions.

An **airway** is a defined corridor of airspace that forms a 'highway in the sky' for aircraft. Airways are often created between ground-based navigation aids (NAVAIDS). An aircraft's flight plan typically describes its routing by naming the airways used.

Traffic is a term used to describe other aircraft in the vicinity of a flight that have the potential to conflict with (be in the way of) an aircraft's flight path.

Separate flight rules are in place depending on whether a flight is in good weather conditions (**VMC: visual meteorological conditions**) where a pilot can see and avoid traffic or poor weather conditions (**IMC: instrument meteorological conditions**) where a pilot requires assistance from an ATCO to maintain safe separation from traffic. For weather conditions to be VMC, there must be minimum **visibility** (distance that can be seen), **ceiling** (height of cloud layer above the surface), and minimum flight path **distance from clouds**.

Heading is the direction that an aircraft is pointed (on the ground) or moving (while in the air) described in relation to degrees of a compass. *Track* describes the expected path of an aircraft.

Squawk is a term used to describe a pilot entering a code on the transponder (e.g., if an ATCO instructs the pilot to 'squawk 7121').

Air Navigation Service Providers

Around the globe, there are many agencies that provide ATC services. These organizations are called air navigation service providers (ANSPs). Some service providers are operated by local governments, some are private companies, while others are multinational organizations. ICAO specifies the following five categories of service providers:

1. State agencies;

2. State-owned self-financing corporations;

3. privatized ATM service providers;

4. regional ATM service providers; and

5. independent private sector ATM service providers of ground- and space-based CNS/ATM (communication, navigation, surveillance / air traffic management) services.

The Civil Air Navigation Services Organisation (CANSO) is a group representing the providers of global air traffic management (ATM) and is a great resource for learning more about ANSPs. The data set out in Table 4.2 below, which details information on the ANSPs from ICAO Council States, comes from CANSO.

Notice that although controllers probably comprise the most visible employee group within an air navigation service provider, they do not make up the majority. A large group of managers, legal and administrative staff, air traffic safety electronic personnel (ATSEP) to support ground-based hardware and software, and other professionals are involved in the work of an ANSP.

4.3 EUROCONTROL

The number of European countries within a relatively small geographic area results in a highly complex air navigation system.

EUROCONTROL is the European Organisation for the Safety of Air Navigation. It is an intergovernmental organization, responsible to its 41 member States, with more than 1900 employees in four countries.

Its goal is to help members achieve efficient, safe, and environmentally friendly air traffic management throughout Europe. A point of confusion for many is that EUROCONTROL is not primarily an ANSP, although it does provide ATM in northern Germany, Luxembourg, Belgium, and the Netherlands from the Maastricht Upper Area Control Centre.

EUROCONTROL is responsible for several activities:

• managing the ATM network, working closely with ANSPs, airports, airspace users, and the military;

• billing, collection, and distribution of aviation charges;

• supporting the European Commission, European Aviation Safety Agency (EASA), and individual States in their regulatory activities;

• contributing to the SESAR Joint Undertaking (discussed later in this chapter) through research, development, and validation; and coordinating civil–military activities in Europe.

Table 4.2 Air navigation service providers for ICAO Council States

State	Air navigation service provider	Facilities (as of 2015)
Australia	Airservices Australia is a government-owned corporation responsible for ATM services.	4204 employees (1054 ATCOs) 2 air traffic service centres 4 terminal control units 29 control towers aviation fire stations at 26 of the country's airports
Brazil	Brazil's Department of Airspace Control (DECEA) is a government group, subordinate to the Brazilian Air Force, responsible for ATM services.	12 000 employees (3512 ATCOs) 5 area control centres 42 approach controls 58 control towers 900+ NAVAIDs
Canada	NAV CANADA is a private sector organization responsible for ATM services in Canadian domestic and western North Atlantic airspace.	4832 employees (1917 ATCOs) 7 area control centres 42 control towers 58 flight service stations 7 flight information centres 41 maintenance centres 50 community aerodrome radio stations with weather information for Canada's North 1000 ground-based NAVAIDs (including 45 radar and 15 ADS-B sites)
China	The Chinese regulatory authority, Civil Aviation Administration of China (CAAC), is responsible for civil ATM.	Unavailable
France	The Direction des services de la Navigation aérienne (DSNA), which is the agency responsible for ATM in France, is operated by France's civil aviation authority, Direction générale de l'Aviation civile (DGAC).	7846 employees (4319 ATCOs) 5 area control centres (12 regional divisions) 86 control towers
Germany	Deutsche Flugsicherung GmbH (DFS) is a state-owned company, operated under private law, responsible for ATM in Germany. DFS took over ATM responsibility from the Federal Administration of Air Navigation Services in 1993.	5938 employees (1716 ATCOs) 4 area control centres 16 control towers
Italy	ENAV S.p.A. is responsible for Italian ATM. ENAV is a joint-stock company completely controlled by the Ministry of Economics and Finance under the Ministry of Transport through the Italian Civil Aviation Authority ENAC (L'Ente Nazionale per l'Aviazione Civile).	4196 employees (1787 ATCOs) 4 area control centres 41 control towers

State	Air navigation service provider	Facilities (as of 2015)
Japan	*Japan Air Navigation Service* (JANS) is a government agency within the Ministry of Land, Infrastructure, Transport and Tourism. JANS is a branch of the Civil Aviation Bureau of Japan (JCAB) providing ATM services for the Fukuoka flight information region (FIR).	4087 employees (1860 ATCOs) 4 area control centres 33 aerodrome control service (tower) 16 approach control service/flight information centres 34 remote aerodrome flight information service centres
Russian Federation	The Main Air Traffic Management Centre	Unavailable
United Kingdom	*National Air Traffic Services* (NATS) is the largest ANSP in the UK. NATS was formed in 1962 and evolved to become a public–private partnership in 1998 – where the Airline Group holds 42% ownership, NATS staff hold 5%, UK airport operator LHR Airports Limited holds 4% and the government holds 49%.1 (Although NATS is the largest, there are more than 60 other service providers operating in the UK).	4252 employees (1467 ATCOs) 2 area control centres 14 UK airports 10 Spanish airports (contracted through joint venture FerroNATS)
United States	The *Federal Aviation Administration* (FAA) is a federal government agency, within the United States Department of Transportation, responsible for ATM.	34 911 employees (18 001 ATCOs) 21 area control centres 512 control towers

[1] NATS, 2016
Source: CANSO, 2015

Communication, Navigation, and Surveillance

There are three key components of ATM that are of paramount importance to the work of ATCOs: 1) communication, 2) navigation, and 3) surveillance. Each of these three elements involve unique technologies, infrastructure, and regulations, and all three must be employed seamlessly by controllers to safely and efficiently separate aircraft.

Communication

Communication is a crucial element of effective air traffic management. Using a variety of communication tools, ATCOs deliver instructions and clearances, and otherwise interact with their flights. Within a specific position, a controller will be designated one or more radio frequencies to monitor. Most ATCOs wear a wireless headset and microphone so that their hands are free and they can move around their workspace. They will also have access to a telephone to communicate with controllers in other units.

International agreements specify radio frequency bands reserved for aeronautical communications, existing primarily in high frequency (HF), very high frequency (VHF), and ultra high frequency (UHF) spectrums. Most aviation radio communications are conducted in the VHF spectrum.

More recently, communication technology has advanced beyond radios with the implementation of *data link* systems. A controller–pilot data link communication (CPDLC) system allows aircraft to transmit and receive voice, text, and pictorial information. In many ways, this technology can be thought of as a text messaging system between pilots, controllers, and airline dispatchers. An example of a data link system is the aircraft communications addressing and reporting system (ACARS). A key feature of ACARS is the automatic detection and report of each major flight phase (out of the gate, off the ground, on the ground, and into the gate) through aircraft sensors. The system automatically transmits 'pings' from the aircraft to ground communication stations.

Standard Phraseology

To avoid miscommunications, pilots and controllers use standard terms and phrases that are unlikely to be misunderstood.

When a pilot makes an initial radio call to ATC, it includes

1. the controller's identification;

2. the aircraft identification; and a message.

For example, a pilot making initial contact with the tower might transmit 'Airport Tower, Aircraft 123, holding at point B2.' This call would indicate the pilot of Aircraft 123 is holding short of the runway at a point on the airport called 'B2' and is awaiting take-off clearance from the tower controller.

When a controller contacts a pilot on the radio, the transmission includes

1. the aircraft's identification;

2. the controller's location and type of service; and a message.

An example of this transmission would be 'Aircraft 123, Airport Tower, cleared for take-off.' For subsequent back-and-forth communications, it is common to stop including the controller's identification and use only the aircraft identification. In the following example, the aircraft identification is underlined:

Pilot: Airport Tower, *Aircraft 123*, holding at point B2.
ATC: *Aircraft 123*, Airport Tower, behind landing Boeing 737, line up runway 09, behind.
Pilot: Behind landing Boeing 737, line up runway 09, behind, *Aircraft 123*.
ATC: *Aircraft 123*, runway 09, cleared for take-off.
Pilot: Cleared for take-off, *Aircraft 123*.

Notice how the pilot repeats crucial information from the controller's message. For those who are new to radio communication, this repetition may seem redundant, but it is critical to ensure mutual understanding and avoid errors.

As radios can transmit only one message at a time, it is possible for transmissions to be 'stepped on' in busy airspace. This means that two transmissions were sent simultaneously, making both inaudible. To make other aviators and controllers aware of the occurrence, it is helpful to point it out over the radio. For example, a pilot may state 'Airport Tower, this is Aircraft 123, last transmission blocked.'

Did You Know?

ICAO has designated English as the international language for global ATC communication.

In an emergency situation where immediate assistance is required, transmissions begin with 'Mayday, Mayday, Mayday'. Urgent messages that do *not* require immediate assistance are prefixed 'Pan-Pan, Pan-Pan, Pan-Pan'. Emergency calls are generally transmitted on the radio frequency currently in use, but if that is not possible the pilot will call on the emergency frequency of 121.5 and squawk the transponder to emergency code 7700.

Did You Know?

In the language of ATC, *squawk* means to select or assign a transponder code to aid in radar identification – generally, ATC will assign a code to a pilot ('Squawk 1234'), which the pilot will then enter into the transponder in the aircraft and use that code while in flight. In the case of emergency, hijacking, or radio failure, a pilot will squawk a distinct code to alert ATC of their situation.

Case Study: A Deadly Miscommunication – KLM Flight 4805 and Pan Am Flight 1736[1]

On 27 March 1977, when a bomb exploded in the passenger terminal of Las Palmas de Gran Canaria Airport in Spain, many inbound aircraft were diverted to the nearby Los Rodeos (Tenerife) Airport, which was too small to properly accommodate them.

When Las Palmas Airport was reopened, the diverted aircraft (with many displaced passengers), were anxious to complete their flights. Two Boeing 747 aircraft (KLM flight 4805 and Pan Am flight 1736) initiated take-off procedures at Los Rodeos; however, other aircraft were parked on and blocking the main taxiway, so the KLM and Pan Am flights were instructed to backtrack on the only runway to get to take-off position. KLM was to taxi down the runway to take-off position and make a 180 degree turn and wait for clearance to take off. Pan Am was given permission to taxi down the runway behind the KLM, take exit C-3, and wait there to be the next in line to take off.

A dense fog made it impossible for the aircraft to see each other, or for ATC to see the runway, the taxiway, or the exits. Without ground radar, ATC relied on radio position reports from the pilots, which made clear communication crucial.

The KLM pilots received a clearance from ATC with departure procedures (but the flight was not yet cleared for take-off). The first officer read back the clearance, which the Captain interrupted with 'We're going.' ATC responded, 'OK . . .' This non-standard terminology may have supported the KLM Captain's incorrect belief that the departure clearance was in fact a take-off clearance.

The controller probably intended 'OK' to mean 'I acknowledge you are in take-off position' because he subsequently transmitted, 'Stand by for take-off, I will call you.' Unfortunately, at the exact same time as the controller transmitted his message to stand by, Pan Am transmitted 'We're still taxiing down the runway, the Clipper 1736.' The two simultaneous transmissions caused a loud squeal over the radio, blocking both crucial messages to the KLM crew.

The KLM flight released brakes, increased engine power for take-off, and began its roll down the runway. The Pan Am was still taxiing down the centre of the runway, directly towards the departing aircraft.

ATC instructed Pan Am to 'report the runway clear' to which Pan Am crew responded, 'OK, we'll report when we're clear.' The KLM flight engineer heard this and expressed concern to the Captain ('Is he not clear, that Pan American?') to which the Captain replied emphatically, 'Oh yes', and continued the take-off. The KLM flight was in the middle of take-off rotation – the nose was beginning to rise up, but the landing gear was still on the ground – when it struck the Pan Am flight, causing 583 fatalities and making this accident the deadliest in aviation history. Only 61 people on the Pan Am flight survived the disaster, while all crew and passengers aboard the KLM flight were killed.

An investigation determined it was communication confusion that led the KLM Captain to initiate take-off without proper clearance, and that contributing factors included interference from simultaneous radio transmissions and foggy weather conditions.

Note

1 Comision de Accidentes, 1978; ICAO, 1978

Phonetic Alphabet

As the quality of radio transmissions can vary, it is important that information be pronounced clearly to ensure understanding. Because many English letters and numbers rhyme, it is easy to confuse them. To prevent this type of confusion, the aviation industry uses code words, which make up the *phonetic alphabet*, for each letter and number. The phonetic alphabet was first adopted by the International Commission for Air Navigation – ICAO's predecessor – in 1932 and has evolved over the years.[4] All who use the air navigation system must learn and use the phonetic alphabet.

Did You Know?

There are several websites and apps that stream pilot–ATCO radio communications (e.g., www.liveATC.net). These are great tools for those just beginning to learn about radio communication procedures.

Although the spelling of some terms in the phonetic alphabet may seem unusual, it is intentional to support pronunciation by non-native-English speakers. For example, the spelling of *Alfa* does not use PH because speakers of other languages may not understand that it should be pronounced as an F. Likewise, *Juliett* ends with two T's because in French, a single T at the end of a word may not be pronounced at all.

Table 4.3 The phonetic alphabet

A	Alfa (AL-FAH)	N	November (NO-VEM-BER)	1	One (WUN)
B	Bravo (BRAH-VOH)	O	Oscar (OSS-CAH)	2	Two (TOO)
C	Charlie (CHAR-LEE)	P	Papa (PAH-PAH)	3	Three (TREE)
D	Delta (DELL-TAH)	Q	Quebec (KEH-BECK)	4	Four (FOW-er)
E	Echo (ECK-OH)	R	Romeo (ROW-ME-OH)	5	Five (FIFE)
F	Foxtrot (FOKS-TROT)	S	Sierra (SEE-AIR-AH)	6	Six (SIX)
G	Golf (GOLF)	T	Tango (TANG-GO)	7	Seven (SEV-en)
H	Hotel (HOH-TEL)	U	Uniform (YOU-NEE-FORM)	8	Eight (AIT)
I	India (IN-DEE-AH)	V	Victor (VIK-TAH)	9	Nine (NIN-er)
J	Juliett (JEW-LEE-ETT)	W	Whiskey (WISS-KEY)	0	Zero (ZEE-RO)
K	Kilo (KEY-LOH)	X	X-ray (ECKS-RAY)	Decimal	DAY-SEE-MAL
L	Lima (LEE-MAH)	Y	Yankee (YANG-KEY)	Hundred	HUN-dred
M	Mike (MIKE)	Z	Zulu (ZOO-LOO)	Thousand	TOU-SAND

Surveillance

To safely separate aircraft from one another, controllers must have an accurate and reliable way of determining each aircraft's position in space. This is accomplished primarily through surveillance radar, although other types of surveillance equipment (such as ADS-B, discussed below) are becoming increasingly common.

Before radar, controllers separated air traffic using data related to distance, time, and altitude communicated via radio reports, or by visually observing aircraft in the airport area. Although this method, known as *procedural control*, is still used in some areas (such as over oceans), three to four times greater separation between aircraft is required when procedural separation is used rather than radar. This increased separation is required for safety, but it reduces the efficiency of the entire system. Because surveillance systems provide controllers with an accurate picture of aircraft in their airspace, separation using radar leads to improved efficiency over procedural control.

Surveillance Radar and Collision Avoidance

Radar systems are classified as *primary* and *secondary*; the two systems work independently of each other.

Primary surveillance radar systems send invisible electromagnetic waves of energy in specific directions. A radar system at an airport has an antenna that spins continually, sending out electromagnetic energy waves. When some of those waves are bounced back – reflected off the metal surface of an aircraft – the system receives this 'echo' and recognizes the presence of an object. The location of the object is determined based on the direction and time delay from the original transmission.

ATCOs use primary radar signals to identify aircraft in airspace – with aircraft represented as an electronic dot on their radar screen. Therefore, primary radar tells controllers that *something* is in the airspace, but not what that something is. Controllers must use other information (radio calls, flight plans, and so on) to determine which dot represents which aircraft. Moreover, it is possible for radar signal to return 'clutter' from non-aircraft sources including ground-based structures, rain or snow, or birds.

Secondary surveillance radar provides additional information beyond that sent by primary radar signals. Secondary radar uses information from an aircraft's on-board equipment – called a *transponder* – to add to the information controllers receives on their radar screens, including aircraft call sign, altitude, or other information. These data can greatly enhance a controller's situational awareness as they are required to do much less mental work to match flight information to the electronic dots on their radar screens. Signals from secondary radar are stronger and larger than those from primary radar and they are less affected by clutter.

Secondary radar works through an aircraft's transponder, which transmits a radio signal to a ground-based receiver. A controller gives a pilot a unique four-digit transponder code (saying 'squawk code 1234') to dial into the transponder. When the ground-based system receives the unique signal from the aircraft's transponder, it compares the code with the flight plan data in the flight management system and presents this information to the controller as the aircraft's call sign.

Transponders may also possess *Mode-C capability*, which enhances the transmitted signal by including the aircraft's altitude (in addition to the call sign) on the controller's radar screen. A computer-calculated ground speed can also be transmitted and read on the controller's screen.

Primary and secondary surveillance radar systems have been in use since the 1930s. Automatic dependent surveillance (ADS), is a newer aircraft-based surveillance technology that references satellites; it has been suggested that ADS will eventually replace radar.

ADS references satellite navigation systems to determine aircraft position and can be received by ATC as a replacement for secondary radar. There are two types of ADS: ADS-A, which transmits information to ATC or other aircraft upon request, and ADS-B, which transmits information automatically to ATC and other aircraft in the vicinity. The ADS-B technology is a key feature in future air traffic management initiatives, covered at the end of this chapter. The features of ADS-B are set out below:

- automatic: it sends information without any action from the pilot or controller;

- dependent: position and speed information rely on the Global Positioning System;

- surveillance: it allows for the precise tracking of aircraft; and broadcast: it sends data to anyone with receiving equipment.[5]

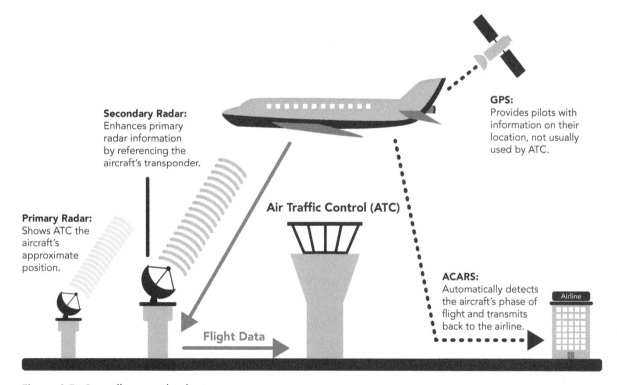

Figure 4.5 Surveillance technologies

Source: Image adapted from BBC News.com, 'How do you track a plane?', 17 March 2014, http://www.bbc.com/news/world-asia-pacific-26544554

Figure 4.6a Radar tower (photo)
Source: "Photograph kindly provided with permission for use by NATS"

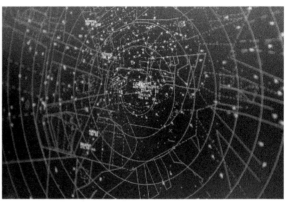

Figure 4.6b Radar screen (photo)
Source: "Photograph kindly provided with permission for use by NATS"

Navigation

Airspace is the term used to describe air overlying a given section of land or water. Each country regulates the use of the airspace overlying its land as well as 12 nautical miles – about 22 kilometres – from its coastlines (based on the Chicago Convention). Oceanic airspace is international, although certain countries take responsibility for traffic in these areas through international agreements.

Air navigation requires a consistent organization and structure of airspace on a global scale. For that reason, the world's airspace is separated into large geographic areas called flight information regions (FIRs). FIRs are then further divided into smaller segments, named Class A through G, depending upon the location of airports and high-traffic areas. Each class of airspace has specific rules associated with the type of operations allowed (IFR, VFR, or both) and whether it is controlled or uncontrolled.

Flight Information Regions

Global FIRs, as designated by ICAO, are geographic areas with clear borders in which a country is assigned responsibility for air navigation services. FIRs usually follow country borders, but there are exceptions (such as over the ocean). Larger countries may be divided into several FIRs, while smaller countries be combined into one FIR. When an FIR crosses international boundaries, international agreements are negotiated to determine how responsibilities for ATM are assigned. Typically, one area control centre (ACC) has responsibility for one FIR.

Controlled and Uncontrolled Airspace

Controlled airspace is a defined sector of airspace where ATC services are provided. Controlled airspace is monitored and managed by air traffic controllers. Pilots require a clearance to enter (i.e., pilots must request and receive ATC permission before entering controlled airspace). Typically, controlled airspace is associated with a high volume of traffic (such as the airspace surrounding airports) or secure areas (such as rocket launches or military operations). Flight in controlled airspace is a combination of IFR and VFR, depending on the airspace classification (A through G).

Uncontrolled airspace refers to sectors where traffic is not supervised by ATC. Therefore, no clearance is required to enter uncontrolled airspace. ATC may provide support services to aircraft through radio communication, workload permitting. Aircraft in uncontrolled airspace typically operate under VFR and pilots are solely responsible for maintaining separation from other aircraft and terrain.

VFR and IFR

The terms *visual flight rules* (VFR) and *instrument flight rules* (IFR) are used to describe two different types of flying and the two sets of regulations that govern them.

VFR refers to flights in which the weather conditions allow pilots to visually separate themselves from other aircraft and terrain. There are regulated minimum weather requirements for VFR flight, called *visual meteorological conditions* (VMC). The precise requirements vary based upon airspace, country, and time of day, but generally dictate a minimum visibility (distance the pilot is able to see unobstructed by fog or other conditions) and horizontal and vertical separation from clouds.

IFR refers to flights in which pilots lack visual reference (i.e., they cannot see other aircraft or the ground, perhaps because of fog or clouds). These flights are conducted primarily by the pilot referencing cockpit *avionics* (a collective term for the cockpit instruments that pilots use during flight navigation). IFR flights may be conducted in either VMC or *instrument meteorological conditions* (IMC), which describes poor weather conditions. As pilots cannot see other aircraft, safe separation is the responsibility of ATCOs. However, the pilot holds the ultimate responsibility for flight safety and has the right to refuse ATC instructions if necessary. Airline flights always fly IFR whether in VMC or IMC weather conditions.

Airspace Classes A–G

ICAO defines airspace classifications in Annex 11; however, States are permitted to assign classes within their airspace as appropriate for their needs. Therefore, while the definitions of the classes are consistent worldwide, the usage of the classes varies by country. For example, Canada has all seven classes of airspace (A through G), Sweden uses only Class C (for all controlled airspace) and Class G (for all uncontrolled airspace), and Australia uses Classes A, C, D, E, and G. Upper airspace (above 18 000 feet) is generally considered Class A, but it may be designated Class G in some areas of the world where air traffic control service is lacking.[6]

In this case, and in other classes of airspace, pilots are responsible for their own separation, using a shared radio frequency to transmit position reports.

The following are the criteria for each airspace class:

- Class A: Only IFR flights are permitted. All flights are provided with air traffic control service and are separated from each other.

- Class B: Both IFR and VFR flights are permitted. All flights are provided with air traffic control service and are separated from each other.

- Class C: Both IFR and VFR flights are permitted. All flights are provided with air traffic control service, and IFR flights are separated from other IFR flights and from VFR flights. VFR flights are separated from IFR flights and receive traffic information for other VFR flights.

- Class D: Both IFR and VFR flights are permitted and all flights are provided with air traffic control service. IFR flights are separated from other IFR flights and receive traffic information for VFR flights; VFR flights receive traffic information for all other flights.

- Class E: Both IFR and VFR flights are permitted. IFR flights are provided with air traffic control service and are separated from other IFR flights. All flights receive traffic information as far as is practical. Class E shall not be used for control zones (the area around an airport controlled by the tower).

- Class F: Both IFR and VFR flights are permitted. All participating IFR flights receive an air traffic advisory service and all flights receive flight information service if requested.

- Class G: Both IFR and VFR flights are permitted and receive flight information service if requested.

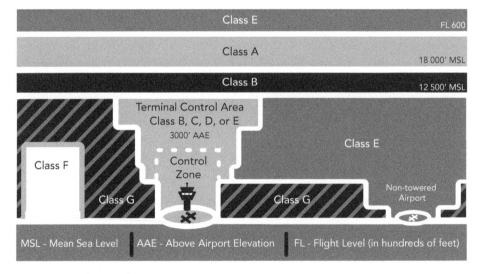

Figure 4.7 Classes of airspace

Did You Know?

The term *upside-down wedding cake* is used to describe the inverted-tier–shape of terminal airspace. Terminal airspace is shaped this way because airspace is controlled above and outward from an airport to support large aircraft which must remain in controlled airspace as they transition to the cruise portion of their flight.

Although the classification of airspace can seem daunting – some call it an 'alphabet soup' because of the many letter designations – its purpose is clear. The intent is to define sections of airspace where different aviation activities can safely occur. The reality is that certain activities don't go well together – high-speed airliners coming in and out of a major airport, skydiving, rocket launches, and student pilot flight training don't mesh. The airspace classification system exists to offer layers of protection that increase safety and efficiency.

4.4   Accident Report: Aeronaves de Mexico 498 and Piper PA-28-181[1]

On August 31 1986 an Aeronaves flight 498 was a DC-9 aircraft on an IFR scheduled passenger flight heading to the Los Angeles International Airport, under radar contact with the Los Angeles terminal ATC.

Simultaneously, a Piper aircraft was flying from Torrance, California to Big Bear, California under VFR and was not in radio contact with any ATC facility.

The Piper inadvertently entered the Los Angeles terminal control area. Skies were clear with a visibility of about 22.5 kilometres (14 miles). ATC did not observe the Piper on the radar, so no traffic advisory was given to flight 498.

The two aircraft collided mid-air two kilometres (1.2 miles) above sea level. Both aircraft fell to the ground, with falling wreckage destroying five houses and damaging seven. The following were fatally injured: 58 passengers and six crew members on the DC-9, one pilot and two passengers on the Piper, and 15 people on the ground.

The investigation of the accident revealed two causes:

1) The entry of the Piper pilot into controlled airspace. This action robbed both aircraft of the protection that the airspace was designed to provide. The Piper's entry into prohibited airspace created a risk that should never have existed.

2) The flawed notion that pilots can visually 'see and avoid' collisions. Follow-up research in a simulator revealed that before the pilots could have been guaranteed to spot the other aircraft, the DC-9 would have been only two seconds from collision and the Piper would have been five seconds from collision. This reinforces the importance of ATC support in providing safe separation between aircraft.

Note

1 NTSB, 1987

Navigation Technologies

So far, we have discussed the ways in which global airspace is divided into sectors and how different rules for navigation apply in these sectors. Just as important are the tools of navigation – the ground-based, aircraft-based, and satellite-based systems and technologies that support aircraft navigation.

Navigation Aids

Aircraft navigation involves on-board avionics – the instruments in the cockpit – which reference both ground-based navigation aids (NAVAIDs) and global navigation satellite systems (GNSS). That is to say, the on-board avionics send and receive data to and from NAVAIDs and/or GNSS. Both NAVAIDs and GNSS refer to groupings of technologies that support navigation. Learning about navigation technologies can be a daunting task for students as it involves a long list of acronyms. A quick reference guide to navigation terms and acronyms is set out here.

Table 4.4 Air navigation technologies

Technology	Description	Category
FMS (flight management system)	Refers to an on-board system that provides multiple sources of information to pilots, including a flight management computer (FMC), aircraft navigation system, automatic flight control/flight guidance system (AFCS or AFGS), and electronic flight instrument system (EFIS).	Avionics
Magnetic compass	Points to magnetic north to ensure that pilots have accurate information about their direction of travel. The aircraft's magnetic compass reading (after being corrected for variation and deviation) will be entered into the pilot's heading indicator (HI) and serve as the primary reference for direction of travel.	Avionics
RNAV (area navigation)	Allows aircraft to fly a unique flight path, supported by ground- or space-based navigation aids, within specific limits. Navigation is based on waypoints, which are specific geographic coordinates, and may be more direct than traditional navigation. An RNAV system may be included as part of an aircraft's flight management system (FMS). Pilots enter a destination and the RNAV system calculates a flight path by connecting a series of waypoints. The flight path is then displayed to the pilot on a cockpit navigation display. Note that RNAV systems may reference VOR, DME, and GPS (see below) for location information.	Avionics
DME (distance measuring equipment)	A ground-based transponder that provides the pilot with a precise *slant-range distance* – the distance, on a slant – from the aircraft in the air to the DME on the ground. DME is usually co-located with a VOR and called VOR-DME. DME can be thought of as similar to secondary radar used by ATC, but in reverse; the aircraft receives the signal from the ground-based DME transponder rather than sending the signal to the ground.	NAVAID

Technology	Description	Category
ILS (instrument landing system)	A precision approach navigation aid that uses two radio beams, offering vertical and lateral guidance. The *glideslope* presents the pilot with an ideal descent angle while the *localizer* provides directional guidance aligned with the centreline of the runway.	NAVAID
INS (inertial navigation system)	A system on board an aircraft that senses the aircraft's movements and rotation to provide speed, altitude, and dead reckoning position to the FMS. The FMS compares INS data against NAVAID and GNSS inputs.	NAVAID
NDB (non-directional beacon)	A land-based radio transmitter using low to medium frequency (LF to MF) or UHF bands that sends a signal of equal strength in every direction. The signal follows the curvature of the earth, so is more accessible to aircraft at a lower altitude, and for a greater distance, than signals from line-of-sight technologies that can be blocked by obstacles (e.g., trees or mountains). NDB signals can be used for non-precision approaches as pilots with direction-finding equipment can determine their bearing to the beacon and track towards or away from it.	NAVAID
TACAN (tactical air navigation system)	A military navigation system that provides slant-range distance from the aircraft to a ground-based station. TACAN is often paired with a VOR (called a VORTAC). TACAN provides DME capabilities for civilian aircraft while military aircraft can use additional features to support non-precision approaches and en route navigation.	NAVAID
VOR (VHF omni-directional range)	A ground-based electronic navigation aid that uses VHF signals. The signals extend in 360 degrees from the station. Aircraft with receiving equipment can determine the magnetic bearing (called *radial*) from their current location to the VOR station. Signals are commonly used to navigate airways or for non-precision approaches.	NAVAID
GPS (Global Positioning System)	All satellite-based navigation systems that provide location, time, and velocity that can be used for non-precision approaches.	GNSS
GLONASS (Global Navigation Satellite System)	GPS was developed by the US, GLONASS was developed by the Russian Federation, and Galileo is being developed by the European Union.	
Galileo		

Navigation Systems

To fully understand navigation, it is important to have a basic understanding of arrival and departure procedures, as well as en route navigation procedures.

Terminal Area Navigation: Arrivals and Departures

To manage incoming aircraft, airport terminal airspace is typically controlled, and pilots and controllers must follow specific procedures when aircraft take off and depart the airport area and when they are approaching to land. These procedures standardize the flow of traffic so that the locations of arriving and departing aircraft are predictable and organized.

When taking off from a major airport, aircraft often follow published *standard instrument departure* (SID) procedures. These SIDs provide a standard expectation between pilots and controllers with respect to managing a departure, separating departing traffic from landing aircraft, and simplifying pilot–controller communications. The SIDs specify the heading, radio communication frequencies, and minimum altitudes the aircraft adhere to after take-off.

For aircraft approaching to land at a controlled airport, a *visual approach* (i.e., a non-instrument approach) may be used if the visibility is good. For a visual approach, a pilot can see the runway and is therefore able to land using visual reference without needing additional support from ATC. However, under poor weather conditions (or at the pilot's choosing), an *instrument approach procedure* (IAP) is used. IAPs are divided into four stages: 1) initial approach, 2) intermediate approach, 3) final approach, and 4) missed approach procedures (if the landing is aborted).[7]

IAPs can be classified as non-precision or precision approaches. *Non-precision approaches* are supported by navigation aids (e.g., VOR or NDB NAVAIDs), which provide the direction to fly towards the runway. During an approach, the pilot may be in IMC (poor weather) and unable to see the runway. For a non-precision approach, the NAVAIDs provide only a heading (direction) to fly by; they do not provide specific information about elevation or approach angle. Therefore, pilots need precise direction on how low they can safely descend without encountering ground-based obstructions. Therefore, a minimum descent altitude (MDA) is identified: to stay safe, the pilot is not allowed to descend below the MDA until the runway can be seen.

If the pilot gets close to the airport, but the weather is too poor for a safe landing, the pilot may reach a predetermined *missed approach point*. This is a specific distance from the runway where it is expected the pilot should be able to see the runway. If the pilot reaches the missed approach point and has no visual contact, he or she must initiate a missed approach: climb, circle the airport, and attempt the landing again (or divert to an alternate airport). As long as the pilot can see the runway before reaching the missed approach point, the aircraft can descend and land at the airport.

Unlike a non-precision approach, a *precision approach* is a descent to landing where the pilot has both lateral (side-to-side) and vertical (up-and-down) guidance. Ground-based navigation aids provide the following guidance to pilots:

- A *glideslope gives* precise information about the ideal descent angle to the runway.

- A *localizer* gives lateral guidance aligned directly with the runway centreline.

This system allows the pilot to follow a precise angle of descent towards the airport on a heading aligned with the runway, while looking out the cockpit window to spot the runway. When the aircraft reaches a predetermined decision height (DH), the pilot must be able to see the runway in order to continue the landing. If the runway cannot be seen, the pilot initiates a missed approach.

As additional guidance is available, the precision approach allows pilots to safely descend lower, as compared to a non-precision approach, before having to declare a missed approach thus increasing the likelihood of landing. However, the tools required for a precision approach are more expensive and are not available at all airports.

4.5 Wake Turbulence

Wing tip vortices are spinning air currents formed as the air flows over an aircraft's wings. Vortices extend from the aircraft's wing tips and trail behind and below the aircraft's flight path. Wing tip vortices create wake turbulence in the air similar to the wake that forms behind a boat travelling through water. Wake turbulence constitutes a challenge at major airports, where there are continual landings and take-offs.

The strongest wake turbulence is caused by heavy aircraft travelling at slow speeds – during take-off and landing – and is most dangerous for small aircraft. If a small aircraft encounters wake turbulence from a heavy aircraft that is landing, the turbulence can cause a rolling force or an uncommanded descent that may be impossible for the pilot to recover from. Wake turbulence is a dangerous condition that has been a causal factor in a number of accidents and serious incidents.

Air traffic controllers play an important role in preventing accidents related to wake turbulence – they enforce separation criteria between IFR aircraft and issue advisories to VFR aircraft. The separation criteria range from two to three minutes, with longer separation time being necessary when there is a significant size differences between aircraft (e.g., a light two-seat aircraft landing behind an Airbus A380). A few minutes of separation allows wake turbulence to decay so that it will not endanger other aircraft; however, it also creates a capacity limitation at busy airports as it slows the rate of departures and arrivals.

Figure 4.8 Wake turbulence

En Route Navigation

Airways are the highways in the sky created to organize global air traffic. Many airways are based on VHF omnidirectional range (VOR) signals. VORs are ground-based NAVAIDs that transmit very high frequency signals in 360 degrees, creating 360 *radials* that extend in straight lines from the VOR. A pilot, after selecting a VOR's frequency on the aircraft's navigation equipment, can navigate towards or away from the VOR on a specific radial. Some aircraft also have distance measuring equipment (DME) to indicate the distance between the aircraft and the station transmitting the signals. VORs are *line-of-sight* instruments, meaning that signals are interrupted by ground-based obstructions. Because of this limitation, a large network of VORs are required throughout various regions.

VORs play a key role in the global air navigation system because their radials provide the structure for airways. Radials that extend between two VORs can form an airway, with an indicated name, specified dimensions, and rules. When two airways cross over each other, they form intersections, which are also given names for navigational purposes.

Low-altitude navigation routes are called *Victor airways* – *Victor* is the phonetic word for V in VOR – while high-altitude routes are called *jet routes*. In Europe, high-altitude routes are called *upper air routes*.

With traditional navigation resulting in pilots flying NAVAID to NAVAID, the routing tends to form somewhat of a zigzag pattern. There are major disadvantages to this approach, including increased flight time (and resulting fuel burn) and congestion around NAVAIDs. As traffic volumes continue to increase, more ANSPs are adopting *performance-based navigation* (PBN) using *area navigation* (RNAV). RNAV is defined by ICAO as

> a navigation system which permits aircraft operation on any desired flight path within the coverage of station-referenced navigation aids or within the limits of the capability of self-contained aids, or a combination of these.[8]

RNAV systems provide pilots with flight plan management (using waypoints), navigation, and guidance information. RNAV systems reference both land- and satellite-based navigation aids to set a direct flight path towards a destination. ICAO Doc 9613 was published in 2008 with extensive descriptions of system and aircraft equipment and performance required to support PBN. A set of globally compatible navigation specifications is also included in the document.

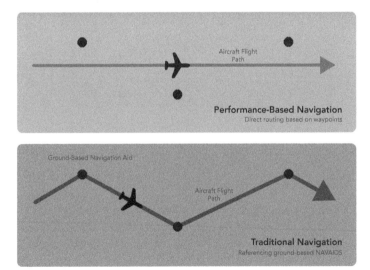

Figure 4.9 A comparison of performance-based navigation and traditional navigation

Land-based navigation has capacity limitations, which cause challenges given the ever-increasing amount of global air traffic. With this in mind, ICAO initiated a Special Committee on Future Air Navigation Systems (FANS) in 1983 to study then-current communication, navigation, surveillance/air traffic management (CNS/ATM) systems and to plan for the future of CNS/ATM. The Committee concluded that global navigation satellite systems (GNSS) would be a key factor in the development of CNS/ATM to support seamless, efficient, and safe global air traffic management.

With the introduction of satellite-based navigation, the capacity of global en route navigation systems increased as pilots were no longer limited to airways based on ground-based NAVAIDs or procedural control. Satellite-based navigation makes it possible to navigate precisely and directly between any two points, rather than following zigzagging highways in the sky. It also allows for precise surveillance in oceanic regions that lack radar coverage.

However, there are challenges associated with GNSS. States traditionally have responsibility for navigation services within their territories, but satellite-based navigation systems complicate this issue. The United States provides the world with its Global Positioning System (GPS) free of charge, while the Russian Federation offers the Global Navigation Satellite System (GLONASS) and Europe has launched the civilian-controlled Galileo system.[9] How, then, does the aviation industry resolve financial, legal, and liability issues associated with this type of navigation? Some States, particularly those that do not control satellites, have expressed concerns about the global navigation system relying upon GNSS as the only means of navigation. To ease global concerns, ICAO has issued a legal framework related to certification, liability, administration, cost recovery, financing, and operating structures for GNSS.[10]

The Future of Air Traffic Management – Global Initiatives

As the aviation industry prepares for increasing air traffic, several innovative ATM projects have been launched. At their core, these projects look to incorporate new technology in ways that improve safety, airspace capacity, and flight efficiency. Some applications target communication improvements, new usage of satellite navigation, improved aircraft surveillance, and controller automated–decision support tools, among others. Although there are several projects ongoing around the globe, two will be highlighted here: NextGen and SESAR.

NextGen

The Next Generation Transportation System (NextGen) is an American Federal Aviation Administration (FAA) initiative to evolve the airspace system through the year 2030. This system is intended to increase the capacity of the airspace through implementation of

- ADS-B satellite-supported aircraft surveillance technology;
- collaborative air traffic management technologies (CATMT), which provide ATC decision support and data-sharing improving collaboration between controllers;

- data communications (DATA Comm) to allow controllers to send digital messages and clearances directly to a display in aircraft cockpits;

- national airspace system voice system (NVS) to improve voice communications technology;

- NextGen Weather, which distributes weather information; and

- a system-wide information management (SWIM) network that distributes NextGen digital information through the airspace system.

SESAR

With European air traffic expected to grow from 9.5 million flights in 2012 to over 14 million in 2035, the region requires a modern and efficient ATM system. In 2004, the Single European Sky (SES) initiative was launched by the European Commission to reform the complicated European ATM system. A key project within SES is the Single European Sky ATM Research (SESAR) collaboration. The goal of SESAR is to modernize ATM throughout Europe by

- tripling the capacity of the airspace system;

- increasing safety;

- reducing environmental impact per flight by 10 per cent; and

- reducing the costs of ATM by 50 per cent.

Now entering the deployment phase, SESAR activities are focusing on large-scale production and implementation of a variety of infrastructure, equipment, and technologies (including Galileo, the European GNSS). A few examples of initiatives include:

- moving to a 4D trajectory management approach, allowing for flight paths to be adjusted in real time based on predicted demand. (4D refers to the three physical dimensions plus time.) A 4D management approach provides flights with a specific arrival time, and in compensation for accepting the timeslot, aircraft are routed directly without deviations. Data has shown a 100 per cent reduction in holding, 6 per cent reduction in total distance flown, 68 per cent reduction in conflicts, and 11 per cent less fuel burnt;

- providing advanced air traffic services through traffic synchronization to improve arrival and departure management, through optimal traffic sequencing;

- improving ATM network services through better information-sharing within a common operational environment; and

- integrating airports into ATM to support collaborative decision-making and improving runway throughput and surface movement management.[11]

Conclusion

Throughout this chapter, we have reviewed key issues in global air traffic management (ATM). The roles of air traffic control officers (ATCOs) have been introduced, along with the air navigation service providers (ANSPs) where ATCOs work. We have detailed the regulations and technologies that support the three key considerations in the safe separation of air traffic: 1) communication, 2) navigation, and 3) surveillance. Finally, two examples of future ATM initiatives – NextGen and SESAR – were discussed.

The goal of this chapter was to provide insight into the many interconnected elements that must work together seamlessly to make global navigation possible.

Key Points to Remember

1. The importance of ATCOs was recognized in the early days of aviation, after a mid-air collision in 1922 demonstrated that pilots cannot always safely separate themselves from other aircraft.

2. ATCOs are responsible for the safe separation of aircraft from: 1) other aircraft, 2) obstructions on the ground, and 3) airspace boundaries. They separate using

   - visual reference (looking out the window to see aircraft);

   - surveillance radar (using precise location information provided by radar); and

   - procedural separation (estimating aircraft position based on rules and pilot reports).

3. Controllers work in four different job roles. *Aerodrome controllers* work in an airport tower and are able to see aircraft on airport surfaces and those landing on or taking off from the active runway. *Approach controllers* manage the arriving and departing aircraft to and from an airport terminal area. *Area controllers* separate aircraft within their airspace during the cruise portion of a flight, and *oceanic controllers* separate aircraft as they navigate across the ocean. ATCOs must earn and maintain a licence to act as a controller. They need ratings to move to a new role (e.g., from an approach controller to an aerodrome controller), and a *validation* for each section of airspace or aerodrome they work in. ATCOs are employed by air navigation service providers (ANSPs), which are the agencies responsible for air traffic management.

4. Broadly, the work of air traffic management can be organized into the categories of communication, navigation, and surveillance.

5. Communication refers to how ATCOs share information with aircraft. ATCOs use radio frequencies to speak to pilots as well as newer data link systems that transmit text-based messages. A globally standardized phraseology is used to

minimize miscommunication in aviation communication – this includes typical phrases, emergency messages, and the phonetic alphabet.

6. To safely separate aircraft, controllers must know their position in space. This is primarily accomplished through surveillance radar. Primary radar sends out energy waves that bounce off metal aircraft and are displayed as a dot on a controller's radar screen. Secondary radar enhances primary radar by giving the dot additional information (such as the aircraft's call sign and altitude).

7. To support the safe navigation of aircraft, the world's airspace is broken down into different categories (with different rules for entering and travelling through each one). ICAO designates seven classes of airspace (A–G), and States can choose which classes meet their needs.

8. Aircraft navigation is supported by various technologies, including aircraft on-board avionics, ground-based navigation aids (NAVAIDs), and global navigation satellite systems (GNSSs).

9. To organize the flow of traffic into, out of, and through busy areas, controllers and pilots abide by the regulations associated with terminal area navigation (including standard instrument departures (SIDs), visual approach procedures, and instrument approach procedures (IAPs)) and en route navigation. Modern navigation approaches include performance-based navigation, which offers a more direct routing for aircraft using area navigation (RNAV).

10. Flight information regions (FIRs) are large sections of airspace aligned with geographic areas (usually following State borders). Each FIR may contain controlled and/or uncontrolled airspace. In controlled airspace, ATCO services are provided and pilots require a clearance to enter, while uncontrolled airspace has no ATCO supervision.

11. Pilots may operate under instrument flight rules (IFR), which means that they fly primarily by referencing their avionics, or under visual flight rules (VFR), which means that they fly with visual reference to the outside world (i.e., what they see out the cockpit window).

12. The future of air traffic management includes NextGen and SESAR. NextGen is a American programme designed to increase airspace capacity through a variety of technologies (including ADS-B, data communications, new voice systems, and system wide information management). The Single European Sky ATM Research (SESAR) programme is a similar initiative that focuses on tripling the capacity of the airspace system in Europe using new trajectory management approaches, traffic synchronization programmes, a common operational environment, and decision-making support for ATCOs.

Table 4.5 Acronym rundown

ACARS	aircraft communications addressing and reporting system
ACC	area control centre
ACU	approach control unit
ADS-B	automatic dependent surveillance – broadcast
AFCS	automatic flight control system
AFGS	automatic flight guidance system
ANSP	air navigation service provider
ATC	air traffic control
ATCO	air traffic control officer
ATM	air traffic management
ATSEP	air traffic safety electronic personnel
CAA	civil aviation authority
CANSO	Civil Air Navigation Services Organisation
CATMT	collaborative air traffic management technologies
CNS	communication, navigation, surveillance
CPDLC	controller–pilot data link communications
DH	decision height
DME	distance measuring equipment
EASA	European Aviation Safety Agency
EFIS	electronic flight instrument system
EFPS	electronic flight progress strip
FAA	Federal Aviation Administration
FANS	future air navigation systems
FIR	flight information region
FMC	flight management computer
FMS	flight management system
FPS	flight progress strip
GLONASS	Global Navigation Satellite System
GNSS	global navigation satellite system
GPS	global positioning system
HF	high frequency
HI	heading indicator
IAP	instrument approach procedure

(Continued)

Table 4.5 (Continued)

ICAO	International Civil Aviation Organization
IFR	instrument flight rules
ILS	instrument landing system
IMC	instrument meteorological conditions
INS	inertial navigation system
LF	low frequency
MDA	minimum descent altitude
MF	medium frequency
NAVAID	navigation aid
NDB	non-directional beacon
NVS	national airspace system voice system
PBN	performance-based navigation
RNAV	area navigation
SES	Single European Sky
SESAR	Single European Sky ATM Research
SID	standard instrument departure
SWIM	system-wide information management
TACAN	tactical air navigation system
UHF	ultra high frequency
VFR	visual flight rules
VHF	very high frequency
VMC	visual meteorological conditions
VOR	very high frequency omnidirectional range
VORTAC	VOR paired with TACAN

Chapter Review Questions

4.1 Which ATCO role do you think would be the most demanding? Explain your answer.

4.2 What is the Civil Air Navigation Services Organisation (CANSO)? How did it come to exist? What is its role in today's aviation landscape?

4.3 Provide three advantages of using secondary and primary radar together as opposed relying solely on primary radar.

4.4 Describe three ways that air navigation is essential to the future of aviation Explain how airlines and air navigation service providers need each other to be viable industries.

4.5 Why is ATCO training and licensing so rigorous? Is such rigor absolutely necessary?

4.6 Consider the use of verbal communication in the aviation industry, and answer the following:

- In what ways is the industry dependent on verbal communication?

- Why is standard phraseology important?

- Do you agree that English should be the international language of aviation? Why doesn't (or shouldn't) each State use its own official language?

- What cultural and linguistic challenges may negatively impact verbal communication within the industry?

- Can you imagine any technologies that may improve communication? If so, how would they do so?

4.7 Why is global airspace organized into so many classes? Is this necessary? How might this system of organization be influenced by States having sovereignty (ownership) over their airspace? Explain.

4.8 Would you expect more accidents to occur under VFR rules or IFR rules? In controlled or uncontrolled airspace? Why do you think so?

4.9 Name an ANSP in your State. How does it operate?

4.10 For an ATCO, which do you think would be more challenging to control: a precision or non-precision approach? Traditional or performance-based navigation? Justify your choice with evidence.

MALAYSIA AIRLINES FLIGHT 370 – A MODERN AVIATION MYSTERY[1]

On 8 March 2014, Malaysia Airlines flight MH 370, a Boeing 777–200ER aircraft, disappeared from ATC radar after departing from Kuala Lumpur, Malaysia. The flight was a scheduled passenger service heading to Beijing, China with 227 passengers and 12 crew members.

The timeline leading up to the crash, as far as can be pieced together, is as follows:

8:41 UTC: Flight takes off from Kuala Lumpur International Airport in Malaysia.

9:07 UTC: MH 370's aircraft communications addressing and reporting system (ACARS) transmits its final message from the flight crew.

9:19 UTC: A radio communication (believed to be from the co-pilot) was transmitted: 'Good night, Malaysian three seven zero.' At this point, the Kuala Lumpur controller expected the pilots to switch their communications to an en route controller in Vietnam.

9:21 UTC: The aircraft's transponder (used to transmit location to ATC ground radar) stopped transmitting. The aircraft did not contact the Vietnamese en route ATC. The ACARS was turned off, such that no more manual messages could be sent; however, the ACARS continued to send automatic 'pings' in the background.

10:15 UTC: Malaysian military radar detects the aircraft flying west over the Indian Ocean, a nearly 90 degree heading change from its course to Beijing.

16:11 UTC: MH 370's ACARS system sends its last automatic ping to satellites. These pings were received every hour between take-off and 16:11. Note that ACARS is a *communication* rather than *surveillance* system, so the hourly pings cannot be used to determine the location of the aircraft, only that it was airborne and functional at the time.

16:19 UTC: MH 370 ACARS transmits one final 'partial handshake' with satellites, eight minutes after the final hourly scheduled ping. This partial handshake – a failed login attempt from the aircraft to the satellite – was the final transmission from the aircraft, and probably indicates the time of the crash.

This incident caused a global media frenzy. For the next month, it was impossible to turn on the news or visit a website without seeing updates, questions, and speculations from aviation experts. All over the world, people were consumed with how it was possible for a large aircraft to simply disappear.

An extensive air and sea search was launched, and the clock was ticking because the battery life of a transmitter on an aircraft black box (flight data recorder or cockpit voice recorder) is only about 30 days. The search was not successful.

After a period of weeks, satellite signal analyses of the hourly pings and final partial handshake led investigators to discover that MH 370 had flown for more than six hours after the final contact with pilots, and had entered the airspace above the Southern Indian Ocean.

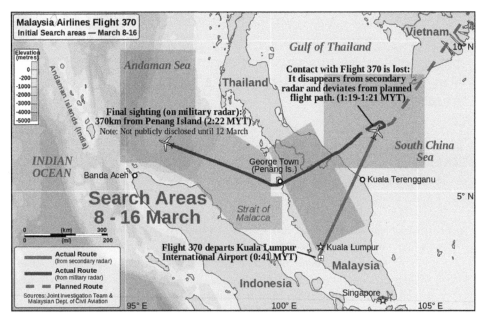

Figure 4.10 MH370 search area (map)

Source: Andrew Heneen [Attribution or CC BY 3.0 (http://creativecommons.org/licenses/by/3.0)], via Wikimedia Commons

In an immense effort to locate the aircraft, the Australian Transport Safety Bureau accepted responsibility from the Malaysian government to lead a search of the Indian Ocean seabed – about 800 kilometres (roughly 500 miles) off Australia's western coast. The search was conducted in two phases:

1. ships, equipped with multibeam sonar, gathered high-resolution images of the seabed; and

2. an underwater search was performed using special equipment including an autonomous underwater vehicle with mounted sonar (a *water drone*).

As of April 2016, 95 000 square kilometres (37 000 square miles) of seafloor had been searched, with the intention of eventually searching 120 000 square kilometres (46 000 square miles). One goal of the search is the recovery of key aircraft components, specifically the cockpit voice recorder and flight data recorder, which will assist tremendously with the Malaysian investigation.

To date, a small wing fragment has been the only confirmed evidence from MH 370. It was found July 2015 on Réunion Island in the Indian Ocean, 3700 kilometres (2300 miles) away from the main search site. This was the first proof that the plane had crashed into the ocean.

Debate is still raging about whether the flight was under human control until the end, pointing to a deliberate action (perhaps by the pilots) as the cause of the accident, or whether humans were somehow incapacitated causing the aircraft to fly on autopilot until it ran out of fuel. Further debate has arisen about pilot access to on-board surveillance technology: sophisticated surveillance technology on an aircraft is rendered useless if pilots are able to turn it off. A wide variety of conspiracy theories related to MH 370 have circulated, ranging from the plausible to the ridiculous. Some have suggested that terrorists are to blame, some believe that it was extraterrestrials, and still others point to the American military, North Korea, or Vladimir Putin. Further theories allege that the crash was predestined, that the plane was switched with another flight that was shot down (and the Réunion Island evidence planted), or that it was all part of a complicated insurance scam.

Setting aside wild conspiracy theories, consider the following: the aircraft's transponder ceased transmission and the ACARS was turned off immediately following the last pilot radio call to the Kuala Lumpur controller (when the aircraft should have been handed over to an en route controller in Vietnam). Simultaneously, the aircraft made a sharp left turn off course.

Note

1 Ashton, et al., 2015

Case Study Questions

Given the evidence and what you have learned in this chapter, make informed guesses about the following questions:

4.11   Could a massive emergency, such as an on-board fire, have been responsible for systems shutting down and for the left-hand turn? Why wouldn't an emergency have been declared over the radio? How could the aircraft continue to fly for the next six hours?

4.12　Could a loss of cabin pressure have caused incapacitation of the crew and passengers, causing the plane to remain airborne with no one at the controls?

4.13　Could a hijacking have occurred? How much aviation knowledge would the hijacker have required to execute this attack? (Note that the pilots did not indicate hijacking using the transponder code for hijacking, the radio, or the ACARS.)

4.14　Is it possible that one of the pilots deliberately diverted or hijacked the aircraft?

References

Aireon, 2016. *It's just ADS-B.* [Online] Available at: http://aireon.com/resources/its-just-ads-b/

Andrade, A. A. L., 2001. *The global navigation satellite system.* Aldershot: Ashgate Publishing Limited.

Ashton, C., Bruce, A. S., Colledge, G. & Dickinson, M., 2015. The search for MH370. *The Journal of Navigation,* 68, p. 1–22.

Boulton, T., 2013. *Why people on planes and ships use the word "mayday" when in extreme distress.* [Online] Available at: www.todayifoundout.com/index.php/2014/01/planes-ships-used-word-mayday-distress/

CANSO, 2015. *CANSO members.* [Online] Available at: www.canso.org/canso-members

Comisión de Accidentes, 1978. *Joint report: KLM–PAA.* Madrid: Ministerio de Transportes Y Comunicaciones.

ICAO, 1978. *Circular 153-AN/56.* Montreal: International Civil Aviation Organization.

ICAO, 2001. *Annex 10 to the Convention on International Civial Aviation: Aeronautical telecommunications, 6th ed.* Montreal: International Civil Aviation Organization.

ICAO, 2008. *Performance-based navigation (PBN) manual, Doc 9613.* Montreal: International Civil Aviation Organization.

Kearns, S. K., Mavin, T. J. & Hodge, S., 2016. *Competency-based education in aviation: Exploring alternate training pathways.* Surrey: Ashgate.

NATS, 2016. *From 1920s to today.* [Online] Available at: www.nats.aero/about-us/our-history/

Nolan, M. S., 2004. *Fundamentals of air traffic control.* 4th ed. Belmont, CA: Thomson Learning.

NTSB, 1987. *AAR-87-07: Collision of Aeronaves De Mexico, S.A. McDonnell Douglas DC-9-32, XA-JED and Piper PA-28-181, N4891F.* Washington, DC: National Transportation Safety Board.

The New York Times, 1922. Americans die in French air craft. *The New York Times,* 8 April, p. 1.

Notes

1　Kearns, et al., 2016

2　NATS, 2016

3　The New York Times, 1922

4　ICAO, 2001

5　Aireon, 2016

6　Nolan, 2004

7　Nolan, 2004

8　ICAO, 2008, pp. 1–xix

9　Andrade, 2001

10　Andrade, 2001

11　www.sesarju.eu/solutions

1. SARPs that impact airport operations are included within Annex 11 to the Chicago Convention.
   a. True
   b. False

2. Which sector of the airport is secured, beginning with the security checkpoint and extending to the airport perimeter fence?
   a. Landside
   b. Airside
   c. Apron
   d. Movement area

5. In general, most of airports have very similar operational and business models.
   a. True
   b. False

3. Parking fees are allocated within the _____ category of airport revenue.
   a. Aeronautical
   b. Non-aeronautical

4. Airport operators must strategically balance several (often competing) factors, such as security with passenger comfort, segregating groups of passengers with limiting walking distances, etc.
   a. True
   b. False

Learning science suggests that thinking through a few questions before you begin studying new material, even if you answer incorrectly, results in improved learning and retention.
Give it a try!

CHAPTER 5

Airports

CHAPTER OUTCOMES

At the end of this chapter, you will be able to . . .

- Summarize the history of airports including the evolution and role of Airports Council International.
- Describe international regulations that apply to airports.
- Explain the organizational structure of airports, specifically key considerations and design elements of airside and landside operations.
- Apply what you have learned to an analysis of Southwest Airlines flight 1248, which was a runway overrun accident.

Introduction

Around the world, there are 41 788 airports used for military, airline, and general aviation; of those, 3883 airports support scheduled commercial airline flights.[1] Airports represent much more than a patch of land. They have important societal and human impacts in facilitating air transportation and fostering local economies.

Airports create direct job opportunities. Globally, 450 000 people work for airport operators, while another 5.5 million work in jobs directly linked to airports (such as retail outlets in the airport, customs and immigration, and catering, among many others).[2] Yet the economic impacts of airports extend beyond transportation-related jobs. Airports also create indirect jobs associated with infrastructure development and the required supply-chain (tourism, ground transportation, fuel, and logistics). More broadly, airports serve to link local economies to international markets.

Did You Know?

The general public is familiar with the term *airport*, yet many have not heard the term *aerodrome*. In the aviation industry, the two terms are used interchangeably – regulators (including ICAO) tend to favour *aerodrome* while industry operators more commonly use *airport*.

Aerodrome is a broad term that describes any location used for take-offs and landings of aircraft (on land, water, or even a mobile platform on a ship). Airports, a subcategory of aerodromes, are typically used for international traffic and have predictable services associated with customs, immigration, and public health.

This chapter will explore the regulations, associations, and operational considerations involved in airport management around the world. Note that security, environmental issues, and safety are all crucial considerations in airport management; however, as these topics are covered in Chapters 6, 7, and 9, respectively, they are only mentioned briefly in this chapter.

Brief History of Airports

In the earliest days of aviation, airport locations were chosen out of convenience. Aviators would take off and land at a beach or from a farmer's field as long as it was flat and located nearby. However, as aviation evolved during the World Wars, it became clear that airports with predictable resources (such as available fuel) and reliable infrastructure (such as strong, durable runway surfaces) were a necessity.[3]

5.1 Early Airports

1909   College Park Airport in Maryland, US began operations; it is believed to be the oldest continuously operated airport in the world.

1916   Amsterdam's Airport Schiphol opened for military traffic; civil operations began in 1920.
Rome's Ciampino Airport began operations.

1919   Paris's Le Bourget Airport began operations; in 1927, it became the landing site of Charles Lindbergh's famous solo Atlantic flight.
Hounslow Heath Aerodrome near London, UK was the first to begin international commercial services.

1920   Sydney Airport in Australia began operations; it is now one of the oldest continuously operated airports.[1]

Note

1 ACI, n.d.

In the 1950s, commercial aviation was an increasingly popular mode of transportation, and airports had to evolve. During this era, airports built their first *passenger terminals* – buildings where passengers transition from ground-based transportation to air-based transportation.

The 1960s marked the beginning of the jet age, which resulted in larger aircraft that could travel greater distances. Airports adapted by building longer runways and jet-bridge systems that allowed for the loading and unloading of passengers from the terminal building to an aircraft without going outside.[4] This new airport infrastructure carried heavy costs, and at that time most airports were managed by government entities – their operations and maintenance costs were paid with public funds.

From the 1970s through the 1990s, many governments around the world moved to privatize airports, establishing commercialization and concession agreements with the private sector. As of 2017, a mixture of public and private approaches to ownership exist around the world: some airports remain government owned and operated; some are privately owned and operated; still others are managed through a hybrid model whereby governments rent airports to non-profit airport associations. Today, 46 per cent of the world's busiest airports have some type of private sector involvement.[5]

Airports Council International

In 1948, only shortly after the 1944 Chicago Convention, 19 airport representatives from the United States gathered to address mutual operational and regulatory issues. This led to the formation of the Airport Operators Council, which evolved to become the Airport Operators Council International (AOCI) based in Washington, DC. Over the following years, two other airport associations were formed in Europe: the Western European Airports Association (WEAA) established in 1950 and based in Zurich, and the International Civil Airports Association (ICAA) established in 1962 and based in Paris.[6]

To more effectively represent airports internationally, the three associations agreed to work together and, in 1970, become the Airport Associations Coordinating Council (AACC) based in Geneva. AACC was granted observer status at ICAO in 1971, giving airports a collective voice associated with international standards development.

Did You Know?

ACI, ICAO, and IATA partner to offer the Young Aviation Professionals Programme (YAPP). YAPP provides an opportunity for young, talented professionals with advanced degrees to participate in each organization's work programmes related to safety, air navigation capacity, economics, and aero-political issues.

In 1991, Airports Council International (ACI) was established as a non-profit organization that succeeded the AACC and the original organizations that it comprised. Today, ACI has its global headquarters in Montreal and five regional offices around the world:

- Africa (Casablanca, Morocco);

- Asia–Pacific (Hong Kong, China);

- Europe (Brussels, Belgium);

- Latin America–Caribbean (Panama City, Panama); and

- North America (Washington, DC, United States).[7]

ACI is the voice of the world's airports, working to represent their collective interests and promoting professional excellence in airport management. ACI serves nearly 600 members who operate more than 1800 airports in 173 countries.[8]

International Airport Regulations

The ICAO Council first adopted SARPs for aerodromes in 1951,[9] which evolved into Annex 14 to the Chicago Convention. Annex 14 is unique among the annexes in the breadth of subjects that it covers – from airport planning, civil engineering, lighting, and search and rescue equipment requirements to methods of preventing wildlife and bird strikes.[10] With such a wide array of issues to cover, Annex 14 is lengthy and is revised regularly as aviation technologies evolve and airports must adapt. For example, newer models of aircraft (such as the Airbus 380) require wider taxiways and Annex 14 had to be updated to reflect this necessity. In fact, Annex 14 is one of the most frequently updated annexes.[11] In 1990, following 39 amendments to Annex 14, the document was split into two volumes:

- *Volume 1 – Aerodrome Design and Operations* contains SARPs and guidance materials on planning, design, operation, and maintenance of aerodromes; and

- *Volume 2 – Heliports* contains SARPs for the design, planning, and operation of heliports.[12]

Aerodrome Certificates

To ensure regulatory compliance, an aerodrome used for international flights must be certified and granted an *aerodrome certificate* (AC) by its civil aviation authority (CAA).[13] To earn an AC, an applicant must submit an aerodrome manual to the CAA for approval. The aerodrome manual will describe how the facility meets regulatory requirements related to operational procedures, management, services, equipment, facilities, and safety management.[14]

Based on an airport's facilities and infrastructure, its AC will specify the type of aircraft that the airport can serve (e.g., the maximum take-off weight or the maximum number of passenger seats). Because ICAO SARPs represent minimum standards, some international variability in ACs will exist. All aerodromes with international flights must meet ICAO standards, but some States may choose to impose stricter airport standards. Note that many States have aerodromes that do not require certification – these facilities are used for general aviation rather than international flights and therefore fall outside the scope of international standards.[15]

Airport Codes

Travellers may know that the aviation industry uses unique letter codes to identify airports. These airport identifiers are commonly used on travel booking websites and are printed on checked luggage tags. However, there are actually two different identifiers given to each airport, one from ICAO and another from the International Air Transport Association (IATA). ICAO codes are primarily used for international standardization in aeronautical flight plans and by air navigation service providers (ANSPs) around the globe. IATA codes, on the other hand, are the codes that travellers are more familiar with – they are used to support airlines, and used by airlines for reservations, timetables, and bag tags. The codes provided in Table 5.1 are IATA codes.

Table 5.1 Airports in ICAO Council States with the most movements, passengers, and cargo, 2015

State	Total movements[1]	Total passengers	Total cargo
Australia	Sydney (SYD): 335,001	SYD: 39,915,674	SYD: 447,149
Brazil	São Paulo (GRU): 295,030	GRU: 39,213,865	GRU: 526,012
Canada	Toronto (YYZ): 443,958	YYZ: 41,036,847	YYZ: 434,777
China	Beijing (PEK): 590,169	PEK: 89,938,628	Shanghai (PVG): 3,275,231
France	Paris (CDG): 475,810	CDG: 65,766,986	CDG: 2,090,795
Germany	Frankfurt (FRA): 468,153	FRA: 61,032,022	FRA: 2,076,734
Italy	Rome (FCO): 315,217	FCO: 40,422,156	Milan (MXP): 511,191
Japan	Tokyo (HND): 438,542	HND: 75,573,106	Tokyo (NRT): 2,122,314
Russian Federation	Moscow (SVO): 265,040	SVO: 31,612,402	SVO: 198,851
United Kingdom	London (LHR): 474,103	LHR: 74,989,795	LHR: 1,591,637
United States	Atlanta (ATL): 882,497	ATL: 101,491,106	Memphis (MEM): 4,290,638

[1] Note that total movements comprise take-offs and landings; total passengers (each counted once) include those arriving, departing, and in direct transit (connecting); and total cargo is measured as cargo loaded and unloaded in metric tonnes.

Source: ACI, 2016a and ACI, 2016b

International Standardization – Time and Location

Airports must rely on international standards associated with time and location, as local time zones and location conditions vary. To ensure consistency and harmonization, international aviation operations use Coordinated Universal Time (UTC) and the location references of *latitude*, *longitude*, and *mean sea level*.

As there are many time zones around the world, airport time is standardized to UTC. UTC has a long history stretching back to 1883 when the railway industry adopted Greenwich Mean Time (GMT) – the legal time in the United Kingdom – as a standard time. With the advent of aviation, GMT was adopted by aviators.[16] In 1928, astronomers suggested GMT be renamed as Coordinated Universal Time (UTC) as it was being used for astronomical observation of star transits (hence the word *universal*). UTC is now the international standard for coordinated civil time.

Airports also require standards related to location. The horizontal location of an airport (i.e., its location on the surface of Earth) is expressed using latitude and longitude, geographical coordinates relative to the equator and the prime meridian.

5.2   Lines of Latitude and Longitude[1]

The equator, which marks the centre of Earth, has a latitude of 0 degrees. Lines of latitude run parallel to the equator, extending north and south to the poles, where they reach 90 degrees.

The prime meridian, which runs through Greenwich in the UK, has a longitude of 0 degrees. Lines of longitude run parallel to the prime meridian, extending east and west. Unlike the equator, which indicates the actual centre of Earth, the prime meridian is an arbitrarily chosen marker, agreed to by the international community in 1884.

Note

1 Nelson, et al., 2001, p. 38

As the surface height of the earth is variable – from low points at ocean level to high points on mountain peaks – airport position must also be expressed vertically. Airport elevation is expressed in feet above mean sea level (MSL). Aircraft altimeter readings, which inform the pilot of altitude, also reference MSL elevations. For example, when a pilot is at an airport with an elevation of 800 feet MSL, the altimeter will read 800 feet while the aircraft is still on the ground. After take-off, the pilot will have to subtract 800 from the altimeter reading to calculate height above ground level.

5.3 Laser Beams – A Growing Concern

As pilots approach an airport for landing, they rely heavily on the visual cues around them. To land successfully, they must make visual contact with the airport. In poor weather conditions, pilots intensely scan their surroundings looking for airport surface lights.

Because a pilot relies on his or her vision, lasers beams have become a significant hazard to aviation. They produce light at an intensity that can permanently damage the retina of the eye instantaneously, even from over 10 kilometres (6.2 miles) away. Lower intensity lasers may not cause tissue damage, but they do disrupt vision. When a pilot's vision is disturbed by a laser beam, his or her view of the outside world is completely disrupted – the windshield turns opaque (illuminated by the light of the laser) and any night vision adaptation is lost.

Lasers used for entertainment purposes are increasingly popular. In the United States alone, there were nearly 200 laser strike incidents per day in 2015.[1] There have also been European reports of air traffic control towers hit by lasers.[2]

To protect pilots and their flights, several protected zones around airports have been established to limit the use of visible laser beams. Laser-beam free flight zones (LFFZs), laser-beam critical flight zones (LCFZs), and laser-beam sensitive flight zones (LSFZs) are established in the proximity of aerodromes to limit the use of visible lasers.[3]

Notes

1 Esler, 2016
2 Esler, 2016

3 ICAO, 2016; ICAO, 2003

Airport Operations

In discussing the structure of airports, it is essential to differentiate between *landside* and *airside* operations.

- The *landside* of the airport is an unrestricted area open to the public. It includes roadways for vehicle traffic, parking lots, and the parts of the terminal building used for airline check-in and baggage drop-off. The landside of the airport ends at the security checkpoint within the terminal building.

- The *airside* of the airport is a secured area that begins at the security checkpoint in the terminal building and extends to the perimeter fence around the entire airfield. Elements of the airside include the secure part of the terminal building, the apron where aircraft park for loading and unloading, and the taxi and runway surfaces. When arriving passengers pass through customs or immigration, they leave the secured airside of the airport and transition to the landside. Note that only individuals with boarding passes or an airport security identification card (ASIC) are permitted airside.[17]

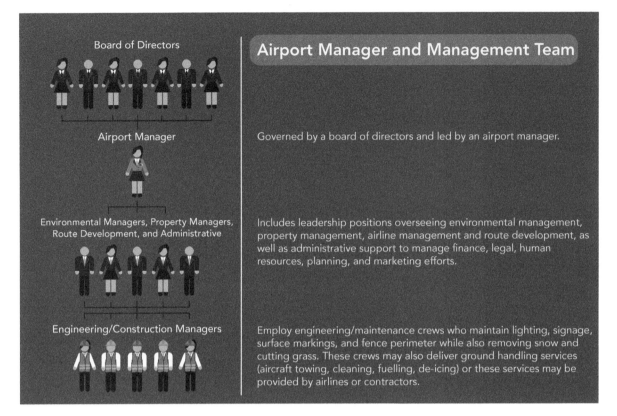

Figure 5.1 The airport community

The airport community includes many people and agencies that collaborate to handle the operations of such a complex facility.

Airport Design

Airports must be strategically designed with consideration given to security and regulatory requirements, passenger flow, clear signage, minimal walking distances, minimal passenger cross-flows, and people movers such as shuttles or moving sidewalks. These design considerations apply to the landside interface between ground transportation and the terminal building, the terminal building itself, and the airside interface between the terminal building and aircraft.

Landside Operations

When arriving via ground transportation, a passenger will first enter the landside of the airport. The landside includes roadways, parking facilities and walkways, and the unsecured part of the terminal building. In landside design, a key consideration is how passengers transition from ground to air transportation and their movements into the terminal building. This includes the ease of access by automobile, short-term and long-term parking, signage, and passenger wayfinding into and within the terminal building. The terminal itself must be designed in a way that takes into account the fact that passengers and greeters may spend

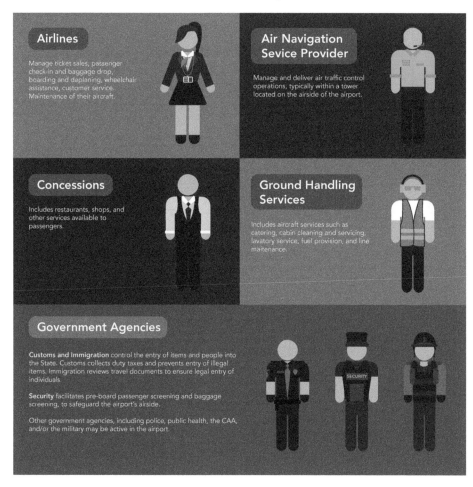

Figure 5.1 The airport community (Continued)

time a long time waiting – therefore, space, seating, and various needs and wants (including food, restrooms, shopping, and so on) must be considered.

Airside Operations

The airside component of an airport includes the secured section of the terminal building, gates, the apron (aircraft parking area), aircraft movement areas (taxiways and runways), the control tower operated by the ANSP, hangars and other buildings, and extends to the perimeter fence that encompasses the entire airside of the airport. The primary activity on the airside is the landing and taking-off of aircraft, but there is a complex support network of vehicles, pedestrians, and machinery used to facilitate and expedite this process.

As noted above, passengers transition from the landside to the airside of a terminal when they pass through security; from that point on, they must navigate the airside of the terminal to reach their departure gate and wait to board their aircraft.

Figure 5.2 A passenger at an airport gate

Apron and Gates

Looking out the window of the terminal building, the passenger will see aircraft parked on what is known as the airport's *apron*. The apron is the surface area that surrounds the terminal building.

5.4   De-icing and HOT

Aircraft must have clean wings for take-off. Because frost and ice are considered *contaminants*, de-icing fluid is often applied to aircraft on the ground before take-off. When de-icing fluid is applied, that aircraft's holdover time (HOT) begins. HOT is the estimated time that the de-icing fluid provides protection from contaminants. If an aircraft has not taken off before its HOT elapses, it must return to the apron for another de-icing.

The terminal has several *gates*, each of which allows an aircraft to connect to the terminal for passenger boarding and deplaning, usually via a jet-bridge.

Turn-around refers to the process of an aircraft arriving at a gate, deplaning, being serviced, boarding new passengers, and then leaving the gate area to taxi to the runway. Airlines and airports collaborate to turn around aircraft as quickly as possible to maximize the efficiency of operations. During the turn-around, while passengers deplane and others board, a variety of services are carried out on the aircraft. These often include baggage unloading and loading, fuelling, toilet service, galley service, cabin cleaning, line maintenance, and de-icing/anti-icing among others (see Figure 5.3).

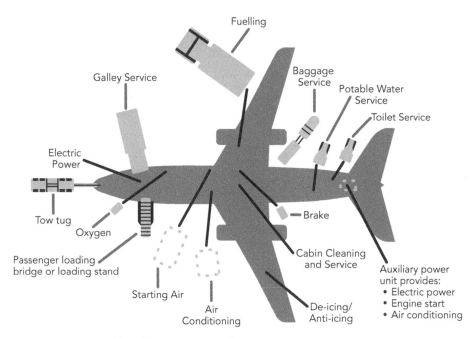

Figure 5.3 Ground handling services at the gate

Source: Adapted from ACI, 2017a; ACI, n.d.

There are a variety of layouts used for the design of terminal buildings and aprons and each layout will have different requirements associated with lighting, marking, and signage. Aprons designed for a passenger terminal are configured differently than those used for general aviation or cargo. In designing terminals and aprons, consideration is given to ensuring the following:

- terminal design does not result in passengers having to walk long distances to reach their gates;

- aircraft can manoeuvre safely close to the terminal building (i.e., wing tip clearances are considered);

- security can be maintained;

- ground-handling service vehicles (trucks, baggage carts, and so on.) can move safely around aircraft with separate service roads;

- the entire area can be lit effectively for night operations; and

- the surface is made of durable material and has a slight slope to allow water to drain off the manoeuvring area.[18]

Depending on specific needs, the above objectives are met through a variety of terminal–gate–apron designs (see Figure 5.4).

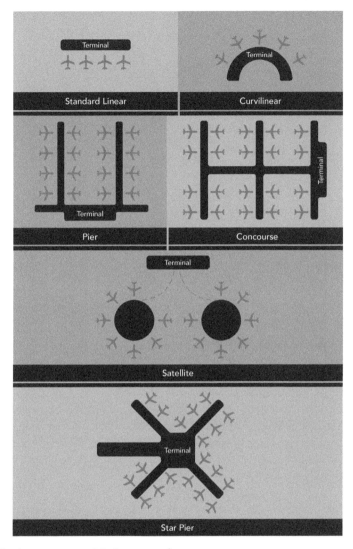

Figure 5.4 Airport terminal & Apron configurations

Manoeuvring Areas

After turn-around, an aircraft must make its way to the runway for take-off. Most aircraft are unable to reverse under their own power, and therefore require a *tow tug* to push them backwards out of the gate area. Once it is facing forward, the aircraft will taxi under its own power on the taxiway and eventually take off from the runway. The term *manoeuvring area* describes the entire network of taxiways and runways at an airport. The surfaces of all movement areas are regularly inspected for foreign object debris (FOD) and for any problematic surface conditions, which could impair aircraft operations. Annex 14 outlines specific criteria for the design and maintenance of manoeuvring areas. These requirements, along with key information about taxiways and runways, are summarized in Table 5.2.

Table 5.2 The essentials of manoeuvring areas

	Taxiways	Runways
Purpose	a series of interconnected 'roads' used by aircraft and vehicles to move around the surface of the airport	surfaces used for the landing and taking-off of aircraft
Markings	Yellow centrelines are painted on taxiways.	White lines are painted on runways. Marking requirements vary depending on whether the runway is used for instrument landings or visual landings. Markings generally include threshold markings, the runway name (numbers), centreline, side stripes, and aiming points to assist pilots during landing.
Lighting	blue (may include edge, centreline, holding stop bar, and obstruction lights)	white (may include edge, touchdown, centreline, and end lights)
Controller	ground controller (located in tower)	tower controller with responsibility for landings and take-offs on the active runway
Naming	Taxiways are named with a letter of the alphabet (A–Z). If a large airport has more taxiways than single letters, then double letters (AA, AB, AC . . .) are used as well. Pilots, controllers, and airport employees use the phonetic alphabet to describe taxiways: taxiway Alpha, taxiway Bravo, etc.	Runways are named based on their orientation in relation to the 360 degrees of a compass (north = 360, east = 090, south = 180, and west = 270). An east–west runway is called runway 09/27. An east-facing aircraft on the end of the runway would take off from runway 09. If the wind was blowing from the opposite direction, the aircraft would take off from the opposite end of the runway (facing west) and would be said to be taking off from runway 27.
Orientation	To minimize fuel burn of taxiing aircraft, taxiways are designed to provide as direct a route as possible between the runway and terminal building.	Aircraft take off facing into the wind, so runways are oriented with consideration given to prevailing winds, land topography and obstacles, fog potential, other air traffic routes, and the types of aircraft using the airport. Multiple runways at an airport may be configured in a variety of patterns: two parallel runways (L and R, for left and right, are added to runway names); an intersecting X-configuration; or a V-configuration, among others.

(Continued)

Table 5.2 (Continued)

	Taxiways	Runways
Signage	Yellow letters or numbers (e.g., A or 09) on a black background are used to mark actual location (i.e., which runway or taxiway an aircraft is currently on). Black letters or numbers on a yellow background provide directional information for navigation. These signs will show a taxiway or runway name with an arrow (for example, B with an arrow pointing to the location of taxiway Bravo).	
Hold short information	Painted ground markings designate where a taxiway ends and a runway begins, so that aircraft and airport vehicles do not inadvertently cross an active runway without permission from a controller. Red signs with white numbers (e.g., 09/27) indicate a runway ahead, and signal that aircraft and vehicles must *hold short* (i.e., wait until clearance is given to proceed).	
Intersections	Named intersections exist where taxiways and runways cross. For example, the B 09/27 intersection occurs where taxiway Bravo crosses runway 09/27, and clear signage is required. Naming intersections helps pilots and vehicle drivers communicate their location to ground controllers and one another.	

Source: Adapted from ACI, 2017a and ICAO, 2016

Did You Know?

Taxiways are named by letter, but not all 26 letters of the alphabet are used. I and O are not used as they could be confused for one and zero, and X is not used because this symbol designates closed sections of the airport.[1]

Note

1 ACI, 2017a

In addition to the manoeuvring areas, the airside of an airport houses a variety of support buildings, including storage and maintenance buildings for aircraft equipment, hangars for aircraft parking and storage, the ANSP's tower, rescue and firefighting facilities, and fuel facilities (such as large fuel storage tanks, fuel trucks, and fuel hydrants).[19]

Airports must have designated safety areas adjacent to where aircraft operate to reduce accident risks in the unlikely event of a runway excursion. Natural hazards must be removed and surfaces must be free of foreign object debris (FOD).

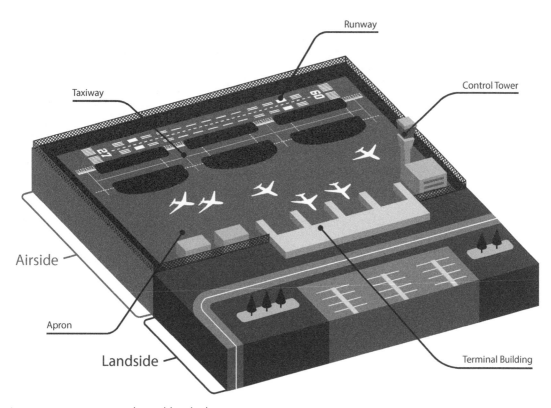

Figure 5.5 Airport airside and landside

Case Study: Foreign Object Debris – Air France Flight 4590[1]

The Concorde was the first *supersonic* aircraft (able to fly faster than the speed of sound) used for scheduled commercial service. Whereas a typical flight from London to New York would take about 8 hours, the Concorde could make the crossing in about 3.5 hours.

The Concorde was used by both Air France and British Airways for commercial scheduled service and had a reputation as one of the safest aircraft in the world. That changed on 25 July 2000 when a Concorde, operating as Air France flight 4590, ran over runway debris at Charles de Gaulle Airport in Paris, France. The debris ruptured a tyre, caused it to disintegrate, and sent tyre fragments into the wing. The impact of the fragments caused a shock wave, which ruptured a fuel tank. The leaking fuel caused *flame-outs* (loss of power) on engines 1 and 2. The pilots, going too fast to stop the aircraft on the runway, decided to take off and circle the airport to land. However, without engines 1 and 2, the aircraft did not have sufficient power from the other two functional engines to remain airborne. The aircraft crashed into a hotel near the airport, fatally injuring all 109 people on board and 4 people in the hotel.

Investigation revealed a metal strip on the runway that was later matched to the rupture mark on the Concorde's tyre. Only five minutes before the Concorde's take-off, the metal wear strip had fallen off a Continental Airlines DC-10 that had taken off for Newark. It was an unfortunate tragedy that on such a large runway surface, the small piece of FOD was precisely aligned with the wheels of the Concorde as it took off.

As a result of the accident, all Concordes were grounded pending investigation. Concorde operations resumed in November 2001 but the aircraft's reputation never fully recovered, and the Concorde was permanently retired in 2003.

This accident illustrates the disastrous potential of FOD. Even small items on the runway can cause devastating outcomes. FOD can cut through tyres, get sucked into engines, become lodged in flight control surfaces, or affect aircraft in many other ways. In this case, the metal strip fell off another aircraft, but FOD can also come from the airport infrastructure (degraded pavement, loose lighting or signs) or the natural environment (wildlife, ice, trees).[2]

To ensure safe operations, airports implement FOD prevention programmes whereby all movement areas, aprons, airside roads, and grassy areas are routinely inspected and audited. Various levels of inspection occur:

- Level 1 — routine daily inspections (ICAO recommends a minimum of every six hours, particularly at dawn, morning, afternoon, and dusk);
- Level 2 – more careful detailed inspection, by foot or at very low speeds; and
- Level 3 – management inspections and audits.[3]

Detailed records of inspections and audits must be kept according to the requirements of the CAA for a minimum of five years.[4]

Notes

1 BEA, 2002
2 ACI, 2017
3 ACI, 2017
4 ACI, 2010

The entire airside of the airport must be enclosed with perimeter fencing – and because an airport typically covers a large area, the fencing needs are extensive. The purpose of the fence is to maintain airport security, by preventing unlawful acts and the entry of wildlife (that could present a hazard to flight operations). Controlled access gates are installed around the perimeter to facilitate secured entry and exit of airport vehicles.[20]

For the purposes of safety and security, airside operations must also include rescue and firefighting services (RFFS), the requirements for which vary based on the types of aircraft that operate at an airport. There are 10 categories of airport, with the highest categories used for airports that support the largest aircraft. Annex 14 sets out the minimum RFFS for each of these 10 categories and includes criteria such as response time, number of required fire-fighting vehicles, and so on.[21]

Airport Economics

As noted earlier in this chapter, airports were historically operated by the government of the State. However, as infrastructure needs increased and the financial burden became too great for public funds to support, many airports shifted to operating models comparable to commercial operations. Today, airports are operated in various ways, including ownership and operation by government, private for-profit corporations, and not-for-profit airport authorities.

Regardless of whether they are operated publicly or privately, all airports must carefully balance revenue with costs.

Revenue

Airport revenue falls into two categories: aeronautical and non-aeronautical. Aeronautical revenue refers to funds collected from airlines, which include

- *landing fees* charged based on an aircraft's maximum take-off weight (to compensate airports for the use of the runway);

- *terminal fees* charged based on an aircraft's seat capacity (to compensate airports for use of terminal facilities); and

- *other fees*, which may include air-bridge fees, tie-down fees, noise fees, security fees, and ground handling fees.

Aeronautical revenue is subject to economic regulation, which places limits on fees to ensure they don't become too high, and airlines continually pressure airports to limit or reduce these charges.

Non-aeronautical revenue refers to funds collected from services not directly associated with air operations, and these are not subject to economic regulation. Based on global averages, the largest sources of non-aeronautical revenue are retail concessions (28 per cent); automobile parking (22 per cent); and property and real estate rent for such things as office space, farming, and golf courses (15 per cent). The remaining 35 per cent comes from a variety of other sources including advertising.[22]

In 2014, the combined revenue for airports worldwide was US$142.5 billion (55.5 per cent aeronautical, 40.4 per cent non-aeronautical, and 4.1 per cent from non-operating sources).[23]

Airport revenue is directly linked to the strength of the economy and the likelihood that people will choose to travel by air. When airlines cancel flights in poor economic conditions, the airport's aeronautical revenue is reduced. To succeed, airport management must explore creative methods of increasing non-aeronautical revenue, so they can keep airline fees low – thereby encouraging airlines to choose their airport as a base of operations – while still generating enough revenue to cover expenses.

Expenses

For most airports, expenses include

- *operating expenses* – all costs associated with running an airport. The largest operating expense is employee costs (34 per cent) followed by contractors, administrators, communications, utilities, maintenance, and waste management costs, among others.[24]

- *capital costs* – large one-time expenses associated with major projects (such as purchasing land, building a new terminal, or investing in de-icing equipment). Most airports carry debt associated with these large investments, and the debt servicing costs represent an ongoing expense.

In 2014, each passenger who moved through an airport generated US$11.78 in aeronautical and $8.58 in non-aeronautical revenue for the airport. The expenses, averaged out on a per passenger basis, were $16.82, giving the airport industry a net profit margin of 16 per cent.[25]

Airports are highly regulated, and regulatory changes can create additional (sometimes unanticipated) expenses. For example, new regulation may result in the need to modify an aspect of a runway or to purchase equipment. Overall, airports are organizations with high fixed costs (associated with building, land, and equipment) and variable revenue sources.

5.5 The Language of Airports[1]

An aircraft's altitude, measured in feet, is expressed on an altimeter in **mean sea level (MSL)**. As the elevation of the ground below an aircraft can vary significantly, because of valleys and peaks, MSL measures altitude compared to sea level to provide an international standard. **Above ground level (AGL)** altitude, conversely, is the height of an aircraft, in feet, above the surface directly below it.

Aerodrome is a broad term for any area on land or water used for aircraft movements. An *airport* is a type of aerodrome that can be used for international traffic and has predictable services (fuel, customs and immigration, etc.).

An **apron** can be considered the parking area for aircraft – it is where aircraft park at a gate, load and unload passengers or cargo, refuel, and undergo line maintenance.

FOD (foreign object debris) describes any inanimate object that is not intended to be on a movement area and can cause harm to aircraft.

Runway **incursions** and **excursions** are both hazardous incidents. An incursion occurs when an unauthorized aircraft or vehicle enters the runway. An excursion happens when an aircraft improperly exits the runway in some way, possibly because of a rejected take-off, the inability to stop, or a slip off the side of the runway.

Movements refer to the number of take-offs and landings at an airport and **movement slots** refer to the maximum number of take-offs and landings a runway can accommodate. Movement slots can be thought of as the runway's capacity.

The **manoeuvring area** includes sections of the airport where aircraft taxi, take off, and land (taxiways and runways), but does not include aprons. **Movement areas** include all manoeuvring areas (taxiways and runways) as well as aprons.

A **hot spot** is a section of a movement area with a high risk of collision or runway incursion. Pilots and drivers must exercise extra caution in these areas.

The runway's **threshold** is the first part of the runway that is usable for landing. The **touchdown zone**, beyond the threshold, is where landing aircraft should first contact the runway surface.

Note

1 Adapted from ICAO, 2016 pages 1–2 through 1–9

Customer Service

Most of the workers within an airport are not employed by the airport itself. An airport can be thought of as a micro-city, which houses a variety of agencies and companies that collectively provide services and amenities to passengers as part of the airport community (see Figure 5.1). Because the airport experience often represents a traveller's first experience in a new State, it is not unfair to say that this first impression can affect a State's reputation; therefore, customer service at the airport has real significance.

All agencies within the airport community collaborate to form an interconnected *service delivery chain*.[26] Agencies serve passengers either directly (such as in a restaurant) or indirectly (such as through ground handling service). Although most of the people who work for these agencies are not employees of the airport, the reputation of an airport is often at stake when a passenger has a negative experience. Consider, for example, encountering a rude server in a restaurant or enduring long wait times to collect baggage because a baggage cart broke down – airport management does not play a direct role in either of these events. Yet airport managers are responsible for this customer service challenge as their goal is to create a positive experience for passengers along the entire service delivery chain.

To create a positive experience, the entire airport community needs to consider important *touch points* with passengers – each is an opportunity to leave a positive impression. However, a plan for *service recovery* is also important. Service recovery describes a predetermined response that is initiated when something goes wrong (such as lost luggage).

Airports develop customer service standards to optimize many aspects of airport operations, including facility design, condition and cleanliness, wayfinding and signage, employee behaviour, and terminal design.[27] Key performance indicators (KPIs) are established to track customer service, recognize and incentivize employees, and clearly communicate important information (such as flight status) to passengers.

Taking a passenger-centric perspective helps airport management to understand the entire experience from booking travel to arriving at the airport, checking in with an airline and moving through security, to boarding, flying, and arriving at a destination. The airport must consider that passengers have a variety of needs depending whether they are business or leisure travellers, on domestic or international flights, flying direct or connecting to another flight. Moreover, many individuals will have specific needs based on personal circumstances.[28]

Positive passenger experiences lead to word-of-mouth marketing, a strong reputation within a community, and increased non-aeronautical revenues (as happy passengers tend to shop more in the airport).[29] Good customer service can also improve safety and security as it can decrease air rage incidents (see the discussion of security in Chapter 6).

Types of Airports

There is a common expression in the aviation industry: *if you've seen one airport . . . you've seen one airport*. The point is that there is so much variability between airports that it's impossible to make broad generalizations about their operations.

For insight on how airports vary, it is helpful to look at the ways they are categorized. ICAO has established aerodrome reference codes, which are assigned to airports based on the size of aircraft that can be accommodated. Airports are coded by number and letter (e.g., a 1B or 2D airport). These codes, detailed in Figure 5.6, directly impact which types of aircraft can be served, as well as the airport design itself.

Code #	Aeroplane Reference Field Length	Code Letter	Wingspan	Wheel Span
1	< 800 meters (m)	A	< 15 m	< 4.5 m
2	≥ 800 m but < 1200 m	B	≥15 m but < 24 m	≥ 4.5 m but < 6 m
3	≥ 1200 m but < 1800 m	C	≥24 m but < 36 m	≥ 6 m but < 9 m
4	≥1800 m and above	D	≥36 m but < 52 m	≥ 36 m but < 52 m
		E	≥ 52 m but < 65 m	≥ 52 m but < 65 m
		F	≥ 65 m but < 80 m	≥ 65 m but < 80 m

Figure 5.6 Aerodrome reference codes
Source: Adapted from ICAO, 2016, p. 1–12
Notes: (1) ICAO provides all measurements in metres. One metre equals 3.3 feet; (2) Wheel span refers to the distance between the outside edges of the main gear wheels.

However, beyond categorization based on the size of aircraft the facilities can support, airports may also be defined by route network, purpose, traffic characteristics, or business model.[30]

Route Network

Airports can be structured as airline hubs or relievers.

- *Hub airports*, like the hub of a wheel connected to many spokes, are centrally located and linked by air routes (like the spokes of the wheel) to many smaller destinations. These airports, with consolidated services, are characterized by higher load factors, but they can face challenges with daily waves of arrivals and departures, and experience heavy congestion at peak times. Hub airports can be severely impacted by an airline's decision to leave. For example, when Northwest Airlines pulled out of Cincinnati (CVG) hub, after merging with Delta Air Lines in 2008, CVG experienced a reduction of traffic by 22 per cent that year, and a further 17 per cent in 2009.[31]

- *Reliever airports* are smaller than hub airports, and can generally accept overflow traffic from a hub (e.g., if the hub airport has reached capacity limits, or is experiencing severe weather, an accident, or a security issue). Reliever airports are less congested than hubs, but passengers are less likely to find direct flights between two reliever airports.

Purpose

Airports vary in their design and configuration based on the type of aircraft traffic supported.

- *International airports* have a proportionately high number of flights to and from international destinations and therefore require customs and immigration services and longer runways for larger aircraft, among other services.

- *Regional airports* primarily serve domestic flights and support smaller populations, such as short flights feeding into international airports. As a result, these airports do not require customs and immigration, have shorter runways, and have facilities to support general aviation activities.

- *Local airports* support general aviation activities within a local community.

Traffic Characteristics

Airports vary in their design and services associated with the type of passenger traffic they handle.

- *Origin–destination airports* are those which serve a majority of passengers (more than 70 per cent) who begin or end their journey at that airport. These airports require more available parking, ticketing, gates, and similar amenities.

- *Transit or gateway* airports are those which serve a majority of passengers (more than 70 per cent) who transition through the airport, transferring from one aircraft to another on their way to another destination. These airports require more ample transit lounges, hotels, food services, and similar amenities. Gateway airports welcome international passengers, process them through customs and immigration, and then transfer them to domestic flights to reach another airport within the State.

- *Alliance hubs* are strategically used as hubs for an airline alliance group and allow passengers and cargo to transfer between partner airlines. Examples of alliance hub airports include Dallas–Fort Worth (serving the **one**world Alliance), Amsterdam Schiphol (SkyTeam Alliance) and Singapore Changi (Star Alliance).[32]

Business Model

Lastly, airports can be categorized by their chosen business model.

- *Cargo hub airports* focus primarily on supporting cargo traffic, both domestic and international. Some cargo hubs offer passenger service as well, while others exclusively serve cargo operations.

- *Business airports* specifically cater to the business aviation sector. These airports will offer a variety of lounges, office space, valet parking, and other premium amenities. Vienna International Airport, for example, developed a VIP and Business Services programme specifically for elite business passengers.[33]

- *Low-cost carriers* (LCCs) focus intently on offering the lowest ticket fares to passengers and will therefore seek airports with the lowest landing and terminal fees. Certain airports cater to these carriers, offering modest terminal amenities that allow for lower fees levied on airlines. An example is the Charleroi Airport in Belgium, which works with Ryanair, a prominent European LCC.[34]

- *Multi-modal ports*, such as India's International Cargo Hub and Airport at Nagpur, integrate a variety of modes of transportation (linking rail, road, air, and sea travel).[35]

- *Destination airports*, a recent phenomenon, offer a variety of amenities – movie theatres, shopping malls, conference facilities (meeting and event spaces), casinos, and museums – that make the airport itself a final destination. The destination model epitomizes the goal of increasing the percentage of non-aeronautical revenue.

Future Challenges

Dramatic growth in international aviation is projected for the coming years.[36] These projections raise questions for airport management teams, which must decide whether to invest in expensive infrastructure upgrades to increase capacity. The risk is that projections can occasionally be wrong, and so uncertainty remains around the choice between launching a construction project to increase infrastructure or putting off the investment and potentially risk reaching capacity limits.

Furthermore, an increase in passenger numbers results in crowding and long lines in airport terminals. Passengers are happiest when they can move freely without delays. When airport terminals become congested, the customer experience suffers with *cross-flows* (i.e., a passenger walking through a flow of other passengers), longer wait times (to check in or to receive checked luggage), customer service breakdowns (as employees are at their service capacity), and lack of comfort (as seating may be fully occupied). Waiting and crowding negatively impact the customer experience.

Overall, capacity management presents one key challenge faced by airport management looking to the future of the industry. Other crucial factors include promoting safety and security and protecting the environment – all of which are covered at length in other chapters of this book.

Conclusion

In this chapter, we have discussed how airports are far more than a patch of land used for aircraft movements. Airports support their local economies, providing direct and indirect employment opportunities, support a global transportation network, and play an important role in establishing a State's reputation.

There are many important considerations in the design and operation of airports. International standards dictate precise requirements for the structure and operation of many elements of an airport. Although these SARPs are created with the best of intentions, some airports can struggle with achieving regulatory compliance as new regulations often require ongoing investments in infrastructure.

Airports represent a place where all stakeholders in the aviation industry come together – airlines, maintenance engineers, pilots, regulators, air traffic management, and airport management must all collaborate to face the upcoming challenges surrounding airports. More than merely a patch of land, an airport is the place where the entire aviation industry intersects.

Key Points to Remember

1. Airports provide employment opportunities. Globally, 450 000 people work for airport operators while another 5.5 million jobs are directly linked to airports.

2. Airports Council International (ACI) is the voice of the world's airports, representing their collective interests and promoting excellence in airport management.

3. ICAO establishes SARPs for airports in Annex 14, which has two volumes: 1) Aerodrome Design and Operations, and 2) Heliports.

   • Aerodromes require *aerodrome certificates* granted by a CAA.

   • Aerodromes use international standards to reference time and location. Time is given in Coordinated Universal Time (UTC), location is given in latitude and longitude, and elevation is given in mean sea level (MSL).

4. Airports are divided into *landside* (the unsecured part of the airport including roadways, parking lots, and terminal check-in areas) and *airside* (the secured part of the airport starting at the security checkpoint, covering the secure part of the terminal, the runways and taxiways, and extending to the airport perimeter fence).

5. Airports can be thought of as small cities where a variety of different agencies work together, including airport management, government agencies, airlines, ANSPs, concessions, and ground handling services.

6. Airport design must consider how passengers flow through the space (including check-in and security, walking distances, signage, and the ease of wayfinding), as well as security and regulatory requirements.

7. The airside of the airport includes

   - the *apron* where aircraft park at gates for the loading and unloading of passengers; and

   - the *manoeuvring areas* (*taxiways*, which are the interconnected roads that aircraft and vehicles use to move around the surface of the airport; and *runways*, the surfaces used for aircraft landings and take-offs). Detailed specifications for the design and operation of manoeuvring areas are outlined in Annex 14.

8. Foreign object debris (FOD) on aircraft movement areas can be extremely hazardous – airports conduct regular FOD inspections to identify and remove it.

9. Airport management must carefully balance revenue and expenses. Airport revenue includes *aeronautical revenue* (fees charged to airlines) and *non-aeronautical revenue* (funds collected from other airport services, such as parking or concessions). Airports seek to increase their non-aeronautical revenue to keep aeronautical revenue as low as possible. Airport expenses include both *operating expenses* (costs of running the airport) and *capital costs* (costs of debt associated with large building projects or purchases).

10. Customer service is a key element of airport management and is linked to the reputation of the airport, which can affect non-aeronautical revenue.

11. Airports vary tremendously in the size of aircraft they can accommodate, their route networks, their primary purpose, traffic characteristics, and business models.

14. Looking to the future, airports face challenges associated with capacity management, safety, security, and environmental protection.

Table 5.3 Acronym rundown

AACC	Airport Associations Coordinating Council
AC	aerodrome certificate
ACI	Airports Council International
ANSP	air navigation service provider
AOCI	Airport Operators Council International
ASIC	airport security identification card
CAA	civil aviation authority
FAA	Federal Aviation Administration
FOD	foreign object debris
GMT	Greenwich Mean Time
HOT	holdover time
IATA	International Air Transport Association
ICAA	International Civil Airports Association
ICAO	International Civil Aviation Organization
KPI	key performance indicator
LCC	low-cost carrier
LCFZ	laser-beam critical flight zone
LFFZ	laser-beam free flight zone
LSFZ	laser-beam sensitive flight zone
MSL	mean sea level
RESA	runway end safety area
RFFS	rescue and firefighting services
SARPs	standards and recommended practices
UTC	Coordinated Universal Time
WEAA	Western European Airports Association
YAPP	Young Aviation Professionals Programme

Chapter Review Questions

5.1 What is Airports Council International (ACI)? How did it come to exist? What is its role in today's aviation landscape?

5.2 What types of SARPs are included in Annex 14? Why is this annex so frequently updated and what challenges does this cause for airport operators?

5.3 Explain why international standardization, including Universal Coordinated Time (UTC), geographical coordinates, and IATA codes are important for international aviation.

5.4 Why is there an important distinction between the landside and airside of an airport? How are operational considerations different between the two?

5.5 What is a turn-around? Why does it matter, both to airports and airlines? Describe three activities that take place while an aircraft is at the gate.

5.6 How are airports important to airlines, communities surrounding airports, and society as a whole? Provide three points of evidence for each.

5.7 What is an aerodrome certificate (AC)? What elements are evaluated by a CAA before one is issued? In your own words, describe how an AC is associated with international aviation safety.

5.8 Why is it difficult, yet important, for airports to maintain high levels of customer satisfaction?

5.9 What type of terminal design does your local airport use? What is one advantage and one disadvantage of this design?

5.10 How does an airport generate revenue? Think of four ideas that might help your local airport generate more aeronautical revenue and non-aeronautical revenue. Describe three expenses that must be managed by your local airport.

5.11 How are airport profits linked to the global economy? Give a real-world example to illustrate your response.

5.12 Referencing the different types of airports, identify three airports in your State that fall in different categories. Explain your choices.

5.13 Customer service in airports is becoming increasingly important. No matter if a restaurant employee is rude or security lines are slow, it is the airport's reputation that takes the hit. How can this be managed? Should airport operations focus more on customer service or bureaucratic efficiency?

CASE STUDY

SOUTHWEST AIRLINES FLIGHT 1248 – A RUNWAY OVERRUN ACCIDENT[1]

On the evening of 8 December 2005, Southwest Airlines flight 1248 was about to complete its flight from Baltimore/Washington International Thurgood Marshall Airport to Chicago Midway International Airport in Illinois, United States. The flight held 98 passengers, three flight attendants, and two pilots.

The pilots anticipated a challenging landing. Their on-board computer calculated the aircraft would successfully come to a stop on the runway with less than 10 metres (30 feet) of runway to spare, as the surface was snowy and slippery and there was a tailwind of eight knots. The aircraft made a successful approach and touched down safely on the runway. Unfortunately, it took the pilots 18 seconds to engage the thrust reversers, which resulted in the aircraft taking too long to slow down (despite both pilots manually applying maximum braking). The aircraft overran the runway, rolling through a blast fence and the airport perimeter fence and onto a nearby roadway where it collided with an automobile and stopped. Tragically, a six-year-old child in the automobile was killed and another occupant injured. Eighteen people on board the aircraft had minor injuries and the aircraft sustained substantial damages.

The cause of this accident was linked to the failure of the airline to deliver clear and consistent training related to landing distance calculations. Another factor was the use of the on-board computer and its failure to include a safety margin in landing calculations that would accommodate unknown variables in the landing.

As a result of the accident, the United States' FAA developed new methods of communicating runway conditions, based on the type of surface contaminant (snow, ice, or water) and its depth, to assist pilots in calculating aircraft braking performance.

One aspect of airport design that can reduce the negative impact of runway overruns (from overshoots or rejected take-offs) is the implementation of a *runway end safety area* (RESA). RESAs provide a clear area – as long a distance as practical – that allow aircraft to overrun a runway without causing harm to people or property. Ideally, RESAs should be built from materials that effectively slow aircraft movement without hindering the movement of fire and rescue services. The length of the required RESA is linked to the airport code (see Figure 5.6) with runways supporting instrument approaches requiring longer RESAs.[2]

Notes

1 Adapted from NTSB, 2006　　　　　　　2 ICAO, 2016, pp. 3–15

Case Study Questions

Applying what you have learned in this chapter, consider the following questions:

5.14 Is it reasonable for airports to bear the cost of mandatory upgrades, such as the construction of RESAs? Consider that new regulations, designed to improve safety, might be prohibitively expensive for airports to implement.

5.15 Should new regulations take into account the cost of implementation? How can a balance be found between the potential benefits and projected costs of new regulations?

5.16 How can regulations support safety operations at an airport without becoming economically burdensome? That is to say, how does the industry balance safety, efficiency, and practicality?

References

ACI, 2010. *ACI airside safety handbook, 4th ed.* Montreal: Airports Council International.

ACI, 2016a. *2015 ACI annual world airport traffic report.* Montreal: Airports Council International.

ACI, 2016b. *2016 ACI airport key performance indicators.* Montreal: Airports Council International.

ACI, 2016c. *ACI airport statistics infographics: Airport economics at a glance.* [Online] Available at: www.aci.aero/Data-Centre/Airport-Statistics-Infographics

ACI, 2017a. *Airside operations course.* [Online] Available at: www.olc.aero/

ACI, 2017b. *Mission, objectives and structure – Airports Council International.* [Online] Available at: www.aci.aero/About-ACI/Overview/Mission-Objectives-Structure

ACI, n.d. *Airport operations diploma program (AODP).* [Online] Available at: www.olc.aero/Courses/Airport-Operations-Diploma-Program--AODP-.aspx

ATAG, 2016. *Aviation benefits beyond borders.* Geneva, Switzerland: Air Transport Action Group.

BEA, 2002. *Accident on 25 July 2000 at La Patte d'Oie in Gonesse (95) to the Concorde registered F-BTSC operated by Air France (Report translation f-sc000725a).* Le Bourget: Bureau d'Enquêtes et d'Analyses pour la sécurité de l'aviation civile.

Esler, D., 2016. *The risk of laser attacks on pilots is real and growing: The strikes are frequent and dangerous, temporarily blinding flight crews.* [Online] Available at: http://aviationweek.com/business-aviation/risk-laser-attacks-pilots-real-and-growing

ICAO, 2001. *Manual on certification of aerodromes, Doc 9774.* Montreal: International Civil Aviation Organization.

ICAO, 2003. *Manual on laser emitters and flight safety, Doc 9815.* Montreal: International Civil Aviation Organization.

ICAO, 2016. *Annex 14 to the Convention on International Civil Aviation: Aerodromes (Volume 1: Aerodrome design and operation), 7th ed.* Montreal: International Civil Aviation Organization.

ICAO, n.d. *ACI – Airports International Council.* [Online] Available at: www.icao.int/secretariat/PostalHistory/aci_airports_international_council.htm

ICAO, n.d. *Annex 14 – Aerodromes.* [Online] Available at: www.icao.int/secretariat/PostalHistory/annex_14_aerodromes.htm

ICAO, n.d. *The Convention on International Civil Aviation: Annexes 1 to 18.* Montreal: International Civil Aviation Organization.

Nelson, R. A. et al., 2001. The leap second: Its history and possible future. *Metrologia*, 38, p. 509–529.

NTSB, 2006. *Runway overrun and collision Southwest Airlines flight 1248, Boeing 737-7H4, N471WN, Chicago Midway International Airport, Chicago, Illinois, December 8, 2005.* Washington, DC: National Transportation Safety Board.

Notes

1 ATAG, 2016, p. 6
2 ATAG, 2016, p. 4
3 ACI, n.d.
4 ACI, n.d.
5 ACI, 2016c, p. 1
6 ICAO, n.d., para. 4
7 ACI, 2017b, para. 9
8 ACI, 2017b, para. 10
9 ICAO, n.d., para. 2
10 ICAO, n.d., para. 6
11 ICAO, n.d., para. 3
12 ICAO, n.d., para. 3
13 ICAO, 2016, p. 1–11
14 ICAO, 2016, p. 1–11

15 ICAO, 2001
16 Nelson, et al., 2001, pp. 38–39
17 ACI, n.d.
18 ACI, 2017a
19 ACI, 2017a
20 ACI, 2017a
21 ICAO, 2016, p. 9–5
22 ACI, 2016c, p. 1
23 ACI, 2016c
24 ACI, n.d.
25 ACI, 2016c
26 ACI, n.d.
27 ACI, n.d.
28 ACI, n.d.
29 ACI, n.d.
30 ACI, n.d.
31 ACI, n.d.
32 ACI, n.d.
33 ACI, n.d.
34 ACI, n.d.
35 ACI, n.d.
36 ATAG, 2016

1. Civil aviation is to be used for peaceful purposes only, but has been an attractive target for criminals as it stands as an example of successful global relations.
   a. True
   b. False

2. Safety and security are terms, with the same meaning, which are used interchangeably within international aviation.
   a. True
   b. False

5. Beyond terrorism, aviation may also be exploited by criminals with profit-motivations, such as drug-smugglers or human traffickers.
   a. True
   b. False

3. Terrorism is an act of unlawful interference that is politically motivated, often taking the form of bombings or hijackings.
   a. True
   b. False

4. The attacks of September 11, 2001 had a tremendous impact on the aviation industry, including the development of he new common security strategy of 'defend the cockpit, at all costs'.
   a. True
   b. False

Learning science suggests that thinking through a few questions before you begin studying new material, even if you answer incorrectly, results in improved learning and retention.
Give it a try!

CHAPTER 6

Security

CHAPTER OUTCOMES

At the end of this chapter, you will be able to . . .

- Discuss the content and application of ICAO's Annex 17, Security: Safeguarding International Civil Aviation Against Acts of Unlawful Interference.

- Describe several international conventions that have created global standards for security.

- Explain how preventative security measures are designed to anticipate unpredictable actions and prevent their occurrence before they impact aviation security.

- Discuss the types of unlawful acts and criminal activities that occur in aviation, including terrorism (bombings and hijackings), drug smuggling, and human trafficking.

- Express how the terrorist attacks of 11 September 2001 affected the aviation industry, specifically in terms of security.

- Outline how modern security initiatives are intended to ensure security, while balancing passenger privacy and the efficiency of passenger and aircraft movements.

- Use your understanding of aviation security to discuss a case study on the 'Underwear Bomber'.

Introduction

The Chicago Convention establishes that civil aviation is to be used for peaceful purposes only and that States are prohibited from using weapons against civil aircraft in flight.[1] Although this principle is globally accepted, the aviation industry represents an attractive target for criminals and terrorists, in part because the international cooperation in the industry results in aviation being a high-profile symbol of global unity. A criminal act, in the aviation industry, can range from drug smuggling or human trafficking to unruly behaviour on an aircraft. Beyond individual criminal activity are acts of unlawful interference (i.e., terrorism), which refer to acts intended to publicize a political agenda through an attack against aviation. As passenger confidence depends on a secure aviation system, and this confidence directly impacts the economic success of the industry, the international community collaborates to create, implement, and enforce security measures.

There is an important difference between safety and security. *Safety* focuses on preventing accidents through the identification and elimination of risk within aviation operations, and will be discussed in Chapters 8 and 9. In this chapter, we will discuss *security*, which focuses on protecting the aviation system from risks associated with intentional wrongdoing and criminal behaviour.

The unfortunate reality of the twenty-first century is that security issues with international implications arise constantly, and these must be identified and addressed. When unlawful acts occur on the ground within a State, that State has the authority to deal with the situation and the offender(s); although there is great variability in how laws are made and enforced around the world, it is generally accepted that the site of the act determines which State has the authority to respond. However, when an unlawful act occurs on a flight between international destinations or in mid-air (perhaps over international waters), many questions arise related to legal authority:

- Which State has the authority to prosecute when unlawful acts occur in flight?

- What authority does the pilot-in-command have in dealing with criminals or terrorists in flight?

- Should States be able to set their own security measures (for example, whether or not they choose to screen passengers and baggage)?

To unify aviation security (AVSEC) activities, all ICAO members abide by the standards and recommended practices (SARPs) published within Annex 17 to the Chicago Convention.

ICAO Annex 17: Security

Annex 17 of the Chicago Convention, titled *Security: Safeguarding International Civil Aviation Against Acts of Unlawful Interference*, sets out the international SARPs that represent the minimum requirements for aviation security.[2]

The need for international security measures was recognized after a series of violent crimes against civil aviation occurred in the 1960s. In response, ICAO held an extraordinary session of the Assembly in 1970 to create an annex to the Chicago Convention that would establish SARPs for international aviation security (with an emphasis on preventing hijackings).

Annex 17 was adopted by the ICAO Council in 1974 with the intent to protect civil aviation activities around the world from illegal actions.[3] Among other things, Annex 17 requires States to create a national civil aviation security programme to enforce aviation security domestically and to coordinate with other States in protecting international civil aviation.

Before 1985, hijacking was considered the biggest security threat to civil aviation. Therefore, Annex 17 established screening systems for passengers, carry-on luggage, and hold baggage (see the case study on the Lockerbie Disaster). In 1989, additions were made to Annex 17 to specify how to deal with 1) items left on aircraft by disembarking passengers, 2) mail and cargo, 3) security for courier services, and 4) passenger–baggage reconciliation.

Annex 17 is supported by guidance material contained in an Aviation Security Manual (Doc 8973), a restricted document available to authorized professionals only, that outlines how security measures should be implemented. Both Annex 17 and Doc 8973 are continually reviewed and updated as necessary by ICAO's Aviation Security Policy (ASP) section of the Aviation Security and Facilitation Office as well as the AVSEC Panel with representatives from several countries and international associations.[4]

The reason that security controls vary between countries is that Annex 17 establishes the *minimum* security standards for international aviation. Some countries implement security systems that extend far beyond these minimums, so global travellers may notice variability as they travel internationally.

Aviation security is a collaborative process that requires involvement of civil aviation authorities (CAAs), airport and aircraft operators, law enforcement agencies, customs and immigration authorities, and air traffic service providers, among others.

Case Study: Pan Am Flight 103 – The Lockerbie Disaster[1]

On 21 December 1988, Pan Am flight 103 had departed London Heathrow airport and was en route at 31 000 feet headed to New York. There were 243 passengers and 16 crew members on board the Boeing 747–121 aircraft. Passengers and crew had no way of knowing that an improvised explosive device (IED), concealed as a Toshiba radio-cassette player, was packed in a Samsonite suitcase in the cargo compartment.

When the IED detonated, it created a large hole in the fuselage and cabin floor, resulting in immediate structural failure of the aircraft. The nose separated from the fuselage within two to three seconds, landing four kilometres (2.5 miles) away. Investigation of the cockpit wreckage showed that all switches and oxygen masks were located consistent with the cruise portion of flight (indicating that the pilots had no time to react to the emergency).

The fuselage disintegrated and fell over an enormous area, with the largest sections falling almost vertically into a residential area in Lockerbie, Scotland.[2] The impact of the fuselage created a large crater, and the jet fuel ignited causing a fireball that reached 3000 metres (10 000 feet) and destroyed several homes.

All 259 people on board the aircraft were killed, along with 11 people on the ground. This accident drew attention to critical gaps in international aviation security.

On November 18 1988, a month *before* the Lockerbie bombing, the Federal Aviation Administration (FAA) in the United States had issued a bulletin describing an IED they had found disguised as a radio cassette player and rigged with a barometric device (which detonates the explosives at a certain altitude, indicating an aircraft was the target). The bulletin cautioned that the IED would be difficult to detect with X-ray inspection.[3] Although the US had a rule prohibiting any carrier from transporting a bag that was not accompanied by a passenger, no such policy existed at Heathrow at that time.

The Lockerbie disaster drew international attention to the danger of IEDs and plastic explosives. The international community recognized that the best way to prevent this type of accident was to ensure that explosive materials could not reach the aircraft in the first place.

Prior to Lockerbie, ICAO had the right to set security standards, but did not have power to enforce these measures due to the risk of infringing on a State's sovereign rights over its own airspace. In the aftermath of the Lockerbie disaster, at the 27th Session of the ICAO Assembly in 1989, ICAO's role in security was enhanced. The organization shifted from merely developing air law to overseeing security implementations among member States.

As a result of this tragedy, international aviation security was reshaped in several important ways:[4]

- requirements were created for the marking of plastic explosives (to improve detection);

- screening systems for detection of trace explosives were modernized;

- a global requirement was established for 100 per cent hold baggage screening; and

- passenger baggage reconciliation systems were developed.

Notes

1 Air Accidents Investigation Branch, 1990 3 Malik, 1998
2 Malik, 1998 4 Ushynskyi, 2009

International Security Conventions

To achieve a secure aviation system, States must aim to meet and (when possible) exceed the safety standards outlined in Annex 17, but there is some flexibility in implementation. For example, Annex 17 requires all passengers and their luggage to be screened, but leaves it up to individual States to decide what screening tools and methods to purchase and implement.

To complicate matters, different countries have different laws associated with how to prosecute aviation criminals. Consider a situation in which one nation's citizens were harmed by aviation criminals in a hijacking – that nation would usually want the right to punish those criminals. Yet if the crime occurred in the air, the State in question may not have access to the offenders who may be in custody in another country.

In the 1960s, it became common for criminals to be harboured and protected by their home countries – many were never brought to justice. Legal loopholes prevented States from capturing and punishing these aviation criminals. To mitigate the issue, the legal team at ICAO facilitated a series of treaties including the 1963 Tokyo, 1970 Hague, and 1971 Montreal Conventions (see Table 6.1). However, challenges remain as States often fundamentally disagree on how offenders should be punished and the treaties are of little help in resolving these international disputes.[5]

The power of these treaties is limited and has led to some countries (particularly those that experience a lot of terrorist activity) to work outside the system by using diplomatic

Table 6.1 Summary of security conventions and protocols

Treaty/convention	Description	Notes
1963 Tokyo Convention: *Convention on Offences and Certain Other Acts Committed On Board Aircraft*	• Established a legal framework for offences committed in flight. • Gave pilot-in-command (PIC) almost absolute power to deal with unlawful acts. (The aircraft was considered a *micro jurisdiction* – essentially its own country referred to as a *State of Nature*.) • Allowed passenger or crew following Captain's orders to restrain offenders without risk of liability. • Established rights and obligations of States in exercising jurisdiction and punishment of offenders, filling the gap for States that had no domestic laws covering AVSEC incidents. • Defined *in flight* as the time between take-off power being applied and the end of the landing roll (beyond which local laws apply).	• Included *Freedom Fighter* clause, which allowed States to refuse to prosecute or extradite an offender whose actions were politically motivated. • Impact limited because several States implicated in hijackings refused to sign on to the Convention.

(Continued)

Table 6.1 (Continued)

Treaty/convention	Description	Notes
1970 Hague Convention: *Convention for the Suppression of Unlawful Seizure of Aircraft*	• Closed the gap that allowed terrorists to find safe haven in States that would not prosecute or extradite them. • Made it an offence to seize or attempt to take control of an aircraft through intimidation or force (and required severe penalties for this offence).	• Included loophole allowing States to use exceptions in their national laws to block extradition of criminals – hijackers still found protection in countries that shared their political views, and ICAO still lacked the power to intervene. • No clear process to settle disputes between States.
1971 Montreal Convention: *Convention for the Suppression of Unlawful Acts against the Safety of Civil Aviation*	• Widened the scope of aviation crimes punishable by severe penalties. • Criminalized any act (or attempted act) of violence against a person on board an aircraft or any act (or attempted act) that destroys or damages an aircraft in service (including placing an explosive device on an aircraft).	• As with previous ICAO treaties, refrained from threatening expulsion from ICAO or revoking of international traffic rights under air service agreements (ASAs).
1978: *Bonn Declaration on Hijacking* (G7 States)	• Banned air services to countries that refuse extradition or prosecution of hijackers or do not return hijacked aircraft. • Invoked in 1981 when 45 hijackers found asylum in South Africa after hijacking an aircraft out of the Seychelles. South Africa initially harboured the hijackers, but relented to diplomatic pressure and charged the mercenaries with hijacking.	• Gave States that had been victimized by hijackings a course of action against States that harbour criminals. • This is a 'declaration' by the G7 States, not a 'convention'. A *declaration* lacks the international status of a *convention* – when convention status cannot be achieved, declaration status is used.

Treaty/convention	Description	Notes
1988 Montreal Supplementary Protocol: *Protocol for the Suppression of Unlawful Acts of Violence at Airports Serving International Civil Aviation*	• Added an accord to expand the list of aviation crimes to incorporate attacks against international airports (in reaction to several bomb explosions at major European airports in the 1980s).	
1991: *Convention on the Marking of Plastic Explosives for the Purpose of Detection*	• Added an accord to dictate the identification, marking, and transport of plastic explosives. • Established International Explosives Technical Commission.	
2010 Beijing Convention: *Convention on the Suppression of Unlawful Acts Relating to International Civil Aviation* and 2010 Beijing Protocol: *Protocol Supplementary to the Convention for the Suppression of Unlawful Seizure of Aircraft*	• Drafted to replace Montreal Convention and amend Hague Convention, expanding the list of aviation offences to include: use of aircraft as a weapon; transport, delivery, or use of weapons of mass destruction on or against aircraft; transport of dangerous explosive, biological, chemical, nuclear, or radioactive materials.	• Not yet ratified or in force
2014 Montreal Protocol: *Protocol to Amend the Convention on Offences and Certain Other Acts Committed on Board Aircraft*	• Adds clauses associated with unruly passengers. • Extends aircraft jurisdiction to include State of Intended Landing and State of Operator. • Specifies minimum behaviours that are considered offences (including physical assault or threats against crew and refusal to follow instructions). • States that airlines have the right to seek compensation for diversions made because of unruly behaviour.	• Not yet ratified or in force

Source: Adapted from Havel & Sanchez, 2014 and IATA, 2015

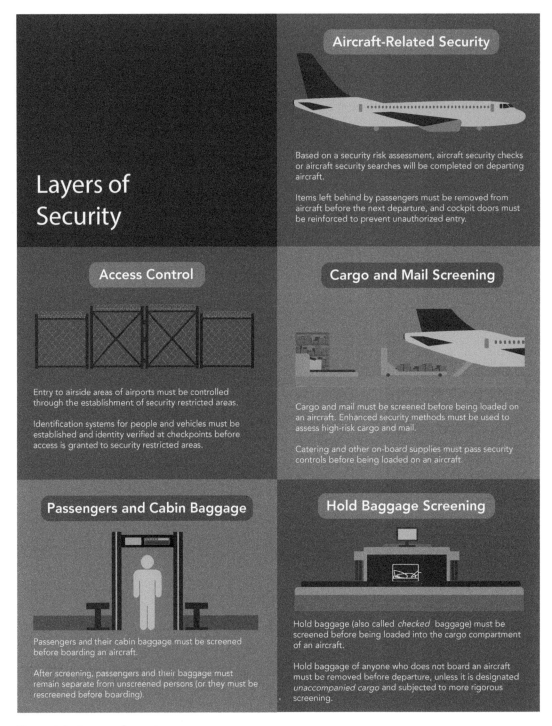

Figure 6.1 Layers of security measures

Source: Adapted from IATA, 2015 and ICAO 2011

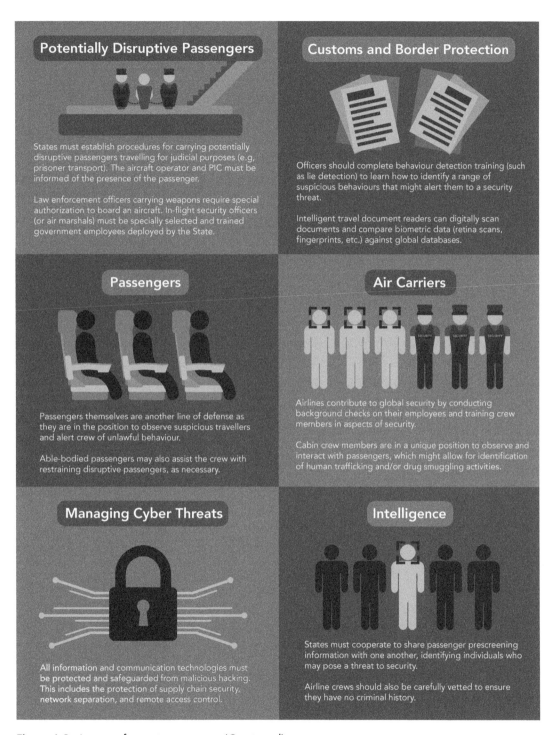

Potentially Disruptive Passengers

States must establish procedures for carrying potentially disruptive passengers travelling for judicial purposes (e.g, prisoner transport). The aircraft operator and PIC must be informed of the presence of the passenger.

Law enforcement officers carrying weapons require special authorization to board an aircraft. In-flight security officers (or air marshals) must be specially selected and trained government employees deployed by the State.

Customs and Border Protection

Officers should complete behaviour detection training (such as lie detection) to learn how to identify a range of suspicious behaviours that might alert them to a security threat.

Intelligent travel document readers can digitally scan documents and compare biometric data (retina scans, fingerprints, etc.) against global databases.

Passengers

Passengers themselves are another line of defense as they are in the position to observe suspicious travellers and alert crew of unlawful behaviour.

Able-bodied passengers may also assist the crew with restraining disruptive passengers, as necessary.

Air Carriers

Airlines contribute to global security by conducting background checks on their employees and training crew members in aspects of security.

Cabin crew members are in a unique position to observe and interact with passengers, which might allow for identification of human trafficking and/or drug smuggling activities.

Managing Cyber Threats

All information and communication technologies must be protected and safeguarded from malicious hacking. This includes the protection of supply chain security, network separation, and remote access control.

Intelligence

States must cooperate to share passenger prescreening information with one another, identifying individuals who may pose a threat to security.

Airline crews should also be carefully vetted to ensure they have no criminal history.

Figure 6.1 Layers of security measures (Continued)

pressure and other statutes. For example, in 1985, Fawaz Yunis hijacked a Royal Jordanian Airlines flight that had several Americans aboard; two years later, American law enforcement lured Yunis to a yacht on international waters under the pretence of a drug deal. Yunis was seized, brought onto US territory against his will, and successfully prosecuted under several American federal statutes that criminalize hijacking and hostage-taking.[6]

Preventative Security Measures

Annex 17 specifies that States should deploy security measures that help achieve a safe civil aviation system. Rather than observing a single security initiative, the civil AVSEC strategy comprises multiple layers of defence based on a philosophy of unpredictability. Effective security requires professionals to continually innovate and vary practices so that they are unpredictable. An unpredictable security system is much more difficult for criminals to study and therefore more difficult to defeat. These unpredictable safeguards are primarily focused on preventing threats from being brought on board aircraft.

Although all the security layers are important, the biggest threat against civil aviation is posed by improvised explosive devices (IEDs). IEDs are homemade explosives designed to cause injury or death. IEDs are commonly concealed within electronics, body cavities, printer cartridges, liquids, and gels. Detecting IEDs and preventing them from reaching civil aircraft is a priority for hold baggage and passenger screening practices.

Hold Baggage Screening

Hold baggage screening (HBS) is the term used to describe the searching of bags for dangerous items before they are placed in the cargo compartment of an aircraft. As of 2006, Annex 17 requires that 100 per cent of hold baggage be screened for weapons, explosives, and other dangerous devices before being loaded onto aircraft. A variety of methods, which vary in efficiency and cost, may be used to conduct HBS searches. Methods range from the use of explosive detection dogs and manual searches by security personnel to sophisticated computed tomography (CT) X-ray machines.

CT X-ray involves baggage being placed on a conveyor and moving through a tube (at rates between 500 and 2000 bags per hour). The CT scanner rotates around the bags, scanning them with X-rays. This process generates a detailed image of the baggage contents, identifying even the specific types of materials within the contents. Automated computer systems analyse the images and flag potentially hazardous materials.

A baggage reconciliation system (BRS) is a secondary process, typically automated, that ensures no baggage is loaded onto an aircraft unless the passenger who checked the bag is already on board. This process ensures positive passenger–baggage matching. The intent is to eliminate the threat of bombers placing IEDs within checked luggage to be loaded onto an aircraft but not boarding the flight themselves, with the intent of detonating the explosives in flight.

Pre-board Passenger Screening

Pre-board passenger screening (PPS) is the process by which passengers are screened for prohibited items. This is accomplished by having passengers proceed through body scanners or walk-through metal detectors (WTMDs) and place their carry-on luggage on a conveyor to be checked by an X-ray machine. The WTMD and X-ray machine are intended to detect metal and electronics that a professional screener could identify as a weapon (such as a gun). This screening is enhanced with explosive detection technology wherein a wand with a small piece of fabric is used to rub a passenger's hands or carry-on baggage to collect a sample, which is then placed in a reader to check for explosive residue. In many airports, full-body scanners have now been incorporated into the PPS process.

Full-Body Scanners (Imaging Technologies)

Imaging technologies may also be used during PPS. Older types of these devices exposed passengers to a type of radiation and measured its reflection (backscatter) while modern systems sense the natural radiation released by the passenger's body (millimetre wave) to identify any concealed devices.[7]

These systems are designed to 'see' under passengers' clothing to identify concealed weapons or explosives, as these devices produce different radiation than the human body. The result is displayed to operators as a digital image that highlights areas of the body where concealed items may be located.

Did You Know?

When first introduced, early imaging was heavily criticized by the media as a privacy violation and even referred to as a 'digital strip search'. The images produced by early technologies clearly showed genitals and breasts. Some claimed these devices were illegal, violating privacy and decency laws including those against child pornography.

Newer imaging technologies have reduced the controversy, as a detailed image of the human form is not shown; instead areas are highlighted on a generic chalk figure outline indicating where screeners should follow up with an in-person search.

Unlawful Acts and Criminal Activities in Aviation

Given the magnitude of the effort invested in security, it is helpful to understand the most common offences and offenders within civil aviation. The term *unlawful acts* can cover a wide spectrum of offences in the criminal codes of various nations, but for the purposes of aviation security, the term refers to those acts identified in the Tokyo Convention: 'acts which, whether or not they are offences, may or do jeopardize the safety of aircraft or of persons or property therein, or which jeopardize good order and discipline on board'.[8] In this chapter, unlawful acts are classified as 1) acts of unlawful interference against civil aviation, and 2) criminal acts.

Acts of unlawful interference against civil aviation refer to terrorist acts, including bombings and hijackings. These acts represent the most pressing security concern in the 21st century. One of the defining elements of a terrorist act is that it is motivated by a political purpose.[9]

A *criminal act*, on the other hand, describes an aggressive, senseless, or violent act that is not terrorism. These include financially motivated activities (such as drug smuggling and human trafficking) as well as acts of individuals, such as assault of a flight attendant during an air rage incident.

6.1 The Language of Security

An **unlawful act** is any act (criminal or not) that jeopardize the safety of civil aviation. There are two subcategories:

- A **criminal act** is one that violates a law in place to protect the public and that includes a penalty for its violation. **Aviation piracy** is a subcategory associated with criminal acts for profit (like drug smuggling).

- An **act of unlawful interference** risks the safety of an aircraft or the people on board, and may include terrorism, bombings, and hijacking. If these acts are politically motivated, they are classified as **terrorism**.

Disruptive passengers are those who don't act appropriately, either in an airport or on an aircraft. Typically, these individuals don't follow instructions and act inappropriately. Alcohol, illegal drugs, and mental health issues contribute to this type of behaviour, which is sometimes called *air rage*.

Human trafficking refers to the moving of people without their consent, through means of intimidation, coercion, or deception, with the intention of exploiting them for forced labour, slavery, sexual exploitation, or the removal of organs.[1] This is different from **human smuggling**, where people choose to participate as they want to seek opportunities in another State.

Aviation security (AVSEC) is the process of protecting civil aviation from unlawful acts. AVSEC incorporates a wide variety of human expertise, screening technologies, regulations, and other strategies such as[2]

- **hold baggage screening** (HBS) – the searching of checked luggage before it is placed on board a flight;

- **pre-board passenger screening** (PPS) – the screening of passengers and carry-on luggage before they enter the airside of an airport; and

- **airport security** – the measures that relate to both landside (building design, patrols, physical measures) and airside (perimeter fence, behaviour detection) security.

Notes

1 United Nations, 2004, p. 42   2 ICAO, 2001, pp. 1–2

Acts of Unlawful Interference

Terrorism is currently the greatest threat to aviation security. Although terrorist acts are rare, they have a significant impact on passenger confidence in air travel and can result in large financial losses for airlines. Terrorists are innovative in planning their attacks and understand that to avoid detection by security personnel, they must constantly vary their methods. Although much global attention has been paid to terrorism through the UN, ICAO, and other organizations, it is difficult to come up with a unified solution simply because there is a lack of consistency in the aims and methods of terrorists.

Although there is no universally accepted definition of *terrorism*,[10] it is generally considered to refer to a politically motivated criminal act. Within the aviation industry, terrorism usually takes the form of bombing or hijacking.

Bombings

There is a long history of bombings within the aviation industry, going back to 1933 when a nitro-glycerine bomb destroyed a United Airlines Boeing 247 in Chesterton, Indiana. The incident is considered the first act of air sabotage in commercial aviation and thought to be connected to Chicago mob activities.

Did You Know?

The word *terrorism* comes from the Latin *terrere*, meaning 'to cause to tremble'.[1]

Note

1 Abeyratne, 2010, p. 185

The history of aircraft bombing incidents includes several famous examples of thwarted attempts. For example, both the 'Underwear Bomber' who attempted to detonate an IED hidden in his underwear on Northwest Airlines flight 253 and the 'Shoe Bomber' who attempted to ignite an IED hidden in his shoes on American Airlines flight 63 made global headlines in 2009 and 2001, respectively.

Yet tragically, many bombing attacks have been carried out successfully. A recent example occurred on 31 October 2015 when Metrojet flight 9268, a Russian airliner, seemingly disintegrated and scattered wreckage over a large section of the Sinai Peninsula. Two hundred and twenty-four people (including seven crew members) were killed, making it the deadliest event in the history of Russian aviation. Terrorist leaders from the Islamic State of Iraq and Syria (ISIS) immediately claimed responsibility for the event through their social media channels and later broadcast images of the IED used in the attack.[11]

Bombing attacks are not limited to aircraft – airports have also been the target of bombings. As you learned in Chapter 5, airports comprise a *landside* (public areas where passengers park, enter the airport, and check in for their flights) and an *airside* (spaces controlled by security checkpoints and accessible only to passengers with boarding passes and personnel with security clearances).

On 28 June 2016, three terrorists used bombs and assault rifles to attack travellers on the landside of Istanbul Atatürk Airport in Turkey (the third busiest airport in Europe). They killed 41 people and injured more than 230 in the attack, which occurred in the parking lot and entrances to the arrival and departure areas. After inflicting as much damage as they could, all three terrorists detonated suicide vests.[12]

Airports Council International (ACI) published guidelines for landside security following the Istanbul attack. Although one might expect the response to have included additional screening inspections at the entrance to the terminal, this course of action would, in fact, further compromise security by resulting in the congregation of people outside the terminal, thus creating new vulnerabilities. Instead, ACI recommended the following landside security measures:[13]

- Building design
  - use blast-proof materials for walls, and shatter-proof glass;
  - incorporate structures to prevent drive-in attacks (e.g., flowerpots, cement obstacles);
  - separate pick-up and drop-off areas from the terminal entrance;
  - rethink space management to reduce large gatherings;
  - reduce areas where a shooter or bomber might access crowds (e.g., terraces);
  - reduce areas where items could be hidden, such as garbage bins;
  - coordinate with city planners with respect to new buildings near airports to analyse security concerns related to balconies, windows, or terraces facing the airport; and
  - consider security measures when constructing new buildings, as it is more expensive to retrofit after they are built.
- Physical measures and process
  - incorporate closed-circuit TV surveillance of public areas; and
  - work with airlines and regulators to identify methods of disseminating crowds.
- Passenger and staff awareness
  - remind passengers to report suspicious behaviour or baggage; and
  - train all airport employees to recognize suspicious activity and to abide by the saying 'see something, say something'.
- Patrols
  - use specialist behaviour detection officers;
  - ensure airport security patrols are highly visible, both for detection and deterrence; and
  - use explosive detection dogs to identify explosives without slowing passenger flows.

- Crisis and incident response

  - establish an emergency response communication and coordination plan, including regular exercises to test the plan;

  - determine how enhanced security measures, in response to a threat, can be removed when threat level returns to normal; and

  - define evacuation and other contingency plans.

Case Study: The Bombing of Air India Flight 182

On 23 June 1985, Air India flight 182, a Boeing 747 aircraft, was headed from Toronto to Delhi with stopovers in Montreal and London. There were 307 passengers (including 82 children) and 22 crew members on board.

After a non-eventful leg from Toronto to Montreal, the flight departed Montreal for London. About 45 minutes before arrival, a rapid decompression of the cabin occurred and the aircraft separated into pieces at 31 000 feet before crashing into the Atlantic Ocean. There were no survivors of the accident, making this terrorist event the deadliest before that of September 11 2001. Only 131 bodies, including those of 30 children, could be retrieved as most were lost to the ocean floor. Several bodies exhibited flail injuries and hypoxia, indicating that they had exited the aircraft alive at a high altitude and perished before impact. Despite a detailed investigation, no evidence of an explosive device was retrieved from the wreckage, although the aircraft structure displayed signs of an explosion.

Just 55 minutes before the Air India flight 182 explosion occurred, an explosive device detonated in Tokyo Narita Airport, killing two baggage handlers and injuring four others. An unaccompanied bag with an IED had arrived in Tokyo from Vancouver and was to be loaded onto Air India flight 301 to Bangkok, which would have led to dual coordinated bombing events had the timing been successful.

Investigation revealed that a passenger calling himself Mr Singh had reserved two flights: CP Air flight 060 from Vancouver to Toronto (connecting with Air India 182 to Delhi) and CP Air flight 003 from Vancouver to Tokyo (connecting with Air India flight 301 to Bangkok). He paid cash for the flights.

Mr M. Singh checked in at Vancouver International Airport for CP Air flight 060 and asked that his bag be transferred to flight 182 in Toronto. An airline agent informed Mr M. Singh that, as he was on standby for flight 182, his bag could not be checked onto that flight. When Mr M. Singh became very aggressive, the agent relented and checked his bag onto flight 182. CP Air flight 060 departed without Mr Singh; however, his bag was on board and was transferred to Air India flight 182 in Toronto.

Meanwhile, Mr L. Singh checked in for CP Air flight 003 from Vancouver to Tokyo and checked a bag. CP Air flight 003 departed with his bag (containing the explosive) on board, but without Mr L. Singh.

The bombing was the project of a Sikh militant group with membership throughout Canada, the United States, England, and India. The alleged mastermind of the bombings, Talwinder Singh Parmar, never faced charges despite significant evidence against him.

After a long investigation by Canadian authorities, at a cost that exceeded CAD$82 million,[1] three conspirators faced charges; however, only one was ever convicted (in 2003) and served time. Inderjit Singh Reyat was given a reduced sentence of five years when he made a deal to testify against his two co-conspirators, Ripudaman Singh Malik and Ajaib Singh Bagri. However, during the trial of his co-conspirators, Reyat lied on the stand and Bagri and Malik were set free due to lack of evidence. For his perjury, Reyat was sentenced to another nine years.

The response by security professionals, aviation safety regulators, and police has been called a 'cascading series of errors'.[2] Canada's security and police agencies were highly criticized for lack of coordination, errors, and turf wars between the groups. Particularly distressing is the fact that the Indian government had warned the Canadian government about the possibility of terrorists placing bombs on board Air India flights in Canada only two weeks before the flight 182 bombing. Furthermore, CSIS (the security agency in Canada) demonstrated 'unacceptable negligence' by erasing hundreds of wiretaps, both before and after the terrorists were the primary suspects in the bombings.

On 23 June 2010, the Prime Minister of Canada publicly apologized on behalf of the country for the institutional failings and treatment of the victims' families. In 2012, the government of Canada issued an *ex gratia* payment as a symbolic demonstration to the families impacted by the event.[3]

Figure 6.2 Air India memorial (photo)

Source: By Artur [CC BY-SA 2.0 (https://creativecommons.org/licenses/by-sa/2.0)], via Wikimedia Commons

Notes

1 Matas, 2002
2 CBC News, 2010

3 Public Safety Canada, 2015

Hijackings

Between 1948 and 1957, 15 hijacking attempts occurred worldwide and received much public attention, which was appealing to terrorists. It has been claimed that aircraft hijackings are a 'contagious' phenomenon because the media coverage of hijacking events provides motivation for this type of attack among other terrorists.[14]

Possibly as a result of media coverage, the number of hijackings rapidly increased, with 385 incidents occurring between 1967 and 1976. In fact, there were 82 hijacking attempts in 1969 alone.[15] Therefore, AVSEC in the 1960s focused primarily on preventing and managing hijacking attempts and on bringing criminals to justice.

In a sampling of 40 hijackings that occurred between 1998 and 2002, a number of trends were noted. Figure 6.3 illustrates some of the characteristics of the hijacking events during this period.

Hijackers range in sophistication from highly organized members of paramilitary groups to desperate and careless individuals seeking escape from a political regime. Organized groups may be waging battle against their governments or trying to draw attention to their political causes. These types of criminals typically establish long-term stand-offs and publicly negotiate demands (such as the release of imprisoned members of their group). Individual hijackers, on the other hand, are usually less organized and their goal is generally more personal – to escape from authorities in a certain part of the world and secure transportation to a country where they may be granted asylum.

Before the events of 11 September 2001 (referred to as 9/11), pilots were taught to deal with a hijacking situation using the *common strategy*: 'accommodate, negotiate, and do not escalate'. Unfortunately, the common strategy was known to both pilots and terrorists alike. Pilots were encouraged to cooperate with demands and land the aircraft safely. If a hijacking occurred on the ground, the goal was to keep the aeroplane on the ground and isolated from other people and operations in an isolated parking position where negotiations could occur safely. Pilots were advised to disable the aircraft, if possible, and airport operators were to bring in obstacles (such as snowploughs or other equipment) to serve as barriers to take-off.[16] If the aircraft was in the air at the time of the hijacking, the strategy was to maintain safe and controlled flight and get the aircraft on the ground as quickly as possible.

The Impact of 9/11

The tragedy of 9/11 had a profound impact on aviation security. On 11 September 2001, 19 terrorists boarded four different commercial flights with the intent of turning the aircraft into guided missiles, and thus defeated the common strategy. (See Figure 6.5 for a summary of the events of that day.) In retrospect, it was noted that terrorists had used vehicles (such as trucks and boats) in other suicide attacks and so the use of an aircraft should not have been entirely unpredictable; however, at that time no one had anticipated this possibility. The common strategy proved to be entirely ineffective in preventing suicide hijackings.

When the hijackings began, there was general confusion from air traffic control (ATC) and no immediate reaction from the military. The standard pilot response to a hijacking is to use the on-board transponder to squawk code 7500, which notifies authorities that an aircraft has been hijacked; however, since the terrorists had aviation training, and this

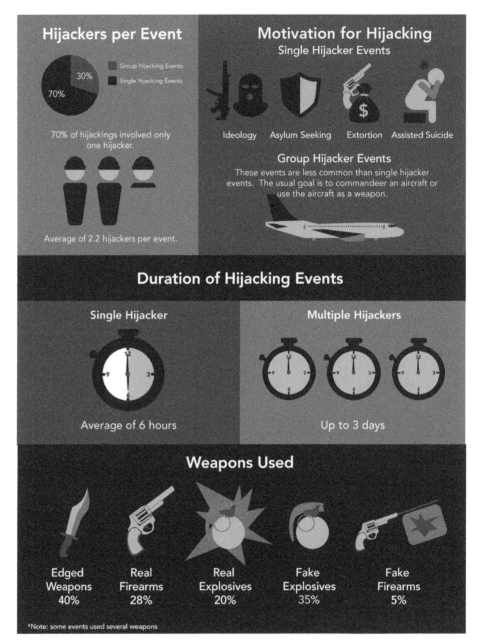

Figure 6.3 Characteristics of hijacking events

Source: Adapted from Williams & Waltrip, 2004

information is freely available in aviation training materials, none of the aircraft squawked 7500 and three of the four aircrafts' transponders were turned off entirely (making it difficult for ATC to track their locations).

After authorities learned what was happening, all airspace in the United States and Canada was closed (with exceptions for military and medical flights). This unprecedented 'ground stop' occurred when SCATANA (Security Control of Air Traffic and Air Navigation Aids), an American emergency preparedness plan, was ordered – all civilian flights were required to land at the nearest airport. There were about 500 international flights inbound for the United States at the time – approximately 250 flights and 40 000 passengers were diverted to land in Canada and other countries, including Mexico. The controllers who coordinated the landings have been highly credited for their actions. National airspace did not reopen until the morning of September 13, when many stranded aircraft and passengers began to make their way to their destinations.

At the time, the aviation industry held an outdated belief that a hijacking would involve a terrorist taking control of an aircraft, ordering pilots to land, and negotiating demands. There was also an assumption that hijackers would be ignorant of aviation practices. The 9/11 terrorists defeated a security system that was based on a 1970s understanding of hijackings. The events of 9/11 changed one of the fundamentals of AVSEC. As a reaction to 9/11, a new common strategy for hijackings was established: 'defend the cockpit, at all costs'.

Figure 6.4 North face of South Tower after plane strike, 9/11

Source: By Robert on Flickr [CC BY-SA 2.0 (http://creativecommons.org/licenses/by-sa/2.0)], via Wikimedia Commons

The 9/11 Commission Report, after thorough analysis, recommended that security intelligence professionals

- take the time to imagine how surprise attacks could be carried out;

- define the most dangerous prospects;

- gather information about those prospects; and

- create defence systems to disrupt the prospects or, at least, produce an early warning.[17]

		American Airlines Flight 11	United Airlines Flight 175	American Airlines Flight 77	United Airlines Flight 93
Boarding	Aircraft	Boeing 767	Boeing 767	Boeing 757	Boeing 757
	Passengers/crew	81/11	56/9	58/6	37/7
	Route	Boston to Los Angeles	Boston to Los Angeles	Washington D.C. to Los Angeles	Newark (New Jersey) to Los Angeles
	Number of hijackers	5	5	5	4
	Airport security contractor	Globe Security	Huntleigh USA	Argenbright Security	Argenbright Security
	Departure time	7:59 a.m.	8:14 a.m.	8:20 a.m.	8:42 a.m. (Scheduled for 8:00)
	Seating of terrorists	3 business class 2 first class	3 business class 2 first class	2 coach 3 first class	4 first class
Hijacking	Estimated time of hijacking	8:14 (Just after 'fasten seatbelt' sign would have been turned off)	8:42–8:46 (after 'fasten seatbelt' sign would have been turned off)	8:51–8:54 (after 'fasten seatbelt' sign would have been turned off)	9:28 (about 45 minutes into the flight)
	Terrorists trained to fly	Mohamed Atta	Marwan al Shehhi	Hani Hanjour	Ziad Jarrah
	Activity in the cockpit	Boston's ATC Center was aware of the hijacking because at 8:25, a terrorist keyed the microphone (presumably trying to communicate with the cabin) and transmitted 'Nobody move. Everything will be okay. If you try to make any moves, you'll endanger yourself and the airplane. Just stay quiet.'	Last transmission from the pilots at 8:42 was their report of a 'suspicious radio transmission' from another aircraft (later discovered to be flight 11). Communication was then cut off.

The aircraft was flown erratically and had several near-collisions with other aircraft.

At 8:58, the aircraft turned towards New York City. | Last routine radio transmission at 8:51. At 8:54, the aircraft deviated from its course and turned south. At 8:56, the aircraft transponder was turned off and at 9:29, autopilot was disengaged.

At 9:34, the aircraft was five nautical miles from the Pentagon and began a descending turn ending at 670 m (2 200'). Throttles were set at maximum power and the aircraft dove into the Pentagon. | Cleveland ATC received calls declaring 'Mayday' followed by shouting 'Hey get out of here – get out of here – get out of here.' At 9:32, a hijacker announced to passengers 'Ladies and Gentlemen: Here the captain, please sit down keep remaining sitting. We have a bomb on board. So, sit.' The hijacker then had the autopilot turn the aircraft to the east.

At 9:39, Cleveland en route controllers heard an announcement that there was a bomb on board. |
| | Activity in the cabin | Two flight attendants and one passenger were stabbed, irritant | Passengers and flight attendants called in reports (from the | Passengers and flight attendants reported terrorists | At least ten passengers and two crew members made a series of calls |

Figure 6.5 Overview of the 9/11 attacks

Source: Adapted from Kean & Hamilton, 2004

Hijacking (Continued)				
(possibly mace) was sprayed in the cabin, and terrorists claimed they had a bomb. Two flight attendants used the airphone to contact American Airlines' reservation office and report the event. At 8:44, flight attendant reported 'Something is wrong. We are in a rapid descent… we are all over the place.' When asked to look out the window, she said, 'We are flying low. We are flying very, very low… Oh my God we are way too low.'	back of the aircraft) that terrorists had used knives, mace, and the threat of a bomb. They reported that flight attendants had been stabbed and that both pilots were killed.	had knives and box cutters and moved passengers to the back of the aircraft.	from airphones to family and friends, learning of the New York attacks and sharing information about their flight. At 9:57, a passenger assault began on the cockpit door. Jarrah rolled and pitched the aircraft to knock passengers off balance, and told another hijacker to block the door. A passenger is heard yelling 'In the cockpit. If we don't we'll die!' Presumably, the terrorists had determined that passengers were only seconds away from entering the cockpit. The controls were turned hard to the right and the aircraft rolled on its back.

Evidence suggests there were no real bombs on board the aircraft, that hijackers had lied to passengers to keep them under control.

Impact					
Crash	8:46: Aircraft crashed into the North Tower of the World Trade Center in New York City. All on board were killed instantly.	9:03: Aircraft crashed into the South Tower of the World Trade Center in New York City. All on board were killed instantly.	9:37: Aircraft crashed into the Pentagon at roughly 850 km/h (530 mph). All on board were killed instantly.	10:03: Aircraft crashed into an empty field in Shanksville, Pennsylvania at 930 km/h (580 mph).
Post-crash	Approximately 50 000 people worked at the Twin Towers. Those who weren't killed in the impact tried to evacuate the 110 storey buildings. At 9:59, the South Tower collapsed, killing all civilians and first responders still inside. At 10:28, the North Tower collapsed, killing many civilians and first responders still inside, as well as many on the ground and in the streets. Many first responders were killed, including 343 from the New York Fire Department, 37 from the Port Authority Police Department, and 23 from the New York Police Department.		125 people inside the Pentagon were killed. 106 people were seriously injured and transported to hospital.	It was determined that the likely target of the flight was the Capitol Building or the White House in Washington, DC. The target was about 20 minutes' flight from the crash location. The passengers are credited as heroes for their bravery and sacrifice, saving an unknown number of lives on the ground by preventing the flight from reaching its target.

In total, 2997 people were killed in the four attacks.

Figure 6.5 Overview of the 9/11 attacks (Continued)

The unfortunate reality is that terrorists invest a lot of time and creativity into their plans to defeat security systems. Security professionals must therefore constantly challenge themselves to imagine the unimaginable, and to innovate their practices to stay ahead of criminal plans.

In November 2001, the Aviation and Transportation Security Act was passed in the US, creating the Transportation Security Administration (TSA), which is now part of Homeland Security in the US. Note that before the existence of the TSA, American airport security was carried out by contractors; now, all airport security in the US is managed by the federal government. Of the US$5.3 billion annual TSA budget, 90 per cent goes towards aviation security, federalizing airport security screeners and deploying new security technology throughout the United States.

The personal, political, and cultural effects of 9/11 have been felt worldwide. In the aviation industry, the consequences of the attacks have also been economic – the 9/11 attacks caused devastating losses for airlines, which exceeded US$12 billion globally in 2001 (including US$7.5 billion in the US and €1.9 billion in Europe).[18] However, the economic repercussions of the 9/11 terrorist attacks continued to be felt for years after 9/11 as the aviation industry was impacted by passengers' new fear of flying. In an effort to prevent airlines from going bankrupt, the US federal government swiftly passed the Air Transportation Safety and System Stabilization Act, which amounted to a $15 billion bailout to airlines. Europe also offered airline support by 1) ensuring airlines access to insurance at pre-9/11 rates, and 2) paying $200 million in aid to European airlines affected by the temporary closure of US airspace.[19]

ICAO responded to 9/11 by revising Annex 17 with security provisions based on the new collective understanding that terrorists could use aircraft as guided missiles to attack ground targets. These new provisions focused on a range of factors including domestic operations, international cooperation (including sharing of threat information), passenger and hold baggage screening, and in-flight security. Probably the most noteworthy change was the reinforcement of cockpit doors to prevent unlawful forced entry.

Criminal Acts

Acts of Individuals

The term *acts of individuals* refers to incidents associated with disruptive passengers, including air rage. Travel can be a stressful process for passengers, with the potential for missed flights, long security lines, or flight delays. These frustrations can lead to a range of responses from passengers from mild irritation to violent air rage that endangers the safety of a flight. Annex 17 defines a disruptive passenger as

> a passenger who fails to respect the rules of conduct at an airport or on board an aircraft or to follow the instructions of the airport staff or crew members and thereby disturbs the good order and discipline at an airport or on board the aircraft.[20]

The Tokyo Convention (see Table 6.1) states that it is against the law to commit any act that jeopardizes 1) the safety of an aircraft, 2) the people and property inside an aircraft, or

3) discipline and good order on board an aircraft, regardless of whether the act is illegal in a given State. It also defines an aircraft as a 'micro jurisdiction' and gives the pilot-in-command (PIC) the authority to deal with an offence, or to instruct crew or other passengers to deal with it, without any risk of liability.

A disruptive passenger may make a variety of poor choices on board a flight. The International Air Transport Association (IATA) provides the following examples of disruptive behaviours:[21]

- consumption of illegal narcotics;

- refusal to follow safety instructions (such as fastening a seat belt);

- verbal or physical confrontations with crew or passengers;

- uncooperative passenger;

- making threats (related to injuring someone or to a bomb on board);

- sexual abuse and harassment; and

- riotous behaviour.

These actions may be the result of a variety of factors including alcohol or drug consumption, irritation with crew or other passengers, frustration with the journey (poor customer service, dirty lavatories, not being able to smoke), mental health issues, or withdrawal symptoms.[22]

Air rage incidents occur regularly and range from the offensive to the bizarre. For example, in 2014, a Thomson flight from Tunisia to Edinburgh had to divert to Gatwick because a 48-year-old woman became heavily intoxicated, loudly demanded cigarettes and a parachute to jump off the plane, and then slapped a girl in a nearby seat. When the cabin crew asked her to calm down, she unbuckled her prosthetic leg and struck them with it, then head-butted a crew member. The woman was eventually handcuffed to her seat and removed by police after landing in Gatwick.[23] An online search will reveal no shortage of further examples of air rage.

Managing disruptive passengers requires the collaborative effort of all airline employees who interact with customers, including check-in agents, lounge staff, gate agents, pilots, and flight attendants. All these people need to feel empowered to take reasonable steps to prevent, handle, categorize, and communicate information about disruptive passengers.

A helpful tool for assessing threats is the four-level categorization of disruptive passengers. ICAO Doc 9811 Manual (Restricted) defines the following framework:

- Level 1 – Disruptive behaviour (verbal): speaking loudly or rudely to crew, not following instructions, using profane language.

- Level 2 – Physically abusive behaviour: physical contact with crew, sexually obscene actions, tampering with or damaging aircraft equipment.

- Level 3 – Life-threatening behaviour: sexual or physical assault, displaying or threatening with a weapon.

- Level 4 – Attempted or actual breach of the flight crew compartment: attempt to sabotage aircraft, gain entry to the cockpit, threat of death or injury to gain entry to cockpit.[24]

Level 1 and 2 behaviours represent safety hazards while Level 3 and 4 behaviours represent security threats to the flight. These threat levels are used by airlines to create standard operating procedures (SOPs) for crew, dictating how to respond to incidents at each of the four levels.

For Level 1 and 2 hazards, airlines should have a warning handout available that describes disruptive behaviours and informs passengers of the legal consequences for these behaviours (ranging from criminal prosecution to financial compensation to the airline if a diversion to a nearby airport is required to remove them from the aircraft). Many airlines also have on-board restraint systems, such as handcuffs or zip ties, so that disruptive passengers can be controlled until landing. Flight crew are permitted to request able-bodied passengers to assist them with restraints, if necessary.

Ultimately, the PIC has authority over the aircraft and has the legal right to take actions necessary to ensure the safety of the flight. However, as the PIC is typically separated from passengers by the reinforced cockpit door, the task of managing and restraining disruptive passengers often falls on flight attendants (who act as agents of the PIC and are legally protected from liability). After an incident, the PIC is responsible for ensuring necessary paperwork and reports are properly completed.[25]

Aviation Piracy

Aviation piracy refers to a category of crime that involves exploiting the aviation industry for profit. Globally, the criminal markets for drug smuggling and human trafficking are highly profitable (see Figure 6.6). Therefore, it is not surprising that the most common acts of aviation piracy are linked to these activities. Strictly speaking, the control of these activities is usually the responsibility of customs and immigration agencies, rather than aviation personnel. However, it is generally accepted that aviation professionals have a role to play in recognizing and preventing these criminal activities.

Did You Know?

When most people think of a pirate, an image of a swashbuckling, seafaring adventurer comes to mind. However, within AVSEC, the term *piracy* refers to the legal term *hostis humani generis* (Latin for 'enemy of all mankind') and is one of the few universal crimes that a State can punish even if the act was not against its own citizens.[1]

Note

1 Havel & Sanchez, 2014, p. 182

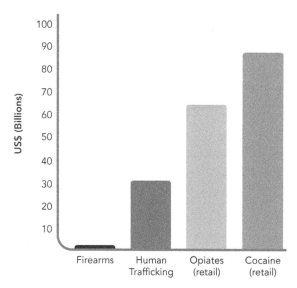

Figure 6.6 Global annual value of criminal markets

Source: Adapted from UNODC, 2010

Drug Smuggling

The production, transport, and sale of illegal substances is a global problem. It is estimated that about 250 million people (one in 20 adults between 15 and 64 years of age) used an illicit substance at least once in 2014.[26]

There are victims at both ends of the drug trade, beginning with the poor, exploited farmers in the developing world who produce the drugs and ending with the addicts (in richer countries) desperate to acquire and consume the substances. However, between these two groups exists a complicated network of criminals seeking profit. According to ICAO:

> The illegal drug trade is an international evil which undermines security, stability, and the health and well-being of persons by stripping them of all capacity and sense of belonging. The phenomenon also weakens the economic, cultural and political foundations of society, and as such, is a major national security problem.[27]

Moving hundreds of tonnes of illegal substances and then returning billions of dollars in cash represents a complicated logistics challenge. Criminals exploit both sea-based and air-based transportation networks to transport their products and profits.

Aviation drug smuggling is most often related to the transportation of cocaine and heroin as these substances are transported long distances between the points of origin and consumption. Substances such as cannabis, methamphetamine, and amphetamine are usually produced and sold locally, and may use aviation transportation for domestic movements, but the focus of this discussion will be on the movements of cocaine and heroin.

Heroin is derived from opium. Of the world's opium, 70 per cent is produced in Afghanistan (supplying neighbouring countries and Europe), 14 per cent comes from Myanmar (supplying East and Southeast Asia and Oceania), while 11 per cent of opium comes from Mexico and Colombia (supplying North American markets).[28]

Cocaine is derived from coca, which is primarily produced in three countries: Colombia, Peru, and the Plurinational State of Bolivia.[29] The global market for cocaine is mostly within North America and Europe.

For both heroine and cocaine, the distance between supply and consumption creates a logistical challenge for criminals, who must establish transportation channels through transit countries to sell their illegal products. Drugs breed corruption and violence. In fact, the highest murder rates in the world lie along main cocaine trafficking routes.

Drug smugglers range from dangerous organized criminals to non-violent individuals seeking extra cash. In rare cases, aviation professionals themselves have participated in illegal smuggling. As airline professionals may receive less scrutiny at checkpoints and border crossings, there is incentive for criminals to seek out and take advantage of these professionals. In 2016, when a flight attendant was flagged for a random security screening at Los Angeles International Airport, she pretended to take a phone call, kicked off her high-heeled shoes, and ran, leaving a suitcase with over 30 kilograms (nearly 70 pounds) of cocaine behind.[30] As smugglers can be of any nationality, gender, or profession, everyone travelling from drug-producing regions to locations that provide a market for narcotics can expect additional scrutiny during security screening.

One challenge in reducing aviation-based drug smuggling is that the law enforcement agency in a country is often separate from the aviation authority. For example, in the United States, the FAA controls pilot licensing and aircraft certification and is not affiliated with the FBI or other law enforcement agencies. When law enforcement was unable to stop repeated drug trafficking by the same pilots and aircraft, the US Aviation Drug-Trafficking Control Act of 1984 required the FAA to invalidate the pilot licence of anyone convicted of violating controlled substance laws and to revoke the registration of any aircraft used to transport controlled substances.[31]

Drug smugglers not only represent a security problem, but also present a safety problem for legally operating aircraft. Smugglers have little consideration for aviation safety while they are transporting their illegal goods. Criminals steal or seize aircraft, violate airspace, land on secret runways, and use phony aircraft call signs and registrations.[32] These operators will commonly fly at night at very low altitudes (to avoid radar contact), with no lights, no radio contact, and no flight plan. This leads to an increased accident rate along drug smuggling routes.

What can be done?

Globally, the United Nations Office on Drugs and Crime (UNODC) is the lead international agency on drug control. In 2009, member States committed to the collective goal of significantly reducing drug supply and demand by 2019.[33] Improvements have been made, as the global interception rates of opiates were about 30 per cent between 2009 and 2014, more than double the interception rates in the period between 1980 and 1997.[34] The global interception rate of cocaine was estimated at between 43 and 68 per cent in 2014.[35]

Security groups within airports are very aware of drug smuggling and use technology to screen for drugs. Law enforcement personnel can apply the concept of *asset forfeiture*, wherein property, proceeds, or instruments of crime are confiscated. Except for the drugs themselves, the confiscated items may be sold and the profits used for good – such as funding anti-drug programmes.

Business aviation professionals and those who charter private aircraft need to be particularly mindful of drug smuggling activities. Many criminals will charter aircraft rather than purchase their own, as it involves less risk. Aviation professionals who offer charter flights should be suspicious of

- unusual requests (such as one-way trips that might indicate the delivery of drugs or money);

- payments made in cash;

- customers who are not concerned with getting a receipt;

- attempts to change or camouflage aircraft tail numbers;

- reluctance to provide name or identification documents;

- unusually large amounts of luggage;

- rental of a hangar, with cash, for a short time;

- desire to use an aircraft that is larger than seems necessary;

- parking in remote areas of the airport;

- security locks on cargo areas of aircraft; and

- modified (enlarged) fuel tanks or antennas.[36]

Human Trafficking

In the twenty-first century, it can be hard to imagine that human slavery still exists. Unfortunately, this violation of human rights is still a reality in our world and the aviation industry is commonly used to transport its victims.

Almost 21 million people are victims of forced labour – 11.4 million women and girls and 9.5 million men and boys around the world. Geographically, the largest number of forced labourers come from the Asia-Pacific region (11.7 million (56 per cent of the global total)), followed by Africa (3.7 million (18 per cent) and Latin America (1.8 million victims (9 per cent)).[37] Victims are typically trafficked from poor to wealthier countries.

Aviation professionals are in a unique position to identify people at risk as they move through the transportation network and to help reduce these human rights violations.

Trafficking versus Smuggling

Human trafficking is defined by the fact that it is *exploitative* and *non-consensual*. This means that human traffickers take advantage of their victims, that victims are enslaved for an indefinite amount of time, and that victims are not being transported of their own free will.

Trafficking is sometimes confused with human smuggling. However, smuggling is quite different because the people transported by human smugglers are participating of their own free will (in violation of laws and international checkpoints) in order to seek opportunities in another country. Unlike trafficking, smuggling situations typically end when the border has been crossed and the smugglers have been paid. The people involved are then free to go.

UN Response to Trafficking

The United Nations created a Protocol to Prevent, Suppress and Punish Trafficking in Persons, Especially Women and Children, which came into force on December 25 2003. This is a global, legally binding instrument that supports international cooperation in investigating and prosecuting human trafficking cases as well as protecting and assisting its victims. Within the document, Kofi A. Annan, then Secretary-General of the UN states:

> I believe the trafficking of persons, particularly women and children, for forced and exploitative labour, including for sexual exploitation, is one of the most egregious violations of human rights that the United Nations now confronts. It is widespread and growing. It is rooted in social and economic conditions in the countries from which the victims come, facilitated by practices that discriminate against women and driven by cruel indifference to human suffering on the part of those who exploit the services that the victims are forced to provide.[38]

Within the UN protocol, States are asked to conduct research, collect information, and launch mass media campaigns to inform the public and combat trafficking in persons. Aviation professionals can play an important role in the identification and reduction of human traffickers. Flight attendants, while carrying out their professional responsibilities, have a unique opportunity to observe and speak with travellers. On a typical flight, flight attendants may observe a traveller that seems suspicious. Airline Ambassadors International (AAI) is a non-profit organization that has grown out of this experience

Nearly 21 million people are victims of modern slavery around the globe. It has been estimated that they generate US$150 billion each year in profits for traffickers (FreeTheSlaves).

Labour Slavery

Manual labour
Domestic work
Maids
Gardeners

79%

21%

Sexual Slavery

Sexual exploitation
Prostitution

Over **1 in 4** of all modern slaves are children.

The **Cost** of Slavery

$40 000 Cost of a slave in the United States in 1850 (in modern dollars).

$90 Average cost of a slave today. If a slave becomes sick or hurt, he or she is often disposed of or killed (Bales, 2004).

Figure 6.7 Modern slavery

Source: Adapted from Bales, 2004; FreeTheSlaves, 2017

among flight attendants. AAI offers flight attendants training on how to spot human trafficking and provides awareness posters for airport and aviation professionals. Table 6.2 lists some of the warning signs that aviation personnel should be familiar with.

Table 6.2 How to spot human trafficking

	How can you identify them?	Who are they?
Victims	• Did not arrange their own travel. May not know where they are going. • May appear to be controlled (look to someone else to answer questions for them) or afraid of their companion. • Responses to questions may seem scripted and unnatural. • Missing or altered travel documents. • May show signs of physical abuse. • May be inappropriately dressed for travel.	• 70 per cent of victims are enticed with promises of affordable vacations, participation in modelling or beauty contests, educational opportunities, or promises of good-paying jobs. • Victims may be of any gender, age, or ethnicity. • Victims tend to be travelling from poor to wealthier regions.
Traffickers	• May pretend to be related to the victim. • Might be observed carefully watching their victims or maintaining physical contact with them (holding their arm). • May answer questions on the victim's behalf. • May not know the name of or personal information about the victim they are travelling with.	• May be of any gender or ethnicity (in one German study 78.1 per cent of human traffickers were male and 21.9 per cent were female). Female traffickers are often used to establish trust more quickly with female victims. • Traffickers can range in age from children to elderly adults. • Almost half of recruiters are someone known to the victim.
	What to do?	What not to do?
	• Be observant of passengers (particularly young women and children). • Notice situations in which it seems that someone is being controlled. • If concerned, start a polite conversation and gather more information. • If you remain concerned, inform the Captain of your suspicions. The Captain should radio ahead to the destination airport and arrange for law enforcement personnel to greet the aircraft and investigate the situation.	• Do not confront the victim or trafficker. • Do not try to rescue the victim or appear visibly upset or alarmed. • Do not do anything to endanger yourself or others.

Source: Portions adapted from AAI, 2015 and UN Global Initiative to Fight Human Trafficking, 2008

Did You Know?

Technology can help in the fight against human trafficking. A free app called TraffickCam allows travellers to photograph their hotel rooms, geotag the images, and upload them to an enormous, constantly updated global database (as of 2016, it contained 1.5 million photos). Referencing the decorations and layout of the room, law enforcement agencies can compare the database to online images associated with human trafficking to recover victims, locate criminals, and investigate trafficking cases.[1] You can help by photographing and uploading a picture the next time you visit a hotel room.

Note

1 PRweb, 2016

Security Initiatives

With a vision towards maximizing security within international civil aviation, several international initiatives are underway. These include ICAO security strategy and audit programmes, and the IATA/ACI Smart Security initiative, among others.

Universal Security Audit Programme – Promoting Global Compliance with Annex 17

The ICAO Comprehensive Aviation Security Strategy (ICASS) incorporates a variety of approaches to addressing new and existing threats, promoting innovation, sharing information, ensuring global compliance with Annex 17 standards, and emphasizing the importance of security among States and stakeholders, and within ICAO.

Following the 9/11 tragedy, ICAO determined that several States were not strictly abiding by ICASS, in that they were not adequately implementing the security standards required by Annex 17. The Universal Security Audit Programme (USAP) was therefore launched to audit, monitor, and prioritize recommendations to reduce risk, and to prescribe corrective actions for States to ensure compliance with Annex 17 standards.

The audit process is a significant undertaking as several security-related agencies in a State must be involved, including the CAA, police, airlines, airports, customs and immigration, military, and foreign affairs, among others. The intent of USAP is to ensure that contracting States have an oversight system guaranteeing positive control and supervision over all civil aviation security activities with the State.

Since USAP was implemented, two full cycles have been completed, which included hundreds of audits and follow-up visits. In the first cycle of USAP (2005–2009) 52.8 per cent of State audits revealed a lack of compliance while follow-up visits showed a reduction to 34.5 per cent, a significant improvement. The second cycle of audits (2008–2013) found a global lack of effectiveness rate of 30.7 per cent.[39] In the future, USAP will be adopting a continuous monitoring approach (CMA), which will regularly monitor each member State's security based on a risk management philosophy, referencing that State's findings from the first two USAP cycles.

Smart Security

Smart Security is a joint initiative between IATA and ACI. The goal of this programme is to develop the next generation of airport checkpoints that maximize security while minimizing inconvenience to passengers. Although the current system works, the industry recognizes that the predicted growth of air traffic represents a challenge for the future. Such an increase will place increased strain on security screening systems and lead to long lines and delays, if the system does not evolve to meet demand.

Smart Security would take a risk-based approach to screening using advanced screening technologies and improving passenger throughput and experience. Smart Security would incorporate the following:[40]

- *Passenger screening* – The walk-through metal detector (WTMD) would be used as the primary measure, with imaging technologies as a secondary measure for those who set off the WTMD. This would allow security stations to process over 800 passengers per hour. (Currently, stations that use imaging technologies as a primary measure have a higher security effectiveness but are less efficient, processing over 300 passengers per hour). In addition, new algorithms will support better detection and decision-making.

- *Cabin baggage screening* – There are limitations to traditional X-ray machines: they are dependent on the operator for detection, and they inconvenience passengers by having them remove certain items from their baggage (e.g., laptops and liquids, aerosols, and gels (LAGs)). When a security operator notices something suspicious, it is flagged for a secondary hands-on search. Smart Security will implement new technologies, including CT and X-ray diffraction (XRD), to provide detailed scans of items without them having to be removed from passengers' bags. In addition, automated target recognition (ATR) systems virtually flag potential threats to support the operator in detecting items.

- *Explosive detection* – explosive trace detection (ETD) systems identify the residue from creating or handling explosives, which is very difficult to remove. ETD systems are portable and highly effective. Explosive detection dogs are also efficient in moving easily through a terminal and screening many passengers. Ultimately, systems must continually evolve in order to remain unpredictable to terrorists.

- *Checkpoint management* – As passenger numbers grow, and capacity constraints are reached, checkpoint management systems (CMS) are a solution for coordinating activities, sharing information, and monitoring checkpoint performance within an airport. Some features include automatic document scanning, configuring lanes to improve throughput, and risk-based screening that references biometrics.

Other Security Considerations

Figure 6.8 details several additional factors relevant to modern AVSEC.

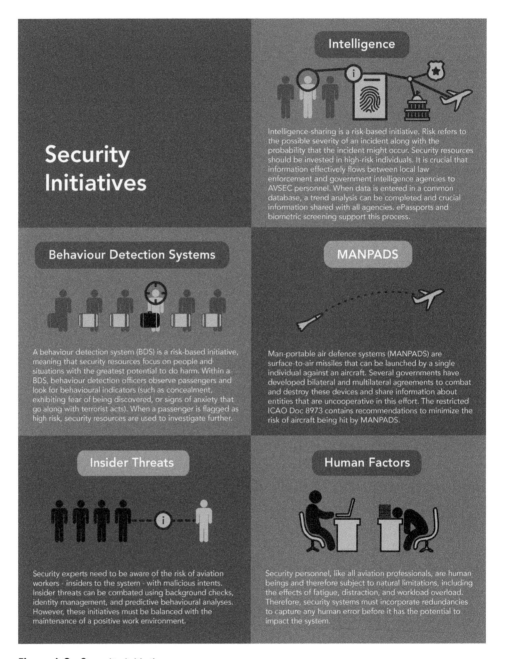

Figure 6.8 Security initiatives

Table 6.3 The security agencies of the ICAO Council States

State	Agency responsible for writing rules	Agency responsible for implementing rules
Australia	Australian government	Private operators
Brazil	Agência Nacional de Aviação Civil	Private operators
Canada	Transport Canada	Canadian Air Transport Security Authority
China	Civil Aviation Administration of China	
France	European Commission (the executive of the European Union)	Direction de la sécurité de l'aviation civile; private operators
Germany	European Commission	Luftfahrt-Bundesamt, private operators
Italy	European Commission	L'Ente Nazionale per l'Aviazione Civile; private operators
Japan	Civil Aviation Bureau	Private operators
Russian Federation	Ministry of Transport of the Russian Federation (Mintrans)	
United Kingdom	Civil Aviation Authority (with guidance from the Secretary of State)	Private operators
United States	Transportation Security Administration	

Source: Brian J. Ho, unpublished research paper, University of Waterloo, 20 March 2017

Conclusion

Aviation security can be a complicated task that requires the coordinated efforts of several groups (AVSEC authorities, security professionals, aircraft and airport operators, local law enforcement, military, customs and immigration, and national security authorities, among many others). To maintain security, aviation professionals are challenged to 'imagine' possible security weaknesses and to plan solutions – as well as to continually innovate practices to ensure unpredictability.

As AVSEC evolves to meet the predicted increase in traffic, risk-based initiatives are key. *Risk* (the potential severity of a specific incident combined with the likelihood that it might occur, as discussed in Chapter 9) must be taken into account as threats are carefully considered and analysed, and security methods are deployed, as necessary. Airports, airlines, and government agencies use risk calculations to strategically determine how to deploy security resources.

The challenge of security is maintaining a balance between the sometimes-conflicting principles of security, privacy, and efficiency. ICAO recommends that security activities should cause only minimal interference or delay to civil aviation activities, as long as the effectiveness of security is not compromised.[41]

Ultimately, it may never be possible to have *perfect* security within such a complex and dynamic system. However, the international aviation community is constantly working to implement innovative risk-based security interventions to predict and prevent threats from endangering aircraft.

Key Points to Remember

1. Civil aviation is to be used for peaceful purposes only, but is an attractive target for criminals. Like the UN, civil aviation requires tremendous international cooperation and stands as a symbol of positive and successful global relations. Terrorists, therefore, consider the aviation industry a setting where they can gain political attention by introducing chaos and anxiety on a world stage.

2. *Safety* and *security* are related, but distinct, concepts. Safety focuses on preventing aviation accidents; security focuses on protecting the aviation system from intentional and criminal wrongdoing.

3. In 1974, ICAO Annex 17 established the standards and recommended practices for AVSEC. Annex 17, along with the supporting documentation in Doc 8973 (the restricted *Aviation Security Manual*), are reviewed and revised regularly to keep up to date with the industry. Other international security conventions have been developed over the years to describe, among other thing, how offenders (typically hijackers of international flights) should be brought to justice; unfortunately, these conventions have some shortcomings.

4. A variety of security measures exist to protect civil aviation, related to airport access control, aircraft-related security, passenger and cabin baggage screening, hold baggage screening, cargo and mail screening, customs and border protection, and the management of potentially disruptive passengers, cyber threats, and intelligence, among others.

5. Hold baggage screening (referring to bags checked in the cargo compartments of aircraft) comprises several search tools and methods including manual search, risk assessment, explosive trace detection, trained dogs, X-ray devices, and baggage reconciliation systems. Pre-board screening of passengers and carry-on luggage uses a variety of devices such as walk-through metal detectors to identify metal and electronics, explosive detection technology to search for explosive residue on passengers' hands and luggage, and imaging technologies to quickly scan passengers' bodies for threats.

6. The term *acts of unlawful interference* refers to terrorism. In aviation, terrorist acts typically take the form of bombings or hijackings.

- Bombings of aviation targets date back to 1933, and include both aircraft and airport targets. Most bombers use IEDs, which represent an ongoing security concern.

- Hijackings became increasingly common in the late 1960s, when hundreds of attempts occurred. The aviation industry responded with the common strategy to 'accommodate, negotiate, and do not escalate' to limit the damage done.

7. The attacks of 9/11 changed everything as terrorists gained access to four aircraft and used them as guided missiles against ground targets. The aviation industry learned that the common strategy was entirely ineffective against suicide missions. This devastating event caused a rapid succession of new security regulations, including mandatory reinforcement of cockpit doors, which supports the new common strategy of 'defend the cockpit, at all costs'.

8. The term *criminal acts*, in aviation, refers to acts of individuals (including disruptive passengers) and aviation piracy, the exploitation of the aviation system by criminals for profit.

- Acts of individuals refer to disruptive passengers, air rage, and mental health incidents on aircraft. The PIC has authority to restrain and take other necessary actions against disruptive passengers to ensure the safety of the flight.

- Aviation piracy refers to illegal acts that are financially motivated, primarily linked to the offences of drug smuggling and human trafficking.

9. Drug smuggling is a global concern that involves the movement of hundreds of tonnes of product and billions of dollars. Within aviation, smuggling generally involves the transport of cocaine and heroin as these substances must be moved great distances between where they are grown and where they are sold. Human trafficking is a US$32 billion criminal market. Almost 21 million people are victims of this practice and many of them are moved through the aviation system. Flight attendants are in a particularly important position to identify and alert authorities of suspicious passengers.

10. As the aviation industry looks to the future, a variety of new initiatives are underway to innovate security technology, to continually consider and assess new security initiatives, and to ensure States effectively implement aviation security standards.

Table 6.4 Acronym rundown

AAI	Airline Ambassadors International
ACI	Airports Council International
ASA	air service agreement
ASP	Aviation Security Policy

(Continued)

Table 6.4 (Continued)

ATC	air traffic control
ATR	automated target recognition
AVSEC	aviation security
BDS	behaviour detection system
BRS	baggage reconciliation system
CAA	civil aviation authority
CMA	continuous monitoring approach
CMS	checkpoint management systems
CSIS	Canadian Security Intelligence Service
CT X-ray	computed tomography X-ray
ETD	explosive trace detection
FAA	Federal Aviation Administration
HBS	hold baggage screening
IATA	International Air Transport Association
ICAO	International Civil Aviation Organization
ICASS	ICAO Comprehensive Aviation Security Strategy
IED	improvised explosive device
ISIS	Islamic State of Iraq and Syria
LAGs	liquids, aerosols, and gels
MANPADS	man-portable air defense systems
PIC	pilot-in-command
PPS	pre-board passenger screening
SARPs	standards and recommended practices
SCATANA	Security Control of Air Traffic and Air Navigation Aids
SOP	standard operating procedure
TSA	Transportation Security Administration
UN	United Nations
UNODC	United Nations Office on Drugs and Crime
USAP	Universal Security Audit Programme
WTMD	walk-through metal detector
XRD	X-ray diffraction

Chapter Review Questions

6.1 Why is an aircraft in flight considered a micro jurisdiction? What rights and responsibilities does this bestow on the Captain (and people acting on his or her directions)?

6.2 What is the difference between aviation safety and aviation security? Which is easier to control, in your opinion?

6.3 Why did hijacking occurrences increase between 1968 and 1985? Was the policy of 'accommodate, negotiate, and do not escalate' appropriate, under those circumstances? Explain.

6.4 What is the difficulty in prosecuting hijackers of international flights? Can you think of an ideal solution?

6.5 It is necessary to strike a balance between privacy, security, and the efficiency (speed) of aviation security. In your opinion, which is most important and why (consider time, cost, risk, and civil rights)? Is modern AVSEC achieving an appropriate balance? Why or why not? What might be done to improve this balance?

6.6 How do you feel about profiling passengers (based on age, gender, and/or ethnicity)? Is this an effective tool to support aviation security? Is profiling subject to prejudice and potential human rights violations? If you or your family members were being profiled, would this change your response?

6.7 Explain three ways in which the terrorist attacks of 9/11 changed aviation. Which of the three will have the most significant impact on your future career in aviation?

6.8 Which of the recommendations for charter companies to recognize drug smuggling activities is the most important, in your opinion? Why?

6.9 What can a cabin crew member do to recognize and stop human trafficking? As a future crew member, what would you do if you were worried (but unsure) if a situation might be human trafficking? Describe fears you might have about alerting authorities to the situation (as well as fears if you chose to ignore it).

6.10 What is the name of the agency responsible for aviation security within your State? What are the biggest security challenges faced by this agency? Through your own research, explain how the agency is evolving to meet these challenges.

THE UNDERWEAR BOMBER – NORTHWEST AIRLINES FLIGHT 253[42]

On Christmas Day in 2009, Umar Farouk Abdulmutallab boarded Northwest Airlines flight 253 departing from Amsterdam for Detroit with 289 people on board.

As the flight approached Detroit, Abdulmutallab (who had spent approximately 20 minutes in the lavatory) returned to his seat and covered his lap with a blanket. Other passengers heard a loud noise that sounded like a firecracker. They found Abdulmutallab in flames, which spread to the carpeting of the aircraft. Passengers restrained Abdulmutallab while flight attendants extinguished the fire. He was relocated to the first-class cabin and held before being taken into custody by the authorities upon landing.

After the foiled plot, Abdulmutallab confessed that he had explosive powder sewn into his underwear. He had worn the explosive-stuffed undergarments for more than two weeks, while he travelled from Yemen to Africa and then to the Netherlands to board the flight. He had wanted to get used to the feeling of wearing the garment, so he had removed it only when showering.

His plan had been to inject a syringe of liquid acid into the explosive powder to form a plastic explosive, that would detonate over the United States and kill all 289 people on the aircraft. However, the explosives were soiled from being worn for more than two weeks, and the degraded explosives caught fire rather than detonating. Abdulmutallab had severe burns on his legs and groin from the fire. This incident was highly publicized in the media and Abdulmutallab was given the title of 'The Underwear Bomber'. He was sentenced to life without parole in 2012 after pleading guilty to all charges, including the attempted murder of 289 people and the attempted use of a weapon of mass destruction.

Although this story has some comical elements, the reality is that if the explosives had not malfunctioned, a tragedy would have occurred. Security screening did not identify the explosives, which raises many questions.

Case Study Questions

Using this case study, and applying what you have learned in this chapter, provide informed responses to the following:

6.11   Hiding explosive materials in sensitive body areas is intended to reduce the likelihood of detection. Terrorists are aware of media reports in which passengers claim 'gate rape', asserting that they were inappropriately touched by security screeners during pat-down. How can security processes balance privacy with detection? Should passengers expect to sacrifice their privacy for a more secure aviation system? Are there any screening processes or technologies that could have prevented the underwear bomber from getting explosive materials on the aircraft?

6.12   Passengers demand 100 per cent security, but express frustration and criticism when their movement through the airport is delayed by security initiatives (such as having to remove shoes and place them in X-ray scanners as a response to the failed shoe-bombing incident). How can security be ensured with the lowest possible impact on passengers' movements through the airport?

6.13   The Underwear Bomber is another in a long line of terrorists willing to sacrifice their lives to accomplish their attacks (consider the Shoe Bomber, the Istanbul Atatürk Airport attackers, and the 9/11 terrorists). With the trend of terrorists adopting suicide tactics, are baggage reconciliation systems still relevant? Reconciliation systems ensure that a passenger is on board before his or her baggage is loaded into the cargo compartment, but these systems do nothing to prevent suicide attacks. Is baggage reconciliation an outdated concept focused on extinct threats, or does it still have value in today's system?

References

AAI, 2015. *Basic human trafficking education.* [Online] Available at: http://airlineamb.org/our-programs/human-trafficking-awareness/aai-presentation/

AAIB, 1990. *Aircraft accident report No 2/90 (EW/C1094): Report on the accident to Boeing 747-121, N739PA.* Aldershot: Air Accidents Investigation Branch.

Abeyratne, R., 2010. *Aviation security law.* Berlin: Springer.

ACI, 2016. *ACI advisory bulletin: Landside security.* Montreal: Airports Council International.

Agnich, L., 2011. Aviation drug-trafficking control act. In: M. Kleiman & J. Hawdon, eds. *Encyclopedia of drug policy.* Los Angeles: Sage Publications, p. 77–78.

Bales, K., 2004. *Disposable people: new slavery in the global economy.* Rev. ed. Berkeley: University of California Press.

BBC, 2016. *Istanbul Ataturk airport attack: 41 dead and more than 230 hurt.* [Online] Available at: www.bbc.com/news/world-europe-36658187

Berlinger, J., 2016. *Flight attendant caught trying to smuggle more than 60 pounds of cocaine.* [Online] Available at: www.cnn.com/2016/03/22/us/flight-attendant-cocaine-smuggling/

CBC News, 2010. *Air India case marred by 'inexcusable' errors.* [Online] Available at: www.cbc.ca/news/canada/air-india-case-marred-by-inexcusable-errors-1.869072

Dearden, L., 2016. *ISIS plane attack: Egypt admits 'terrorists' downed Russian Metrojet flight from Sharm el-Sheikh for first time.* [Online] Available at: www.independent.co.uk/news/world/africa/isis-plane-attack-egypt-terrorists-downed-russian-metrojet-flight-from-sharm-el-sheikh-islamic-state-a6893181.html

FreeTheSlaves, 2017. *Slavery today.* [Online] Available at: *www.freetheslaves.net/about-slavery/slavery-today/*

Harrington, J., 2008. *Drug trafficking.* [Online] Available at: www.ainonline.com/aviation-news/aviation-international-news/2008-09-09/drug-trafficking

Havel, B. & Sanchez, G., 2014. *The principles and practice of international aviation law.* New York: Cambridge University Press.

Holden, R., 1986. The contagiousness of aircraft hijacking. *American Journal of Sociology,* 91(4), p. 874–904.

IATA/ACI, n.d. *Smart Security.* Montreal: International Air Transport Association / Airports Council International.

IATA, 2015. *Guidance on unruly passenger prevention and management, 2nd ed.* s.l.: International Air Transport Association.

ICAO Working Paper, 2013. *Illicit trafficking of narcotic drugs and psychotropic substances by air.* Montreal: International Civil Aviation Organization.

ICAO, 1944. *Convention on Civil Aviation ("Chicago Convention").* Chicago: International Civil Aviation Organization.

ICAO, 2011. *Annex 17 to the Convention on International Civil Aviation: Security: Safeguarding international civil aviation against acts of unlawful interference, 9th ed.* Montreal: International Civil Aviation Organization.

ICAO, 2013. *Security & facilitation: Universal Security Audit Programme. Analysis of audit results: November 2002 to June 2013.* Montreal: International Civil Aviation Organization.

ICAO, n.d. *The Convention on International Civil Aviation: Annexes 1 to 18.* Montreal: International Civil Aviation Organization.

ILO, 2017. *Statistics and indicators on forced labour and trafficking.* [Online] Available at: *www.ilo.org/global/topics/forced-labour/policy-areas/statistics/lang--en/index.htm*

Jedrychowski, E., 2012. Post-September 11 US and European airline industries: Navigating through the bailouts, bankruptcies, liquidations and mergers. *Connecticut Journal of International Law,* 28(1), pp. 177–198.

Kean, T. & Hamilton, L., 2004. *The 9/11 Commission report: Final report of the National Commission on terrorist attacks upon the United States.* Washington, DC: National Commission on Terrorist Attacks upon the United States.

Malik, O., 1998. Aviation security before and after Lockerbie. *Terrorism and Political Violence,* 10(3), pp. 112–133.

Matas, R., 2002. Cost of Air-India probe hits $82-million. *Globe and Mail,* September 19.

National Consortium for the Study of Terrorism and Responses to Terrorism, 2016. *Global Terrorism Database,* s.l.: s.n.

Price, J. & Forrest, J., 2016. *Practical aviation security: Predicting and preventing future threats.* 3rd ed. New York: Elsevier.

PRweb, 2016. *Travelers use exchange initiative's TraffickCam app to fight sex trafficking by uploading hotel room photos to national database.* [Online] Available at: www.prweb.com/releases/2016/06/prweb13497362.htm

Public Safety Canada, 2015. *Remembering Air India flight 182.* [Online] Available at: www.publicsafety.gc.ca/cnt/ntnl-scrt/cntr-trrrsm/r-nd-flght-182/index-eng.aspx

Rodrigues, C. & Cusick, S., 2012. *Commercial aviation safety.* 5th ed. New York: McGraw-Hill.

Senate of the United States Government, 2010. *Attempted terrorist attack on Northwest Airlines flight 253: Report of the Select Committee on Intelligence (Report 111–199).* Washington, DC: US Government Printing Office.

UN Global Initiative to Fight Human Trafficking, 2008. *Workshop: Profiling the traffickers (Background Paper).* Vienna: The Vienna Forum to fight Human Trafficking, February 13–15 2008.

United Nations, 1969. *Convention on offences and certain other acts committed on board aircraft.* Vienna: United Nations.

UNODC, 2004. *United Nations covention against transnational organized crime and the protocols thereto.* Vienna: United Nations Office on Drugs and Crime.

UNODC, 2010. *World drug report.* Vienna: United Nations Office on Drugs and Crime.

UNODC, 2016. *World drug report.* Vienna: United Nations Office on Drugs and Crime.

Ushynskyi, S., 2009. Pan Am flight 103 investigation and lessons learned. *Aviation,* 13(3), pp. 78–86.

Vesty, S., 2014. *One-legged woman attacks holiday flight cabin crew with prosthetic leg in air rage incident.* [Online] Available at: www.dailyrecord.co.uk/news/scottish-news/one-legged-woman-attacks-holiday-flight-3944119#MQwcOWUTdXucVSIY.97

Williams, C. & Waltrip, S., 2004. *Aircrew security: A practical guide.* Aldershot: Ashgate Publishing Limited.

Notes

1 ICAO, 1944
2 ICAO, 2011
3 ICAO, n.d.
4 ICAO, n.d.
5 Havel & Sanchez, 2014, p. 208
6 Havel & Sanchez, 2014, p. 211
7 Rodrigues & Cusick, 2012, p. 290
8 United Nations, 1969, p. 222
9 Havel & Sanchez, 2014, p. 183
10 Havel & Sanchez, 2014, p. 188
11 Dearden, 2016
12 BBC, 2016
13 ACI, 2016
14 Holden, 1986
15 Williams & Waltrip, 2004, Ch. 3, p. 2
16 Price & Forrest, 2016, p. 359
17 Kean & Hamilton, 2004, p. 339
18 Jedrychowski, 2012
19 Jedrychowski, 2012, p. 181
20 ICAO, 2011, pp. 1–2
21 IATA, 2015, p. 13
22 Williams & Waltrip, 2004
23 Vesty, 2014
24 IATA, 2015, p. 28
25 IATA, 2015, p. 37
26 UNODC, 2016, p. ix
27 ICAO Working Paper, 2013, p. 3
28 UNODC, 2016, p. 27
29 UNODC, 2016, p. 35
30 Berlinger, 2016
31 Agnich, 2011
32 ICAO Working Paper, 2013
33 UNODC, 2010
34 UNODC, 2016, p. 27
35 UNODC, 2016, p. 36
36 Harrington, 2008
37 ILO, 2017
38 UNODC, 2004, p. iv
39 ICAO, 2013
40 IATA/ACI, n.d.
41 ICAO, 2011, p. 2–1
42 Senate of the United States Government, 2010

1. Aviation is fundamentally impacted by environmental factors (such as weather) – and activities in this sector have a direct effect on the environment.
   a. True
   b. False

2. _____ describes the scientific study of the atmosphere.
   a. Weather
   b. Climate
   c. Meteorology
   d. Environment

5. All aviation activities account for about ___ per cent of global CO_2 emissions.
   a. 2
   b. 4
   c. 9.5
   d. 14

3. A goal of climate change initiatives is to reduce global warming before annual average temperatures reach two degrees Celsius above the 1850-1900 benchmark.
   a. True
   b. False

4. Annex 16 has two volumes that contain SARPs associated with environmental protection. The topics of the two volumes are:
   a. Pollution and noise
   b. Noise and engine emissions
   c. Engine emissions and fuel disposal
   d. Fuel disposal and waste management.

Learning science suggests that thinking through a few questions before you begin studying new material, even if you answer incorrectly, results in improved learning and retention.
Give it a try!

CHAPTER 7

Environment

CHAPTER OUTCOMES

At the end of this chapter, you will be able to . . .

- Discuss aspects of meteorology – including the atmosphere, air pressure, air masses, wind, and lifting forces – and their impact on aviation operations.

- Describe the changing world climate, including the impact of significant weather events on aviation activities.

- Explain environmental protection initiatives such as 1) the United Nations Framework Convention on Climate Change, 2) ICAO's initiatives for carbon neutral growth from 2020, and 3) ICAO's carbon offsetting and reduction scheme for international aviation.

- Identify the international Standards in Annex 16 associated with environmental protection, specifically related to aircraft noise and engine emissions.

- Describe the global community's collaboration in forecasting weather and disseminating the information internationally.

- Use your understanding of meteorology and its impacts on aviation operations to analyse the Delta flight 191 accident, which was associated with wind shear.

Introduction

Aviation is heavily influenced by environmental conditions. Uncontrollable natural forces, as simple as temperature and the movement of air, have a direct impact on the efficiency and economics of air travel.

The study of weather is a discipline in itself, one which easily fills many excellent textbooks. As an introduction for students of international aviation, this chapter will introduce key aspects of the natural environment that affect aviation, examine the impact that aviation activities have on the environment, and discuss sustainable development initiatives. An overview of how the world collaborates to gather and share weather information, and ICAO's role in this process, will also be provided.

Although sometimes confused, weather and climate are not the same thing. *Weather* refers to the short-term environmental conditions on a specific day that can change quickly (rainy or dry, cold or hot, clear or cloudy), while *climate* can be thought of as an average of the weather conditions over a longer period (months or years). In addition to a discussion of weather and climate, this chapter will provide an introduction to *meteorology*, which is the scientific study of the atmosphere related to forecasting weather.

7.1 The Language of the Environment

Convection describes the vertical movement of air. Convection can trigger weather events that impact aviation, such as turbulence and wind shear.

Visibility, for aviation purposes, is used to describe how far a person can see. It is measured as the distance at which a black object can be seen against a white background (or a light can be seen against a dark background). Many aviation regulations specify a minimum visibility for certain types of operations.

Aviation weather reports use several acronyms. These acronyms are so common that within operations, it is rare for the full description of weather services to be used. Some of the common acronyms are listed below:

- A **METAR** is a *meteorological terminal aviation routine weather report* that describes weather conditions at an airport.

- A **TAF** is a *terminal area forecast*, which describe the weather predicted in the area around an airport.

- An **ATIS** is an *automatic terminal information service* – a pre-recorded description of weather conditions at an airport (broadcast on a continuous loop and updated at regular intervals).

- An **AIRMET** is an *airmen's meteorological*, which refers to information on moderate weather phenomena (mostly of interest to smaller general aviation aircraft).

- A **SIGMET** refers to *significant meteorological* information about weather events that are hazardous to all aircraft.

- A **GAMET** refers to *general aviation meteorological information* of forecast weather conditions for low-level flights.

- A **SIGWX** describes *significant weather* at high levels, such as thunderstorms, turbulence, fronts, and jet streams.

- A **VOLMET** describes regional weather information for aircraft in flight (from the French *vol* (flight) and *météo* (weather)).

- A **PIREP** is a *pilot report* of significant weather that has been observed during a flight.

- A **NOTAM** is a *notice to airmen* of hazardous weather conditions.

Weather

Significant weather events, such as severe storms, heavy rains, and changes in prevailing winds, directly impact aviation operations. These events may cause airlines to change their routes and airports to invest in infrastructure such as improved drainage to manage very heavy rain, or longer runways (as hot temperatures reduce the lifting properties of the air, which requires a longer take-off run to achieve flight). On a broader scale, significant weather events can influence people's decision to travel at all.

The variety of ways in which hazardous weather conditions can directly impact aviation operations are set out in Table 7.1.

Table 7.1 Weather hazards and their effects on aviation

Weather hazard	Description	Impact on aviation
Fog/mist	Fog and mist occur when moist air at ground level cools to its dew point temperature (causing the water droplets to condense and be suspended in the air). Fog and mist are essentially the same phenomenon; they can be thought of as 'clouds on the ground'. This condition is classified as *mist* if visibility is more than one kilometre (0.62 miles) and *fog* if visibility is less than that.	Fog and mist can have a significant impact on airport operations. Take-offs and landings may be delayed until visibility improves. Reduced visibility can lead to runway incursions.

(Continued)

Table 7.1 (Continued)

Weather hazard	Description	Impact on aviation
Icing	Ice and snow that have accumulated on an aircraft must be removed from the wings before take-off. However, icing can also accumulate while an aircraft is in flight. In-flight icing is not caused by ice in the clouds, which would be unlikely to stick to an aircraft, but rather by supercooled liquid water droplets or a cold-soaked airframe. • *Supercooled* means that the water in the atmosphere is below its freezing point, but is still a liquid. Supercooled water droplets freeze on contact with a solid surface (in this case, the leading edge of a wing). • An aircraft can become *cold soaked* (cooled below freezing) if it has been travelling at an altitude with cold air. When the aircraft descends into warm moist air that is above freezing, moisture can freeze on the aircraft causing structural icing.	When ice accumulates on an airframe, it significantly impairs operational performance. Lift and thrust are decreased while drag and weight are increased. Ice can also block ports on the aircraft that disrupt on-board avionics (altimeter, airspeed and vertical speed indicators) and antennas necessary for communications. Many aircraft incorporate de-icing or anti-icing systems to manage or prevent ice accumulation.
Turbulence	*Turbulence* refers to a powerful and unsteady movement of air. Turbulence can result from the wake of another aircraft, convection (such as occurs near thunderstorms), airflow over mountains, and wind shear.	When aircraft fly through air currents that vary in direction or speed, the flight will experience turbulence. Turbulence varies in severity from minor bumps to violent jolts that have the capacity to cause injury to those on board and structurally damage the aircraft.
Volcanic ash[1]	A volcanic eruption can emit a massive quantity of very small solid particles into the atmosphere. This volcanic ash can impact weather for several months following the eruption, as it accumulates within high-level clouds and moves with the wind. As the particles are very small, they are not visible by air traffic control or on-board weather radar.	The primary risk associated with volcanic ash is engine malfunction resulting from the ash melting and then fusing to internal engine components. This can lead to engine surges and flame-outs (loss of power). Ash can also cause windshield and aircraft skin corrosion.

Weather hazard	Description	Impact on aviation
Wind shear	Wind shear describes wind that changes abruptly in speed, direction, or both. Wind shear can occur both horizontally and vertically. Wind shear is typically associated with thunderstorms, wind flow over mountains, jet streams, or air mass fronts.	Wind shear is particularly dangerous during take-offs and landings when aircraft are close to the ground (without sufficient altitude to recover). Horizontal wind shear on landing can result in a switch from a headwind to a tailwind. When this happens, the approach may be too fast and a runway overrun can occur. Downward vertical wind shear on take-off can overpower the pilot's climb and force the aircraft into the ground.
Severe storms		
Thunderstorms	Thunderstorms are turbulent, severe storms defined by the presence of thunder and lightning. Thunderstorms are caused by lifting of warm, moist, unstable air. As air rises, it cools and its moisture condenses to form rain or hail. Thunderstorms contain strong air currents that move violently up and down. They can contain strong downward air currents called *downbursts* that disperse in all directions after hitting the ground, causing strong winds. A downburst affecting an area of four kilometres (2.5 miles) or less is called a *microburst*.	As a rule, all aircraft avoid thunderstorms as they are a significant hazard to aviation. Aircraft can be struck by lightning, which is distressing but rarely affects the safety of the flight. The greater risks of thunderstorms are hail damage, turbulence, and the possibility of downbursts. These hazards are particularly dangerous during the take-off and landing phases while flights are close to the ground (lacking the altitude to recover).
Tornados	A tornado is a violent, but short-lived, rotating column of air reaching down to the surface of Earth. As tornados require strong updrafts to form, they are often caused by thunderstorms. Wind speeds in tornadoes can reach up to 200 knots. Tornados are most commonly found in North America.	It is unlikely that an aircraft in flight would survive an encounter with a tornado. Tornados can cause severe destruction to airports, air traffic control equipment, and aircraft on the ground. After notice of an inbound storm, aircraft are often evacuated (flown to another area out of reach of the storm).

(Continued)

Table 7.1 (Continued)

Weather hazard	Description	Impact on aviation
Hurricanes (*North America*) Cyclones (*Southeast Asia, Northeast Australia, South Pacific, Mexico*) Typhoons (*Japan and the Philippines*)	*Hurricanes, cyclones,* and *typhoons* are all different names for the same type of storm. These are tropical, revolving storms that form over oceans. The storm's rotating winds move inward and upward, forming an eyewall with the most intense rain, encircling a central 'eye' of the storm that has calm conditions. These can be large, long-lasting storms that may range from 160 to 1600 kilometres (100 to 1000 miles) in diameter, reach the full height of the troposphere, and move across water at speeds of 16 to 40 kilometres an hour (10 to 25 miles per hour). When they make landfall, they can cause further severe weather. In 1969, for example, Hurricane Camille launched nearly 100 embedded tornadoes.	These storms have very high damage potential. Violent winds, torrential rains, and severe turbulence make it impossible to travel by air through these storms. They can damage ground-based equipment (airports and air traffic control) and cause fatal injuries to people. Moreover, these storms can have long-lasting effects as it can take years to rebuild infrastructure that was destroyed. As with tornados, hurricanes, cyclones, and typhoons will lead to evacuation of aircraft (if enough advance notice is given).

[1] Skybrary, 2016b

Source: Contents of table adapted from Christopherson, et al., 2016 and Skybrary, 2016a, 2016b

Aviation Weather Services: ICAO and the WMO[1]

Because of the direct, and often severe, impact that weather has on aviation operations, ICAO plays an important role in the distribution of weather information. Consistent and reliable worldwide data is crucial for facilitating the safe navigation of aircraft. Like many aspects of aviation, this requires international cooperation and a standardized approach among all the countries of the world.

However, the need for accurate weather reports predates aviation. The first international meteorological conference was convened in Brussels, Belgium, in 1853 for the purpose of establishing an international strategy for sharing weather observations made by ships at sea.

In 1873, the first International Meteorological Congress was held in Vienna, Austria-Hungary, and the International Meteorological Organization was established. This organization launched a variety of initiatives to standardize weather observations and share the information internationally.

In 1947, at a meeting of the International Meteorological Organization in Washington, DC, States worked to create the World Meteorological Convention with the intent that the World Meteorological Organization (WMO) would succeed the International Meteorological Organization. This Convention came into force in 1950, when it was ratified by 30 signatories.

Figure 7.1a Flag of the WMO
Source: Public Domain, https://commons.
wikimedia.org/w/index.php?curid=547545

Figure 7.1b Flag of ICAO (photo)
Source: Public Domain, https://commons.
wikimedia.org/w/index.php?curid=548538

Specific to aviation, international SARPs associated with weather were developed in Annex 3: Meteorological Service for International Air Navigation, after the Chicago Convention came into force in 1947. The goal of Annex 3 is to establish guidelines, specifically for aviation professionals, for the distribution of weather information.

In 1953, ICAO and the WMO established a partnership. The intent was to secure cooperation and ensure that efforts were not unnecessarily duplicated. The partnership established that ICAO has responsibility for specifying the aeronautical meteorological service needs of pilots and operators while the WMO is responsible for providing the meteorological information. WMO and ICAO continue to maintain a close cooperative partnership, ensuring that the weather service requirements of aviators are met. The WMO also works closely with the International Air Transport Association (IATA) to understand and meet the needs of airlines.

National Meteorological Services

Each of ICAO's contracting States has a domestic meteorological service with the authority to provide weather data to facilitate international air navigation. Although these are domestic services, they must meet the Standards outlined in ICAO's Annex 3 as well as WMO requirements. In some countries, the CAA acts as the meteorological authority, while other States have a separate national meteorological service (NMS).

There are about 70 States that don't have the capability to generate and use climate information and forecasts,[2] and so the WMO is working to improve their access to climate services. The Global Framework for Climate Services, established in 2011, assists the least developed countries, including small island States and other vulnerable countries, to strengthen their national climate and meteorological capabilities.[3] The WMO is also working with partners to help countries protect themselves from climate risks through multi-hazard early warning systems, impact-based forecasts, and risk-informed warnings.

Forecasting weather is a complicated process, which involves collecting a large amount of data about current conditions, compiling this information into a database, and referencing this data against known models of how the atmosphere changes over time. Modern forecasting gathers international data from a variety of sources, including environmental satellites, Doppler radar, and automated weather observing systems (AWOS) among others.

ICAO describes the meteorological requirements for international aviation in Annex 3,[4] which specifies that the services set out in Table 7.3 below must be delivered by contracting States.

Table 7.2 National meteorological services in ICAO Council States

Australia	Bureau of Meteorology
Brazil	Instituto Nacional de Meteorologia
Canada	Meteorological Service of Canada
China	China Meteorological Administration
France	Météo-France
Germany	Deutscher Wetterdienst
Italy	Servizio Meteorologico
Japan	Japan Meteorological Agency
Russian Federation	Russian Federal Service for Hydrometeorology and Environmental Monitoring
United Kingdom	Met Office
United States	National Oceanic and Atmospheric Administration

Source: WMO, 2016a and WMO, n.d.

Table 7.3 Annex 3 Requirements for Weather Services

Global forecasting and weather offices

World area forecast system (WAFS) – If a State takes responsibility for providing weather services for a world area forecast centre (WAFC), it must prepare global forecasts of routine and significant weather, issue forecasts to meteorological authorities, and receive and distribute information about radioactive materials or volcanic ash in the atmosphere.

Aerodrome meteorological office – States must establish at least one aerodrome and/or meteorological office to provide weather services for international aviation.

Meteorological watch office (MWO) – If a State provides navigation services for a flight information region (FIR), it shall maintain an MWO to monitor weather conditions impacting flight operations in the area and to provide SIGMET and AIRMET information.

Volcanic ash advisory centre (VAAC) – A State that accepts responsibility for a VAAC must monitor satellite and airborne data to detect volcanic ash in the atmosphere, forecast movement of the ash 'cloud', and issue advisory information to other weather offices.

Tropical cyclone advisory centre (TCAC) – A State that accepts responsibility for a TCAC must monitor the development of cyclones using satellite and radar data and issue advisory information to other weather offices.

Weather observations and reports

Aeronautical meteorological stations – A State must establish aeronautical meteorological stations at aerodromes within its territory, which make routine weather observations at fixed intervals (typically every hour) and provide ATIS reports (for use by arriving and departing aircraft at that airport) and METAR reports (for distribution beyond the airport). METAR reports will contain location and time of report along with weather data including surface wind direction and speed, visibility, runway visual range, current weather, cloud amount/type/height, air temperature, dew point temperature, and pressure.

Aircraft observations and reports

Aircraft observations of weather – A State must arrange for aircraft in its registry to submit weather observations of international air routes, including both routine observations and special observations of unusual weather conditions (such as severe turbulence, icing, mountain waves, thunderstorms, and volcanic ash). These observations include PIREPs of weather conditions encountered by aircraft in flight.

Forecasts

Aerodrome forecasts – Weather forecasts around airports are issued as TAFs and include the location, time, date, and predicted weather conditions, such as surface wind, visibility, weather, clouds, and any expected changes. Only one TAF can be valid at any given time, with each new forecast cancelling out the previous one.

Area forecasts for low-level flights – In areas with a high proportion of flights below 10 000 feet, routine area forecasts must be issued. These reports include AIRMET, GAMET, and SIGWX phenomena.

SIGMET and AIRMET information and warnings

SIGMET – This information describes the occurrence of significant metrological events (en route weather or atmospheric conditions that may impact the safety of flight operations).

AIRMET – This information describes en route weather phenomena that may impact the safety of low-level flights below 10 000 feet.

Aerodrome warnings – This information describes weather conditions that may impact airport facilities and aircraft on the ground in the vicinity (including those that are parked).

Wind shear warnings and alerts – This information relates to observed or expected existence of wind shear (which may impact landing and departing aircraft) at an airport between the ground and 1600 feet.

Services for operators and flight crew

Meteorological information services – Weather information must be supplied to operators and flight crew for flight planning purposes, both before departure and while in flight. This information should include forecasts, METAR, TAF, SIGMET, volcanic ash, AIRMET, aerodrome warnings, and ground-based weather radar information. The meteorological office must provide briefings to flight crew members upon request.

Source: Adapted from ICAO, 2016a

Meteorology

Meteorology is the scientific study of the atmosphere associated with forecasting the weather. As noted earlier in this chapter, meteorology is a complex science, to which only the briefest introduction can be included in this textbook. Key elements of meteorology relevant to the aviation industry will be discussed, including the composition of the atmosphere, air pressure, temperature, air masses, and wind.

Atmosphere

Earth is surrounded by a mixture of gases (78 per cent nitrogen, 21 per cent oxygen, 1 per cent water and other gases), which we call the *atmosphere*. The atmosphere is warmed by Earth's surface, which is warmed by net radiation (including the sun's energy).

As illustrated in Figure 7.2, the atmosphere is divided into layers, with the lowest being the *troposphere* where weather and clouds exist. At the top of this layer is the *tropopause* at

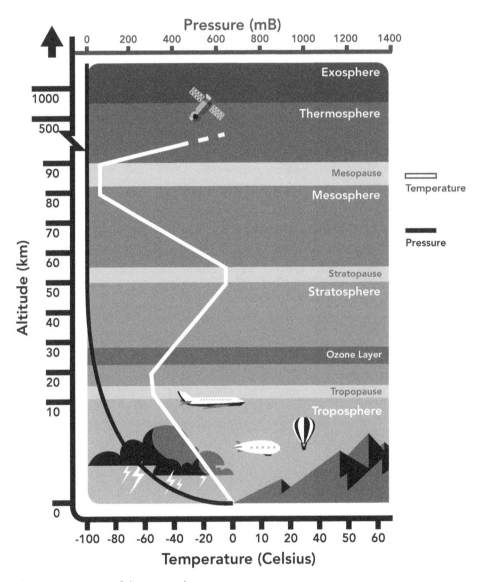

Figure 7.2 Layers of the atmosphere

about 11 kilometres (36 000 feet) above Earth. Here temperatures are very cold and various flight-relevant phenomena take place (including jet streams and clear air turbulence).[5] Thunderstorms can penetrate the tropopause, but are mostly found in the troposphere and stratosphere.

Above the tropopause, the temperature increases throughout the *stratosphere* as a result of the ozone layer. Ozone, a very rare molecule in our atmosphere, plays an important role by filtering ultraviolet light, allowing only a limited amount to reach Earth's surface. Unfortunately, human-produced chemicals (including halocarbons and chlorofluorocarbons (CFCs)) have a depleting impact on the ozone layer and are a significant environmental and health concern (see Textbox 7.2 on the *Montreal Protocol*).

Above the stratosphere, the temperature drops again throughout the *mesosphere*, with the lowest atmospheric temperatures at the *mesopause* (85 kilometres or 280 000 feet), which marks the top of the mesosphere. Above this layer is the *thermosphere*, where the temperature again rises because of solar radiation. Above the thermosphere, the atmosphere transitions to the *exosphere*, which is generally considered *space*.

Although aviation takes place in the troposphere, it is important to understand the composition of the entire atmosphere because what exists above the troposphere directly impacts the weather and the climate that is experienced.

7.2 Montreal Protocol[1]

In the early 1970s, chlorofluorocarbons (CFCs) were commonly used in aerosols such as hairspray, as refrigerants, and in insulation and packing materials. American chemists hypothesized that these CFCs released in the atmosphere rise to the stratosphere, and when combined with solar radiation, release chlorine atoms that destroy large amounts of ozone molecules. Following international investigations into the impact of CFCs these chemicals were eventually banned in Canada, Sweden, Norway, and the United States.

In 1985, a 'hole' in the ozone layer was discovered over Antarctica by the British Antarctic Survey. This finding led the United Nations to establish the multilateral Vienna Convention for the Protection of the Ozone Layer that same year. The power of the Convention was enhanced in 1987 with the legally binding Montreal Protocol on Substances that Deplete the Ozone Layer. The Montreal Protocol is an international treaty that regulates and phases out the use of chemicals that deplete the ozone layer (i.e., CFCs, halons, and other ozone-depleting chemicals [ODCs]).

The amount of stratospheric ozone decreased in both the Arctic and Antarctic regions through the 1990s. Most scientists expect the ozone layer to eventually recover, which is an indication of the tremendous success of the Montreal Protocol.

Note

1 Rafferty, 2011

Air Pressure

Air is composed of molecules that are constantly moving and bumping into each other and other objects, exerting a force measured as *air pressure*. It is air pressure that makes flight possible. Recall from Chapter 2 that the wings of an aircraft are designed with a slight curve to the front edge, causing air to move more quickly above the wing than below it. The subsequent difference in air pressure – high pressure below the wing and low pressure above it – results in lift. However, air pressure continually changes based on the density and temperature of air molecules.

The density of air molecules decreases with height above Earth's surface, resulting in lower air pressure at higher altitudes (see Figure 7.2). Although the atmosphere extends hundreds of kilometres above Earth, it is most dense closest to the surface. Half of the air mass in the atmosphere is below 5.5 kilometres (18 000 feet) while 90 per cent is below 16.2 kilometres (53 000 feet).[6]

Air pressure is also linked to temperature. When air is heated, energy is transferred to air molecules, which causes increasing movement. Heated, faster moving molecules spread out and exert less pressure on their surroundings than colder molecules.

Air pressure directly impacts aircraft performance. Therefore, the current air pressure at an airport is an important element of weather reports. When air pressure is low, the lifting force on the wings will be lower and engines will produce less power. This requires pilots to make adjustments, perhaps by reducing the weight of the aircraft (taking on less cargo or fewer passengers) to compensate for poor performance.

Pressure Altitude

Aircraft altimeters, which give altitude information to pilots, are calibrated to air pressure. However, because the atmosphere is in a continual state of change, an *international standard atmosphere* (ISA)[7] model is used to calibrate altimeters during design, testing, and calibration processes. The ISA is a hypothetical model, agreed upon by ICAO, which assumes air is still and contains no water vapour or dust.

The ISA has a standard pressure of 1013.25 hectopascals (29.92 inches of mercury) and 288.15 degrees Kelvin (15 degrees Celsius) at sea level elevation. Both pressure and temperature decrease with altitude, with air pressure decreasing one inch, and temperature decreasing two degrees Celsius for every 1000-foot gain in elevation in this hypothetical model.[8]

7.3 Standard Pressure Altitude

Standard pressure altitude is a useful tool for pilots. By comparing the conditions at an airport to the standard, they can quickly estimate how their aircraft will perform on a given day. If the pressure that day is above the standard, aircraft performance will be higher (wings produce more lift and engines produce more power). If the air pressure is lower than the standard, then the opposite will occur.

Temperature

After the sun's energy passes through the atmosphere, heat is absorbed by Earth's surface. This varies with season, time of day, and latitude. Keep in mind that the sun's angle in the sky, and Earth's location in its orbit, change constantly. Temperatures around the world vary significantly as a result of air movement, surface conditions, and Earth's rotation.

As air contains moisture, when the air temperature is cooled (often resulting from a lifting force – cooling around two degrees Celsius for every 1000-foot gain in altitude) the dew point temperature can be reached. The *dew point* is the temperature at which the water naturally present in air begins to condense, causing clouds and fog.

Air Masses and Wind

The surface of Earth (along with its temperature, moisture level, and stability of air) impacts the conditions of the overlying air. The combination of temperature, moisture, and air stability can create large air masses, defined by their moisture and temperature characteristics. Classifications include continental (C: dry), maritime (M: wet), polar (P: cold), or tropical (T: hot). The longer an air mass remains still, the more defined these attributes become.[9]

Air masses do not stay still for long, and as they move across the surface of Earth, they directly impact weather. As a CP (dry, cold) air mass moves into a warm area, it brings with it colder temperatures; however, its new location will cause it to slowly warm until the cold air mass eventually dissipates. Warm air masses typically have low pressure, while colder air masses have higher pressure.

Wind is caused by differences in air pressure. High pressure air moves towards areas of lower pressure, and the greater the difference in pressure, the stronger the associated winds.

Did You Know?

As light moves through the layers of atmosphere, longer wavelengths (reds and yellows) pass through while shorter wavelengths (blues) are absorbed and scattered, which causes the sky to appear blue.

Wind has both a speed and a direction. In aviation, wind speed is measured in knots (1 nautical mile or 1.852 km) per hour. Wind direction is given based on the compass orientation from which it comes (not the direction in which it is going). For example, if a wind moves from west to east, it is considered a westerly wind from a direction of 270 degrees.

At airports, it is common to see a windsock that provides pilots with a visual reference of wind speed and direction (see Figure 7.3).

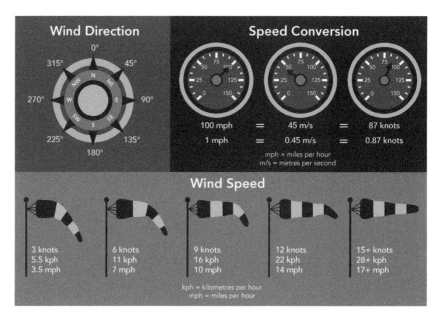

Figure 7.3 Wind direction, wind speed and speed conversion

Lifting Forces

There are several atmospheric phenomena that cause air to be lifted. These include *convergent lifting* as air moves towards low pressure areas; *frontal lifting* as air moves upward along the edge of another air mass; *orographic lifting* when air is pushed upward because of rising terrain (such as a mountain); and *convective lifting* where air is pushed upward by *thermals* (rising currents of warm air caused by surface heat).[10]

With respect to aviation, what's important to remember is that when air is lifted away from Earth's surface, it cools, which has a direct effect on weather. When air cools to its dew point, clouds will form. More aggressive lifting forces – such as a quickly moving air mass causing frontal lifting – can result in violent weather such as a squall line of thunderstorms.

Climate

As noted at the beginning of the chapter, climate describes the average weather conditions over a period of time, typically years. To track and understand the human impact on global climate, the global average temperature each year is compared against the average from 1850–1900 (i.e., before the dramatic increases in emissions associated with human use of fossil fuels in industry). This era is used as a benchmark to understand the impact of human activities on the world's atmosphere. On a global level, international agencies have set a goal of limiting global warming to two degrees Celsius above the 1850–1900 benchmark. The limit of two degrees Celsius is considered the threshold between acceptable and dangerous climate change.[11]

The WMO's 2016 report on world climate indicates that 2015 was the hottest year on record – the world reached the milestone of a global average temperature that was one degree Celsius above the pre-industrial benchmark.[12] That year also witnessed numerous extreme weather events such as heat waves, droughts, floods, and strong tropical cyclones.

Climate change also increases the likelihood of significant turbulence, with stronger turbulence occupying significantly more airspace and directly impacting aviation operations.[13]

Did You Know?

Extreme weather events are responsible for the deaths of tens of thousands of people each year.[1] Beyond the immediate damages these storms can have on buildings and infrastructure, they have longer lasting repercussions: droughts impacting crops (which lead to disease related to malnutrition) and floods (which trigger infectious disease outbreaks and exacerbate the rates of malaria, meningitis, dengue fever, and diarrhoea). Although climate change affects the entire world, the most severe impacts are felt by the most vulnerable populations in developing countries.[2]

Notes

1 WHO/WMO, 2012, p. 4   2 WHO/WMO, 2012

Greenhouse Gases and Global Climate Change

Climate change is linked to greenhouse gases (GHGs) in Earth's atmosphere. GHGs both absorb and emit heat (i.e., thermal radiation), causing a warming process called the *greenhouse effect*. There are four greenhouse gases that occur naturally in our environment: water vapour (H_2O), carbon dioxide (CO_2), methane (CH_4), and nitrous oxides (NO_x). Other GHGs (including CFCs linked to ozone depletion) are human-made. The total greenhouse effect comprises the combination of natural and human-made GHGs.

The presence of natural GHGs in the atmosphere is beneficial to humanity, capturing some of heat from the sun to maintain an average global temperature that supports life. Carbon dioxide, for example, is a natural part of our environment and has been for millions of years. Organic matter (plants and animals), forest fires, and volcanoes all contribute CO_2 to the atmosphere. Through photosynthesis, plants remove CO_2 from the atmosphere and much is absorbed by seawater in the world's oceans. However, in the late 1800s, the burning of fossil fuels and the acceleration of other industrial and agricultural activities resulted in a massive increase in the amount of GHGs, including CO_2, released into the atmosphere. The increase in GHGs in the atmosphere leads to more heat being captured, and the result has been global warming.

The major source of human-caused CO_2 emission is the burning of fossil fuels (although deforestation also increases atmospheric CO_2 levels). Of the natural GHGs, CO_2 has the most impact because it is emitted in far greater amounts than the others. However, methane (a more potent GHG than CO_2) is a growing concern, as it is being released from warmed permafrost (under the sea and on Earth's surface) as well as from livestock such as cattle.

What Is Being Done to Combat Climate Change?

Despite the disturbing trends in climate data, there are many challenges associated with regulations designed to control climate change. Such regulatory controls would limit people's actions, which raises questions about human rights (consider, for example, the public reaction to limiting the amount of automobile use or prohibiting pleasure flying). Regulation could also directly impact a company's operations, which in turn, affects economies. Furthermore, there are those who deny the human impact on climate, believing the changes to be either entirely natural or non-existent.

7.4   Environmental Protection and International Law[1]

Developing international law to protect the environment is complicated. The International Court of Justice describes a general obligation of States to respect the environment of other States; however, there is no global governing body on environmental issues and it has been difficult to get all the countries of the world in agreement.

There are only three examples of environmental *hard law*. These relate to 1) the phasing out of ozone toxins, 2) the International Maritime Organization's (IMO) ban on dumping plastic garbage at sea, and 3) reporting obligations related to GHG emissions, established under the Kyoto Protocol.

Other than these three examples, most environmental agreements are considered *soft law*, which use language like 'to the extent possible', and set vague timelines.

Note

1 Havel & Sanchez, 2014

The United Nations plays a leadership role in managing global climate change. The United Nations Framework Convention on Climate Change (UNFCCC) came into being with the goal of stabilizing GHG concentrations in the atmosphere.

The UNFCCC, sometimes called the Rio Convention as it was adopted at the 1992 Rio Earth Summit, came into force in 2004. This convention required industrialized countries to report regularly on their climate change policies and to measure and submit an annual report of GHG emissions.[14]

Linked to the UNFCCC is the Kyoto Protocol, which was adopted in Kyoto, Japan in 1997 and came into effect in 2005. This multilateral agreement commits States to internationally binding emission reduction targets. The Kyoto Protocol exempted international aviation and maritime emissions as both industries had their own specialized agency to monitor global activities: ICAO and the International Maritime Organization (IMO).[15] However, domestic aviation activities were included. Note that industrialized States have a heavier burden than developing countries under the Kyoto Protocol, as they have been the largest contributors to GHG emissions through years of industrialized activity.

States must meet their emissions targets, but *market-based measures* (detailed below) can be used to make adjustments to these targets.[16]

While ICAO was tasked with exploring how to reduce CO_2 from international aviation in accordance with the Kyoto Protocol, the European Union (EU) took the initiative to develop an emissions trading scheme (ETS) of its own. This market-based measure (MBM) capped emissions from commercial flights operating in the EU. The ETS measured the emissions from an entire flight and levied a relative carbon price on the airline.[17] If an airline exceeded its allotment, it would have to purchase allowances through auction or from another airline.[18] Of course airlines were opposed to this measure as the escalating costs cut into profit margins. However, before the EU policy went into effect, it was decided that flights from nations outside the EU would be excluded, as ICAO was making progress towards a global carbon trading scheme. Further, some countries including the United States, were strongly opposed to the policy.[19]

Table 7.4 Examples of market-based measures within the Kyoto Protocol

International emissions trading	Allows a State that exceeds its GHG emissions limit to trade credits with another State that has excess capacity – this has created the carbon market.
Clean development mechanism (CDM)	Allows a State to undertake an emissions-reduction project in a developing country in return for emission credits, which can be sold, traded, or used to meet Kyoto targets. Example projects include installation of rural solar electricity infrastructure or energy efficient boilers.
Joint implementation (JI)	Allows two States to form a partnership to earn emission reduction credits through an emission removal project. For example, Denmark and the Czech Republic partnered to install a new technology to reduce emissions from a fertilizer manufacturing plant in the Czech Republic. Funds earned from selling carbon credits can be used for green investments.[1]

[1]UNFCCC, n.d.

Source: Adapted from UNFCCC, 2014a

ICAO was successful in establishing several CO_2-related policies, including a voluntary two per cent annual improvement in global fleet fuel efficiency to the year 2050 (resulting in a 50 per cent reduction in net emissions from 2005 levels)[20] but its market-based measure wouldn't be confirmed until 2016 (see the discussion on CORSIA below).

Paris Agreement

In 2016, the Paris Agreement came into effect, enhancing the powers of the UNFCCC. The goal of the Paris Agreement is to commit the world to lower GHGs emissions while boosting clean energy businesses. The specific aims of the Paris Agreement are to

1. hold the global average temperature increase to well below two degrees Celsius above pre-industrial levels;

2. increase the ability to adapt to adverse impacts of climate change and to foster climate resilience and low GHG emissions development without threatening food production;

3. make financial pathways that support low GHG emissions and climate-resilient development. *Climate-resilient development* includes early warning systems, emergency preparedness, risk-assessment and management, among other things.[21]

To meet these three aims, States are to undertake ambitious efforts that demonstrate a progression over time, while recognizing that developing States require support to implement these goals. This agreement also introduces nationally determined contributions (NDCs) to be reviewed in 2023 and every five years thereafter, and recorded in a public registry. Although the Paris Agreement does not impose punishments upon States, the public disclosure of NDCs creates a 'name and shame' system to highlight and encourage governments to take environmental protection actions.

Although international aviation is not included as part of the Paris Agreement, emissions from domestic aviation are to be included in the NDCs. For reference, all aviation activities account for about two per cent of global CO_2 emissions (approximately 0.7 per cent from domestic and 1.3 per cent from international aviation).[22]

International Aviation – Carbon Neutral Growth from 2020

Although aircraft are becoming more fuel efficient, which reduces CO_2 emissions, these gains are offset by projected traffic growth which will result in more total aircraft in the skies. This necessitates a global approach to managing emissions from international aviation. In 2013, the ICAO Assembly resolved to maintain global net emissions at the level expected in the year 2020 (a strategy termed *Carbon Neutral Growth from 2020*). This will be accomplished through improved technologies (lighter airframes and more efficient engines), operational improvements (more direct flight routing), alternative fuels, and market-based measures.[23]

To complement the UNFCCC Paris Agreement, the *Carbon Offsetting and Reduction Scheme for International Aviation* (CORSIA) has been put forward by ICAO as a global market-based measure to address CO_2 emissions from international aviation. CORSIA is to be implemented in phases, starting with voluntary participation by States in the pilot phase (2021–2023) and the first phase (2024–2026), and eventually requiring the participation of all member States (2027–2035). All operators (regardless of whether their State is participating in the phase) are required to gather and submit emissions data to ICAO.

CORSIA offsets aviation emissions by facilitating a reduction in emissions from other sectors (with 'emissions units' representing one tonne of CO_2). This is similar to the UNFCCC's clean development mechanism (see Table 7.4), which occurs within the global carbon market.

Did You Know?

It can be difficult to grasp what a tonne of CO_2 actually represents. Imagine a balloon with a diameter of about 10 metres (33 feet). Filled with CO_2, that balloon would weigh about one tonne.

The average North American family produces about 24 of these balloons each year, through activities such as powering their home and transportation.

After a year or so, about half of those 24 balloons would be absorbed by plants or the oceans while the other half remain in the atmosphere and contribute to global warming (along with about half of that family's emissions from all previous years).[1]

Note

1 Chameides, 2007

Annex 16: Environmental Protection

To foster environmental protection, ICAO publishes international SARPs in Annex 16: Environmental Protection, which has two volumes: Volume 1 – Aircraft Noise, and Volume 2 – Aircraft Engine Emissions.

Aircraft Noise

One of the first environmental concerns to be considered by ICAO was the issue of aircraft noise around airports. Noise, defined as unwanted sound, is an environmental pollutant. Anyone who has spent time around airports understands that aircraft generate a lot of noise, which can be an annoyance to nearby residents.

However, more than a simple nuisance, noise actually has far-reaching health impacts. A community's exposure to aircraft noise (associated with an international airport) has been linked to poor health, an increase in the usage of sleep medication, and a slight increase in cardiovascular disease.[24] Children seem particularly vulnerable to noise, as chronic exposure to aircraft noise in a school is linked to learning and memory impairments among youth.[25]

Therefore, noise pollution with its potential for irritation and harm can be a troubling issue in a community.

The growing public outcry against aircraft noise led to the London Noise Conference in 1966 and a subsequent Assembly resolution in 1968. The initial mandate of this work was to measure aircraft noise, understand the human tolerance to aircraft noise, and create aircraft noise certification and abatement criteria. This work on noise evolved to become Volume 1 of Annex 16 to the Chicago Convention.

Annex 16 defines the maximum noise emission for aircraft types (separating them into a variety of different categories). The four key strategies used to manage aircraft noise include 1) reducing the aircraft noise itself, 2) improving land-use planning and management, 3) employing noise abatement operational procedures, and 4) setting operating restrictions.

Did You Know?

Modern jet aircraft are 75 per cent quieter than the first jets that entered service. New generations of aircraft, with modern airframes and engines, continue the trend towards quieter operations.[1]

Note

1 ATAG, 2016

The aviation community is very aware of the issue of noise and works collectively to measure and manage the noise produced by aviation operations. Aircraft on international trips must carry a *noise certificate* (approved and issued by the State of Registry). Manufacturers support the goal of noise reduction by using innovative materials and technologies to create

Figure 7.4 Qantas aircraft over houses on approach to London heathrow

Source: Public Domain, https://commons.wikimedia.org/w/index.php?curid=3165371

quieter aircraft and engines. Airports do their part by completing land studies to identify ways of managing the noise impact on surrounding communities.

Engine Emissions

In an effort to reduce aircraft engine emissions, Annex 16 also includes SARPs associated with limiting the emissions of nitrogen oxides, CO_2, hydrocarbons, and smoke from engines.

One challenge faced by the aviation industry is related to the long lifespans of aircraft; therefore, the sector can be locked in to the use of older technology for decades. Although aircraft and engine manufacturers continue to innovate to improve fuel efficiency, the tremendous growth projected for aviation would result in increased CO_2 emissions for the sector without improvements in efficiency. [26]

Aircraft emissions (like emissions from other human-made sources) impact the composition of the atmosphere. Environmental impacts of aviation emissions fall generally into two categories:

1. direct emission of GHGs including CO_2, NO_x, hydrocarbons (HC), H_2O, sulfur oxides (SO_x), and non-volatile black carbon (BC); and

2. emissions that trigger the generation of clouds (contrails).[27]

Both types of emissions can lead to a radiative force (heat) in the atmosphere that contributes to global warming. Any changes in Earth's heat balance have far-reaching effects associated with air circulation and weather.

There are several initiatives underway to reduce emissions:

- *More efficient aircraft* – Manufacturers are continually innovating their practices to produce more fuel-efficient airframes and engines. This reduces CO_2 emissions but also results in lower operating costs, which is a win-win situation for operators and environmentalists.

- *Alternative fuels* – There is ongoing research and development related to a variety of alternative fuels, including biofuels derived from natural sources (such as vegetable oil, sugar, or algae). There is also innovative research that seeks to use solar energy to combine water and CO_2 into hydrocarbons to produce a fully synthetic jet fuel.[28]

- *Selective flight levels* – As contrails contribute to global warming, it is possible to restrict aircraft to specific flight levels where known atmospheric conditions would not form contrails. However, this limitation places a capacity restraint on aviation that would increase controller workloads and potential traffic conflicts.[29]

It is challenging to define the precise impact of aviation on the environment. However, work at ICAO continues to identify and implement environmental protection initiatives. Much of this work occurs within the Committee on Aviation Environmental Protection (CAEP).

Did You Know?

Have you ever looked up in the sky and noticed a white trail behind the flight path of an aircraft? Although some people believe this to be smoke or pollution from aircraft engines, the truth is that these condensation trails (*contrails*) are actually clouds.

Engine exhaust produces water vapour, and in some weather conditions this vapour raises humidity and causes clouds to form. Contrails can become long-lasting cirrus clouds, which have radiative (heating) properties that can contribute to global warming.

Figure 7.5 Contrails

Committee on Aviation Environmental Protection

ICAO's Assembly has made environmental protection a global priority for international aviation. CAEP is responsible for completing much of this work on environmental initiatives. CAEP was established by Council in 1983 with the goal of adding or revising SARPs associated with environmental issues, specifically those associated with noise and engine emissions. CAEP is a technical committee, reporting to the Council, which includes members from a variety of government and non-governmental organizations.

CAEP seeks to protect the environment without sacrificing the viability of the international aviation system. Therefore, in addition to environmental impact, they also consider technical feasibility, economic reasonableness, and interdependence of measures.

Key ongoing initiatives of CAEP include:[30]

- *Aeroplane CO_2 emissions standards* – To reduce emissions from air transport, these standards encourage more fuel-efficient technologies in aircraft design and development.

- *Non-volatile particulate matter (nvPM) standards* – Aircraft engines produce very fine particulates (soot or black carbon), which can impact air quality around airports. New standards would define the allowable mass and number of nvPMs produced by aircraft on landings and take-offs.

- *Aviation system block upgrade (ASBU) environmental analysis* – Within the current air navigation system, efficiency will degrade two per cent every decade if improvements are not made. The proposed *block upgrade* will improve airport operations, interoperable systems, capacity and flexible flights, and the efficiency of flight paths to produce fuel savings (and associated CO_2 savings).

- *Noise* – Helicopter noise reduction technologies and unmanned aircraft noise standards are being reviewed.

- *Aircraft recycling* – Development of recommendations on recycling of decommissioned aircraft is an emerging area.

Sustainable Development in Aviation

Although the growth of aviation has had a variety of positive impacts (e.g., economic and employment-related), the resulting increase in total aircraft flying makes it difficult to reduce the CO_2 emissions from the sector.

Rising out of an historic UN summit, 17 sustainable development goals (SDGs) came into force at the beginning of 2016 and apply through the year 2030. The SDGs are meant to apply to all people, to collectively fight inequalities, tackle climate change, and end poverty. The SDGs are intended to be a call to action, to encourage everyone – the poor and the wealthy – to promote prosperity while protecting Earth.

Did You Know?

Environmental protection and economic growth are often competing goals. Sustainable development seeks to achieve both, yet this can be a difficult balance to achieve.

The term *sustainable development* is an important consideration within aviation. It means that development should meet the needs of the present without sacrificing future generations' ability to meet their needs.[31] Sustainable development is built on three pillars of equal importance: 1) economic growth, 2) social inclusion, and 3) environmental protection.

Figure 7.6 United Nation's 17 Sustainable Development Goals (SDGs)

Note: Suzanne K. Kearns, the author of this text, supports the Sustainable Development Goals.

Sustainable development within aviation is of particular importance. As a UN agency, ICAO embraces the SDGs and invests in accomplishing these objectives. The following seven goals are of particular importance to aviation:

- Goal 4 – ensuring inclusive and quality education for all and promoting lifelong learning;

- Goal 5 – achieving gender equality and empowering all women and girls;

- Goal 8 – promoting inclusive and sustainable economic growth, employment, and decent work for all;

- Goal 9 – building resilient infrastructure, promoting sustainable industrialization, and fostering innovation;

- Goal 10 – reducing inequality within and among countries;

- Goal 13 – taking urgent action to combat climate change and its impacts; and

- Goal 17 – revitalizing the global partnership for sustainable development.

Based on these key SDGs, ICAO has established the following global priorities for aviation:

- aviation safety;

- air navigation capacity and efficiency;

- security and facilitation;

- economic development; and

- environmental protection.

A key initiative within this framework is the No Country Left Behind (NCLB) programme through which ICAO offers assistance to States in the implementation of SARPs. The intent is to resolve global safety concerns highlighted by audits to give every State access to the socio-economic benefits of international air transport.[32]

Conclusion

As discussed throughout this chapter, the international aviation industry is fundamentally linked to the environment in which it operates. For this reason, the forecasting and communication of standardized weather information is a necessity.

An ongoing challenge for international aviation is environmental protection. Representatives within the aviation industry have made the argument that the economic importance of aviation is one reason why strict CO_2 regulations are not feasible for this sector. For example, 80 per cent of aviation CO_2 emissions are from flights over 1500 kilometres (932 miles) for which there is no practical alternative form of transportation.[33] However, the industry's emissions have grown almost consistently year-on-year since its emergence.[34]

International aviation has heard this call to action and has responded by creating several initiatives to support the global community's goal of remaining below the two degree Celsius global warming benchmark. Environmental protection initiatives include using new technology to improve aircraft fuel efficiency (including the use of sustainable biofuels), navigating more direct flight paths through air traffic modernization, improving infrastructure, and implementing CORSIA, which is a global market-based measure. These initiatives are intended to support the industry's goal of carbon neutral growth from 2020.

Key Points to Remember

1. Aviation is fundamentally linked to the environment in which it operates – the industry is affected by environmental factors and simultaneously affects the environment

2. *Weather* refers to short-term environmental conditions, and *climate* is the average of weather conditions over a longer period. *Meteorology* is the scientific study of the atmosphere. At least a basic understanding of weather, climate, and meteorology is necessary for those in the aviation industry.

3. ICAO's Annex 3 dictates SARPs for the forecasting and dissemination of weather information by a State's national meteorological service (NMS).

4. The atmosphere can be organized into layers, based on temperature characteristics.

5. *Air pressure* refers to the force that air molecules exert on their surroundings. *Air masses* are sections of air with defined moisture and temperature characteristics. *Wind* is caused by air pressure differences, as high pressure (cold) air moves towards low pressure (warm) areas. When air is lifted, it cools and moisture can condense to form clouds. Aggressive lifting can result in violent weather (e.g., thunderstorms).

6. The world's climate is warming. Annual temperatures are compared against the average from the years 1850 to 1900. The goal is to reduce global warming before annual temperatures reach two degrees Celsius above that benchmark.

7. Global warming results in significant weather events, such as severe storms, heavy rains, and changes in winds. These can produce many weather-related hazards for aviation operations.

8. Global climate change is linked to greenhouse gases (GHGs) in Earth's atmosphere, which both absorb and emit thermal radiation (heat). Of all GHGs, CO_2 is the most impactful as it is emitted in far greater amounts than the others.

9. The United Nations is a leader in managing global climate change through

   - the United Nations Framework Convention on Climate Change (UNFCCC), adopted in 1992 and ratified in 2004, with the goal of stabilizing GHGs in the atmosphere;

   - the subsequent Kyoto Protocol, adopted in 1997 and ratified in 2005, which commits States to emission reduction targets; and

   - the Paris Agreement, ratified in 2016, which introduces publicly disclosed nationally determined contributions (NDCs) for States.

10. All aviation activities account for about two per cent of global CO_2 emissions (0.7 per cent domestic and 1.3 per cent international operations). Although aircraft are becoming increasingly fuel efficient, those gains are offset by projected increases in total aircraft in the skies. However, ICAO has committed to carbon neutral growth from 2020, whereby global net emissions from aviation will be held at the level expected in the year 2020, despite any increases in total aircraft flying.

11. To complement the UNFCCC Paris Agreement, ICAO has introduced the *carbon offsetting and reduction scheme for international aviation* (CORSIA), which is a global market-based measure to address CO_2 emissions from aircraft.

12. ICAO publishes SARPs related to environmental protection in Annex 16, which has two volumes: Volume 1 – Aircraft Noise, and Volume 2 – Aircraft Engine Emissions.

   - Aircraft noise around airports is an environmental concern. Four key noise-reduction strategies include reducing the aircraft noise, improving land-use planning and management, employing noise abatement procedures, and setting operating restrictions.

- Aviation emissions that impact the environment include direct emission of GHGs and those that trigger the generation of clouds (contrails). Emissions can be reduced by flying more efficient aircraft, using alternative fuels, and restricting aircraft to selective flight levels.

13. The Committee on Aviation Environmental Protection (CAEP) conducts much of ICAO's work on assessing environmental impacts of aviation. CAEP seeks to balance four key considerations: 1) technical feasibility, 2) economic reasonableness, 3) environmental benefit, and 4) interdependence of measures.

14. The United Nations has identified 17 sustainable development goals (SDGs) that are meant to apply to all people, to promote prosperity while protecting Earth. These goals balance economic growth, social inclusion, and environmental protection. The SDGs have led ICAO to identify several global priorities for aviation, including the No Country Left Behind (NCLB) initiative through which developing States are offered assistance in the implementation of ICAO SARPs.

Table 7.5 Acronym rundown

AIRMET	airmen's meteorological information of moderate weather phenomena
ASBU	aviation system block upgrade
ATC	air traffic control
ATIS	automatic terminal information service
AWOS	automated weather observing system
BC	non-volatile black carbon
CAA	civil aviation authority
CAEP	Committee on Aviation Environmental Protection
CAT	clear air turbulence
CDM	clean development mechanism
CFCs	chlorofluorocarbons
CH_4	methane
CO_2	carbon dioxide
contrail	condensation trail
CORSIA	carbon offsetting and reduction scheme for international aviation
CVR	cockpit voice recorder

(Continued)

Table 7.5 (Continued)

ETS	emissions trading scheme
EU	European Union
FAA	Federal Aviation Administration
FIR	flight information region
FO	first officer
GAMET	general aviation meteorological information
GHG	greenhouse gas
H_2O	hydrogen dioxide (water)
HC	hydrocarbons
ICAO	International Civil Aviation Organization
IATA	International Air Transport Association
IMO	International Maritime Organization
ISA	international standard atmosphere
JI	joint implementation
LLWAS	low-level wind shear alert system
MBM	market-based measure
METAR	meteorological terminal aviation routine weather report
MWO	meteorological watch office
NO_x	nitrous oxides
NCLB	No Country Left Behind
NDC	nationally determined contributions
NMS	national meteorological service
NOTAM	notice to airmen of hazardous weather conditions
nvPM	non-volatile particulate matter
ODC	ozone-depleting chemical
PIREP	pilot report
SARPs	standards and recommended practices
SDGs	sustainable development goals
SIGMET	significant meteorological information about hazardous weather phenomena
SIGWX	significant weather
SO_x	sulfur oxides
TAF	terminal area forecast
TCAC	tropical cyclone advisory centre
TDWR	terminal Doppler weather radar

UNFCCC	United Nations Framework Convention on Climate Change
VAAC	volcanic ash advisory centre
VOLMET	regional weather information for aircraft in flight
WAFC	world area forecast centre
WAFS	world area forecast system
WMO	World Meteorological Organization

Chapter Review Questions

7.1   What are the differences between weather, meteorology, and climate? Why does the aviation industry need to be cognizant of all three?

7.2   Why is the ISA model important in understanding air pressure for pilots?

7.3   Why is it important that aviators understand wind speed and direction? Explain how runway 'names' are linked to the direction of prevailing winds.

7.4   Describe three ways in which climate change negatively impacts international aviation.

7.5   Which environmental consideration discussed in this chapter has the potential to most significantly impact your future career in aviation (in the short term and long term)? Explain your answers.

7.6   In your opinion, which of the four atmospheric lifting forces is the most dangerous to an aircraft? Why?

7.7   What is ICAO's plan for limiting international aviation's global carbon footprint? Describe two advantages and two challenges of this approach. Can you think of any other methods to control international aviation emissions? Explain.

7.8   Which of the UN's 17 sustainable development goals (SDGs) can be positively impacted by aviation? Is it possible that aviation might negatively impact any of the SDGs? Justify your answers.

7.9   What is the name of the national meteorological service (NMS) in your State?

7.10 Choose one of the meteorological service requirements for international aviation and explain why it is important to local air operators in your State that the information provided be accurate. Consider the safety, economic, and efficiency impacts of incorrect information.

7.11 Explain two ways in which Annex 16 impacts aviation operations in your State.

7.12 Do you think solar-powered planes are a practical long-term solution to carbon emissions in aviation? Explain your answer. If you don't consider solar-powered planes a practical solution, identify another environmental innovation that you believe offers a better approach to dealing with emissions.

DELTA FLIGHT 191 – THE DEADLY IMPACT OF SEVERE WEATHER[1]

On 2 August 1985 at about 6:00 p.m., Delta Air Lines flight 191 (a Lockheed L-1011 aircraft) was on approach to land at Dallas/Fort Worth (DFW) International Airport. It was over 100 degrees Fahrenheit (about 38 degrees Celsius) and the air was moist, providing the ideal conditions for convective thunderstorm development.

A rain shower had begun near the airport but did not concern pilots or controllers in the area. As the aircraft approached the airport, flight 191 was sequenced to land behind two other flights, which landed successfully. Another pilot who observed the weather from the ground said that the weather cell looked harmless, like showers.

At 6:03, the approach controller broadcast, 'we're getting some variable winds out there due to a shower . . . out there, north end of DFW.' As the aircraft approached the runway, the shower intensified. At 6:04, the first officer (FO) said, 'Lightning coming out of that one', to which the Captain responded, 'Where?' The FO replied, 'Right ahead of us.'[2]

The flight encountered the northern gusts of a microburst, associated with a developing thunderstorm at 1000 feet above the ground. This caused the flight to experience a headwind increase that accelerated the aircraft airspeed to 173 knots. The pilots moved throttles to idle in an attempt to preserve the approach speed of 150 knots.[3]

The pilots continued their descent and the CVR caught the sound of rain hitting the aircraft (at 6:05:05), indicating the flight had entered the rain shaft beneath the storm. At 6:05:21, the Captain warned the FO, 'You're gonna lose it all of a sudden; there it is' and then 'Push it up, push it way up.' These words were followed by the sounds of engine power at maximum. The headwind decreased 25 knots and the downdraft increased from 18 to more than 30 feet per second (1800 feet or 550 metres per minute), causing the aircraft to lose 44 knots of airspeed in 10 seconds.

At 6:05:35, the aircraft hit the southern gust front of the microburst, which caused severe variations in wind along all three of the aircraft axes. Within a single second, the airspeed decreased from 140 to 120 knots, vertical wind reversed from a 40-foot per second downdraft to a 20-foot per second updraft, and a severe lateral gust hit the aircraft (causing a rapid roll to the right). Pilots worked vigorously to fly the aircraft, but it was impossible to correct the flight path. At 6:05:46, the flight was 280 feet (85 metres) above the ground in a descent rate of almost 5000 feet (1500 metres) per minute.

Witnesses saw the landing aircraft emerge from the rain shaft and the left engine strike an automobile and light pole on Highway 114, adjacent to the airport. A large explosion followed as the fuel tanks ignited and the aircraft rolled and struck two airport water towers. One vehicle driver on the ground and 134 passengers and crew were killed in the accident. Twenty-nine people on board the aircraft survived (27 with injuries).

Although the pilots were both experienced and qualified, they had not been formally trained in microbursts. The primary risk associated with microbursts is wind shear, which causes the direction and/or speed of the wind to change dramatically within a short distance. Wind shear is particularly dangerous during take-off and landings when aircraft are close to the ground.

CASE STUDY

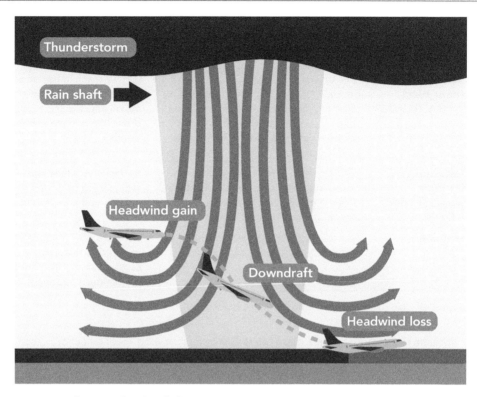

Figure 7. 7 Microburst and Delta Flight 191

One of the causes of wind shear is a downburst from a convective cloud (such as a thunderstorm). The smaller the scale of the downburst, the more dangerous to aircraft as wind changes are localized within a small column that pushes aircraft towards the surface. The term *microburst* refers to a downburst with a diameter of less than four kilometres (2.2 nautical miles).[4]

The Delta flight 191 accident revealed the devastating potential of weather events, but had an important impact on the aviation industry. As a result of this accident, several improvements have been implemented as detailed below:

Table 7.6 An overview of wind shear–related issues and technologies

	At the time of the accident:	*After the accident:*
Airborne weather radar	Weather radar systems were primarily for en route weather avoidance and could not detect wind shear.	In 1988, the FAA mandated that all turbine airline aircraft have an airborne wind shear warning and flight guidance system. New weather radar technologies were introduced in the 1990s to alert crew to the presence of wind shear ahead of their aircraft.

(Continued)

Table 7.6 (Continued)

	At the time of the accident:	After the accident:
Low-level wind shear alert system (LLWAS)	LLWAS was in operation at the time of the accident, using wind sensors around an airport to detect and alert ATC to wind shear. However, these systems had a time delay, which made them ineffective for warning pilots.	LLWAS is still in use as an information tool, but due to the time delay, is still not an effective technology for the prediction and avoidance of severe weather.
Terminal Doppler weather radar (TDWR)	TDWR detects presence of wind shear by measuring direction and velocity of wind-driven precipitation. Technology was mature, but not installed at most airports.	By 1997, ground-based Doppler radars were deployed at 45 high-risk airports in the United States.
Training	Pilot training in wind shear was not mandatory.	Pilots are now required to practise rapid wind shear recognition and escape manoeuvres that use the full performance capability of the aircraft.

Source: All table content adapted from FAA, n.d., paras 20–29

Despite many improvements, there are still limitations on wind shear detection. Clear air turbulence (CAT) and wake turbulence are particularly challenging to detect. CAT is usually associated with storm outflows, the merging of jet streams, or the impact of air flowing over terrain (mountain waves). Wake turbulence (discussed in Chapter 4) also remains a challenge that is undetected by technology.

The Delta flight 191 accident illustrates the challenge of operating aircraft within the inherently uncontrollable environment. It also demonstrates how, in response to such an event, the industry rallies to create and introduce new technologies to make air travel safer.

Notes

1 NTSB, 1986
2 NTSB, 1986, p. 3

3 FAA, n.d., p. para. 11
4 FAA, n.d., p. para. 20

CASE STUDY

Case Study Questions

Using this case study, and applying what you have learned in this chapter, provide informed responses to the following:

7.13   This accident illustrates how environmental factors can present hazards to aviation, even with the concerted effort of a skilled flight crew. Which of the reactions to this event have the greatest potential impact to improve safety (training, air-based technologies, or ground-based technologies)?

7.14 Consider the many professionals involved in ensuring a safe flight – pilots, air-line personnel, air traffic controllers, and aircraft manufacturers. Where do you think the balance of responsibility lies for detecting, managing, and/or avoiding dangerous weather?

7.15 As has been illustrated through this textbook, many safety measures or advance-ments are made as a reaction to an accident or a major event. How can industry professionals challenge themselves to identify risk and implement new safety approaches before a major event occurs?

References

Abeyrante, R., 2014. *Aviation and climate change: In search of a global market based measure.* New York: Springer.

ATAG, 2016. *Aviation benefits beyond borders.* Geneva, Switzerland: Air Transport Action Group.

Bows-Larkin, A. et al., 2016. *Aviation and climate change – The continuing challenge.* London: John Wiley & Sons Ltd.

Butterworth-Hayes, P., 2013. Climate change and aviation; Forecasting the effects. *Aerospace America,* pp. 1–30.

CAEP, 2016. *Informal briefing to the ANC: 10th meeting of the Committee on Aviation Environmental Protection (CAEP/10).* Montreal: International Civil Aviation Organization.

Chameides, B., 2007. *Picturing a ton of CO_2.* [Online] Available at: http://blogs.edf.org/climate411/2007/02/20/picturing-a-ton-of-co2/

Christopherson, R. W., Birkeland, G. H., Byrne, M.-L. & Giles, P. T., 2016. *Geosystems: An introduction to physical geography.* 4th ed. Toronto: Pearson Learning Solutions.

FAA, n.d. *Accident overview: History of flight 191.* [Online] Available at: http://lessonslearned.faa.gov/ll_main.cfm?TabID=1&LLID=32&LLTypeID=2

Franssen, E. A. M., van Wiechen, C. M. A. G., Nagelkerke, N. J. D. & Lebret, E., 2004. Aircraft noise around a large international airport impact on general health and medication use. *Occupational and Environmental Medicine,* 61, pp. 405–413.

Havel, B. F. & Sanchez, G. S., 2014. *The principles and practice of international aviation law.* New York: Cambridge University Press.

Hileman, J. I. & Stratton, R. W., 2014. Alternative jet fuel feasibility. *Transport Policy,* Volume 34, pp. 52–62.

ICAO, 1993. *Manual of the ICAO Standard Atmosphere extended to 80 kilometres, Doc 7488/3, 3rd ed.* Montreal: International Civil Aviation Organization.

ICAO, 2016a. *Annex 3 to the Convention on International Civil Aviation: Meteorological service for international air navigation, 19th ed.* Montreal: International Civil Aviation Organization.

ICAO, 2016b. *Environment: Why ICAO decided to develop a global MBM scheme for international avia-tion.* [Online] Available at: www.icao.int/environmental-protection/Pages/A39_CORSIA_FAQ1.aspx

ICAOa, n.d. *Achieving climate change goals for international aviation.* [Online] Available at: www.icao.int/Newsroom/Presentation%20Slides/Achieving%20Climate%20Change%20Goals%20for%20International%20Aviation.pdf

ICAOb, n.d. *No country left behind.* [Online] Available at: www.icao.int/about-icao/NCLB/Pages/default.aspx

ICAOc, n.d. *ICAO and the World Meteorological Organization.* [Online] Available at: www.icao.int/secretariat/PostalHistory/icao_and_the_world_meteorological_organization.htm

Lester, P. F., 2013. *Aviation weather, 4th ed.* Englewood: Jeppesen.

NTSB, 1986. *Delta Air Lines, Inc., Lockheed L-1011-385-1, N726DA, Dallas/Fort Worth International Airport, Texas, August 2, 1985 (NTSB/AAR-86/05).* Washington, DC: National Transportation Safety Board.

Prather, M. & Sausen, R., 1999. Potential climate change from aviation. In: J. E. Penner, et al. eds. *Aviation and the Global Atmosphere.* Cambridge: Cambridge University Press, pp. 187–197.

Rafferty, J. P., 2011. *Montreal Protocol,* Chicago: Britannica Academic.

Skybrary, 2016a. *Tornado,* s.l.: Skybrary.aero.

Skybrary, 2016b. *Volcanic Ash & Aviation Safety,* s.l.: Skybrary.aero.

Stansfeld, S. A. & Matheson, M. P., 2003. Noise pollution: non-auditory effects on health. *British Medical Bulletin,* 68, pp. 243–257.

UNFCCC, 2014a. *International emissions trading.* Bonn: United Nations.

UNFCCC, 2014b. *United Nations Framework Convention on Climate Change.* Bonn: United Nations.

UNFCCC, n.d. *10 Highlighted JI Projects.* [Online] Available at: http://ji.unfccc.int/index.html

United Nations, 2015. *Paris Agreement.* Geneva: United Nations.

United Nations, 2016. *Sustainable development goals: 17 goals to transform our world.* [Online] Available at: www.un.org/sustainabledevelopment/development-agenda/

WHO/WMO, 2012. *Atlas of health and climate.* Geneva: World Health Organization and World Meteorological Organization.

Williams, V., Noland, R. B. & Toumi, R., 2002. Reducing the climate change impacts of aviation by restricting cruise altitudes. *Transportation Research Part D,* 7, pp. 451–464.

WMO, 2016a. *National services.* [Online] Available at: https://public.wmo.int/en/about-us/members/national-services

WMO, 2016b. *WMO statement on the status of the global climate in 2015 (WMO-No. 1167).* Geneva: World Meteorological Organization.

WMO, 2017. *Global framework for climate services.* [Online] Available at: www.wmo.int/gfcs/

Notes

1 Section adapted from ICAOc, n.d.
2 WMO, 2016b
3 WMO, 2017
4 ICAO, 2016a
5 Lester, 2013
6 Christopherson, et al., 2016, p. 70
7 ICAO, 1993
8 ICAO, 1993
9 Christopherson, et al., 2016, p. 208
10 Christopherson, et al., 2016, p. 209
11 Bows-Larkin, et al., 2016
12 WMO, 2016b, p. 2
13 Butterworth-Hayes, 2013
14 UNFCCC, 2014b
15 Havel & Sanchez, 2014
16 Abeyrante, 2014
17 Bows-Larkin, et al., 2016
18 Havel & Sanchez, 2014
19 Bows-Larkin, et al., 2016
20 ICAOa, n.d.
21 United Nations, 2015
22 ATAG, 2016, p. 7; ICAO, 2016b
23 ICAO, 2016b
24 Franssen, et al., 2004

25 Stansfeld & Matheson, 2003

26 Bows-Larkin, et al., 2016.

27 Prather & Sausen, 1999

28 Hileman & Stratton, 2014

29 Williams, et al., 2002

30 CAEP, 2016

31 United Nations, 2016

32 ICAOb, n.d.

33 ATAG, 2016, p. 7

34 Bows-Larkin, et al., 2016, p. 2

1. Malaysia Airlines Flight 370, an aircraft that was mysteriously lost at sea in 2014, would be categorized as an:
   a. Accident
   b. Incident
   c. None of the above.

2. The global aviation accident rate is about ____ accidents per million departures.
   a. 3
   b. 9
   c. 22
   d. 42

5. Which category of accident is the most common within the worldwide commercial jet fleet?
   a. Landing-related accidents
   b. Approach-related accidents
   c. Controlled flight into terrain
   d. Loss of control – in flight

3. Search and Rescue (SAR) professionals have responsibility for locating an accident site as quickly as possible, aiding survivors, and conducting the accident investigation.
   a. True
   b. False

4. Which country has responsibility for leading an accident investigation?
   a. The country where the accident occurred
   b. The country where the aircraft was registered
   c. The country where the aircraft operator was based
   d. The country where the aircraft manufacturer was based

Learning science suggests that thinking through a few questions before you begin studying new material, even if you answer incorrectly, results in improved learning and retention.
Give it a try!

CHAPTER 8

Accidents

CHAPTER OUTCOMES

At the end of this chapter, you will be able to . . .

- Describe the global accident rate and the related costs of an accident.
- Explain the role of search and rescue.
- Differentiate between *accidents* and *incidents*.
- Discuss how aviation accidents are investigated, referring to accident investigation authorities, safety technologies, and the investigation process.
- Explain the primary causal factors of aviation accidents in various sectors of the industry.
- Apply what you have learned to answer questions about the Swissair 111 case, an accident that resulted in a lengthy and challenging investigation process.

Introduction

Safety is a vitally important factor in the success of aviation around the globe. All aviation professionals share a common respect for protecting the safety of air travellers. Yet safety is not something that happens by accident. Safety is expensive and requires continuous effort, energy, and innovation. The entire aviation industry continually invests resources into understanding risk, implementing technologies and initiatives to improve safety, and distributing safety promotion materials. The walls of operators are plastered with posters with mantras like

- *It's better to be on the ground wishing you were in the air than in the air wishing you were on the ground.*

- *Maintenance personnel get tired too. Proper rest, and he's ready; not enough rest, and nobody's ready.*

- *Managing risk is a big job . . . Yours!*

The unfortunate reality is that perfect safety in such a complex and interconnected system is probably impossible. Therefore, when aviation accidents do occur, the sole objective of investigations is the prevention of future accidents and incidents – not to place blame or liability.[1] Although every annex of the Chicago Convention includes SARPs that contribute to the safety of aviation, there are three annexes explicitly concerned with safety – Annex 12: Search and Rescue, Annex 13: Aircraft Accident and Incident Investigation, and Annex 19: Safety Management. The focus of this chapter will be on the first two of these annexes; safety management will be discussed in Chapter 9.

Global Accident Rate

The industry's constant investments in safety have resulted in aviation being an exceptionally safe mode of transportation. In 2014, ICAO determined the global accident rate to be three accidents per million departures.[2] In the airline sector, there were 10 fatal accidents out of 40.4 million flights in 2016.[3] That means the odds of being involved in an accident were one in 4 040 000 – if you took an airline flight every single day, you might expect an accident in 11 069 years.

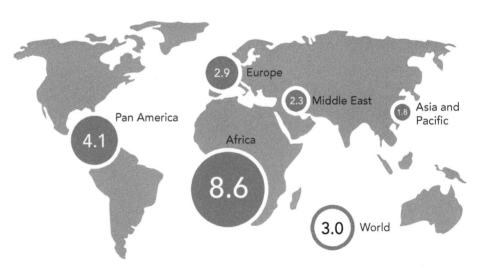

Figure 8.1 Global accident rate (per million departures)
Source: Adapted from ICAO, 2015, p. 9

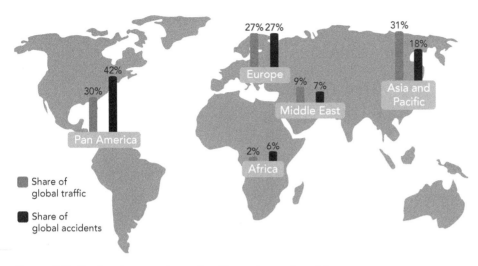

Figure 8.2 Regional comparison of traffic to share of accidents

Source: Adapted from ICAO, 2015, p. 9

Although accident rates are very low, they do vary by geographic region. ICAO groups countries into five *regional aviation safety groups* (RASG) based on geographic proximity rather than any political affiliations: Africa (AFI), Asia and Pacific (APAC), Middle East (MID), Europe (EUR), and Pan America (PA).

As you can see in Figure 8.1, the regional accident rate is highest in the African region and lowest in the Asia and Pacific region; however, the global accident rate is mediated by the fact that Africa has the lowest traffic volume of all the regions (see Figure 8.2).

From a historical perspective, the global accident rate is much lower today than it was in the past. In 1960, the global accident rate for commercial jet fleet was approximately 50 accidents per million departures. This rate dropped dramatically through the 1960s and early 1970s, primarily as a result of improved aircraft manufacturing, systems, and technologies.

However, since the mid-1970s, despite continued safety efforts, the accident rate has decreased only slightly. In Figure 8.3, notice how the accident rate drops dramatically from 1960 to 1970 and then remains relatively flat. This accident rate plateau leads some professionals to argue that the current accident rate is just the cost of doing business, and that further safety innovations are not justified as they are expensive and do not significantly decrease the global accident rate. The problem with this theory is that air traffic volume is increasing, resulting in more aircraft in the sky. The increased traffic volume means that, even if the rate of accidents is stable, the number of accidents that occur in a year will increase.

If the number of annual accidents increase, the resulting media coverage is likely to impact air travellers. Imagine how people would react if aviation accidents were reported every other week – even if the accident rate remains stable and this increase in accidents is simply a reflection of more aircraft flying, it may lead passengers to choose other forms of transportation (or decide not to travel at all).

Safety is, and always will be, a priority within the aviation industry, one that is recognized and respected by all aviation professionals. For these reasons, safety research and innovations are ongoing.

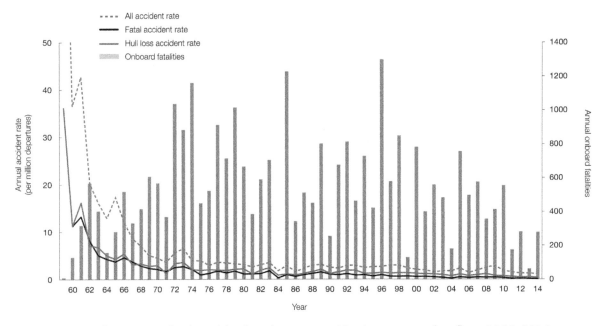

Figure 8.3 Accident rates and onboard fatalities by year: worldwide commercial jet fleet, 1959–2014
Source: Boeing, 2015, p. 16

Cost of Aviation Accidents

There is no escaping the fact that accidents are extremely expensive. Air carriers hold insurance to cover *direct losses* that may result from an accident. Direct losses include the cost of the aircraft itself as well as injury and fatality claims paid out to passengers or their families. Insurance may also cover the cost of a replacement aircraft to join the airline's fleet and cover the planned routes of the aircraft that was lost.

Accidents also cause *indirect losses* – sometimes many times greater than the direct losses – which are not compensated for through insurance. Indirect losses include an operator's tainted reputation after an accident, loss of key employees, and loss of productivity of staff who must manage the aftermath of an accident.

The impacts of an accident can be far-reaching and include effects on communities, small businesses, and society. An Australian report estimated that the cost of each fatality resulting from an aviation accident was AU$2.17 million. This calculation considers the lost *human capital* of that individual, based on life-years-lost and expected earnings, which result in lost contributions to workplace, household, and volunteer efforts.[4]

Not only is aviation safety a crucial consideration from an ethical and moral perspective, the costs of an accident are so great that safety is also good for business. A common saying in aviation is *If you think safety is expensive, try an accident!*

To explore the aviation industry's response to accidents, the remainder of this chapter will discuss search and rescue (SAR), accident investigation, and the leading causes of accidents.

Aircraft Accident Investigation

Aviation occurrences are classified as either an *accident* or an *incident*. These two terms have distinct meanings in the aviation industry, as detailed in the Language of Accidents feature.

8.1 The Language of Accidents

In the aviation industry, an **accident** involves loss of life or loss of an aircraft, and occurs between the time people board the aircraft with the intention of flight until they disembark. An *incident*, on the other hand, does not result in loss of life or loss of an aircraft.

ICAO defines an accident as:

An occurrence associated with the operation of an aircraft which, in the case of a manned aircraft, takes place between the time any person boards the aircraft with the intention of flight until such time as all persons have disembarked, or in the case of an unmanned aircraft, takes place between the time the aircraft is ready to move with the purpose of flight until such time as it comes to rest at the end of the flight and the primary propulsion system is shut down, in which:

a) a person is fatally or seriously injured as a result of:

- being in the aircraft, or

- direct contact with any part of the aircraft, including parts which have become detached from the aircraft, or

- direct exposure to jet blast,

except when the injuries are from natural causes, self-inflicted or inflicted by other persons, or when the injuries are to stowaways hiding outside the areas normally available to the passengers and crew; or

b) the aircraft sustains damage or structural failure which:

- adversely affects the structural strength, performance or flight characteristics of the aircraft, and

- would normally require major repair or replacement of the affected component,

except for engine failure or damage, when the damage is limited to a single engine, (including its cowlings or accessories), to propellers, wing tips, antennas, probes, vanes, tires, brakes, wheels, fairings, panels, landing gear doors, windscreens, the aircraft skin (such as small dents or puncture holes), or for minor damages to main rotor blades, tail rotor blades, landing gear, and those resulting from hail or bird strike (including holes in the radome); or

c) the aircraft is missing or is completely inaccessible.[1]

The ICAO definition of an incident is: 'An occurrence, other than an accident, associated with the operation of an aircraft which affects or could affect the safety of operation.'[2]

A **fatal injury** associated with an aircraft accident is defined as a death that occurs within 30 days of the accident.

The term **hull-loss** describes an aircraft that is destroyed or damaged beyond repair, missing, or inaccessible. This is the aviation equivalent to an automobile insurance company writing off a car when the damages would cost more to repair than the value of the car.

Accidents **investigations** are formal processes that include gathering information, drawing conclusions, determining causal factors, and making safety recommendations. The **investigator-in-charge** (IIC) is the person responsible for leading the accident investigation.

In an accident investigation, both causal and contributing factors may be identified. **Causal factors** are those that led directly to the 'causation chain of events', which led to the accident. **Contributing factors** are those that had the potential to adversely impact flight safety, but did not directly cause the accident.

Notes

1 ICAO, 2010b, pp. 1–1   2 ICAO, 2010b, pp. 1–2

Generally, any occurrence during flight operations that results in a fatality, serious aircraft damage, or a lost or missing aircraft is classified as an *accident*. Accidents *must* be investigated to determine what caused the event.

Sometimes, what is classified as an accident may seem counter-intuitive. For example, if a coffee pot in an aircraft's galley exploded during flight and fatally injured a passenger or cabin crew member it would be classified as an accident and require an investigation. However, if a maintenance professional was fatally injured doing work on the ground, it would not be considered an accident because the aircraft was not conducting flight operations at that time.

An *incident* is an unsafe occurrence that is less severe than an accident. Incidents *may* be investigated. Incidents can be thought of as near-accidents. For example, if two aircraft fly too close together (referred to as a *loss of separation*) this near–mid-air collision would be classified as an incident.

Whatever hazard or risk caused an incident may cause an accident in the future, and therefore investigators can gain valuable insights from such occurrences. Moreover, since no one was fatally injured and no severe damage occurred, people tend to be more willing to share information and less worried about possible legal action. Examples of incidents include:

- fuel quantity issues requiring pilots to declare an emergency;

- aborted take-offs or landings on a closed or occupied runway;

- smoke or fire in the aircraft (extinguished before causing an accident);

- anything requiring the use of emergency oxygen;

- engine failures that did not lead to an accident;

- flight crew incapacitation;

- near-collisions with terrain;

- runway incursions;

- take-off or landing incidents;

- system failures; and unintentional release of an external load (typically from a helicopter).[5]

Incidents that involve aircraft with a mass over 2250 kilograms (4960 pounds) are required to be investigated. Incidents involving lighter aircraft *might* be investigated, at the discretion of the accident investigation authority (AIA). The unfortunate reality is that there are too many small aircraft incidents for authorities to launch investigations into every occurrence; therefore, only those that are of particular interest or relevant to broader safety initiatives will be investigated.

Search and Rescue

When an aircraft goes missing, the first response is to initiate search and rescue (SAR). The goal of SAR is to locate the crash site, rescue any survivors, and salvage the safety technologies (discussed later in this section) from the wreckage to facilitate the subsequent investigation process. Understandably, SAR personnel must react quickly to accomplish these goals.

With respect to international aviation, SAR can present many challenges. Consider for example, a situation where an aircraft registered in one State crashes in a foreign State's territory.

- Could that foreign State *choose* not to deploy SAR?

  - What happens if a State lacks SAR personnel or equipment?

- If the SAR process is lengthy (and therefore expensive), which State has the obligation to fund the efforts?

- If the crash site is on a border between two States, which one is responsible for SAR?

To ensure global harmonization, the 1944 Chicago Convention laid the foundation for global SAR by declaring that States are required to provide assistance in their territory to aircraft in distress and permit the owners of the aircraft, or authorities of the State in which the aircraft is registered, to provide assistance.[6]

Annex 12 to the Chicago Convention outlines SARPs associated with SAR. Annex 12 requires all States to provide SAR services within their territories as well as areas where sovereignty is undetermined, such as parts of the oceans and other areas as determined through air navigation agreements.[7] Key themes of this annex are the communication, cooperation, and coordination of SAR, the preparations required, and the operating procedures in emergency situations.[8]

Coordination with other States is crucial. To avoid delays related to negotiations or permissions, agreements are often established in advance through written SAR agreements, common SAR plans and procedures, pre-authorization of rescue coordination centres, and joint SAR training exercises.[9]

Annex 12 is supported by the International Maritime Organization's (IMO's) *International Aeronautical and Maritime Search and Rescue Manual* (IAMSAR) with three volumes that focus on specific SAR duties: 1) organization and management, 2) mission coordination, and 3) mobile facilities.[10]

The headquarters of SAR activities is called the *rescue coordination centre* (RCC), which leads the communication, cooperation, and coordination of the wide group of professionals and agencies involved in SAR. RCCs must be staffed 24 hours a day, with professionals who can speak English and the language required by that State for radiotelephony communication. Each RCC is led by the *SAR mission coordinator* who is responsible for leading the

- planning and coordination of aerial searches (including dispatching aircraft, flow control over search areas, and safe separation between aircraft);

- maintenance of operational safety;

- provision of emergency medical services to survivors; and

- evacuation of survivors.[11]

Generally, a single State will not conduct SAR independently, as most lack the necessary resources. Cooperation is required as most SAR equipment is borrowed from the military or owners of aircraft, watercraft, and land vehicles. SAR also makes use of specialized personnel (diplomatic, medical, police, military, security, and so on).[12]

SAR response to emergency situations has three phases:[13]

1. The *uncertainty phase* begins when radio contact is lost with an aircraft, or when an aircraft does not arrive at a destination as planned. The RCC may be activated and begin collecting and evaluating reports.

2. The *alert phase* begins when the RCC alerts SAR units and begins further search action.

3. The *distress phase* begins when there is reasonable certainty that an aircraft is in distress. The RCC is responsible for locating and providing assistance to the aircraft as quickly as possible, communicating with stakeholders (aircraft operator, State of Registry, air navigation service provider, adjacent RCCs, and accident investigation authorities). The SAR plan is written up and its execution begins.

After SAR professionals have located an aircraft in distress, and evacuated any survivors, the scene is protected to facilitate the forthcoming investigation. To help facilitate the SAR process, certain safety technologies are built into aircraft.

Safety Technologies

Because aircraft must be located as quickly as possible, and the cause of an accident must be determined by investigators, international aviation law requires installation of two safety devices within aircraft:

1. an *emergency locator transmitter* (ELT) in every aircraft, which alerts SAR authorities when an accident occurs and broadcasts a signal from the crash site; and
2. *black box* flight recorder(s) in large aircraft, which record aspects of the flight to help investigators determine why an accident occurred.

Emergency Locator Transmitter

An ELT is installed in every aircraft. When the ELT senses an impact or water – or is manually activated – it transmits a distress signal on the emergency frequencies 121.5 MHz and 406 MHz. These signals are detected by non-geostationary satellites, which alert SAR authorities. It is crucial that SAR professionals can locate the scene of an accident as quickly as possible, as the lives of survivors often depend on a speedy response.

Black Box Flight Recorders

After SAR professionals have located the crash site and survivors have been evacuated, attention shifts to determining the cause of the accident, which is the role of accident investigators within the AIA. A crucial tool in an investigation is the aircraft's *black box*. Contrary to its nickname, a black box is actually bright orange in colour (to improve visibility in potentially harsh conditions), and its proper name is a *flight recorder*. Flight recorders preserve key pieces of data about a flight so that investigators can recreate the events that led up to an accident.

Figure 8.4 Flight data recorders

Source: By National Transportation Safety Board [Public domain], via Wikimedia Commons

There are usually two different flight recorders on an aircraft: the *flight data recorder* (FDR) located in the tail section to maximize survivability, and the *cockpit voice recorder* (CVR) located near the aircraft cockpit. Two secondary types of recorders, often built into the FDR or CVR, are the *airborne image recorder* (AIR) and a *data link recorder* (DLR).[14]

FDRs are typically about 16 centimetres high, 13 centimetres wide, and 50 centimetres deep (6 inches by 5 inches by 20 inches) and weigh about 5 kilograms (11 pounds).[15] The FDR continually records various aircraft parameters that allow investigators to accurately recreate the aircraft's operations leading up to an accident. *Parameter* is the term used to describe a collectable data point, and the number of parameters recorded varies based on the FDR. At a minimum, parameters must include altitude, airspeed, heading, acceleration, and microphone keying (which helps investigators correlate the FDR data to voice recordings on the CVR).[16] However, modern FDRs record thousands of parameters. As of 2016, all FDRs must be capable of storing at least the last 25 hours of operations.

CVRs are audio recorders that collect the voices of the pilots along with sounds in the cockpit (such as auditory alarms, switches, engine noise, and cabin crew communications, among other things). An average CVR is 16 centimetres high, 13 centimetres wide, and 32 centimetres deep (6 inches by 5 inches by 13 inches) and weighs about 4.5 kilograms (10 pounds).[17] CVRs give accident investigators crucial insight into the actions and behaviours of the pilots immediately before an accident. Older magnetic tape recorders stored the last 30 minutes of flight on a loop (with new recordings continuously overwriting older recordings), while modern recorders store at least a two-hour loop. CVRs are required on all aircraft with a maximum certificated take-off mass over 5700 kilograms (12 600 pounds).

AIRs are video cameras that record the cockpit area. Because of concerns expressed about crew privacy, AIRs, when used, are typically mounted to exclude the head and shoulders of the crew.

DLRs save data link communication messages, equal in recording duration to the CVR. A DLR is needed when an aircraft's flight path is authorized or controlled through use of data link messages, for example if an ADS-B system is in use (see Chapter 4). In this case, all data link messages (both uplinks to the aircraft and downlinks from the aircraft) must be recorded on the aircraft DLR.

These four types of flight recorders are found in most commercial aircraft; however, there are a few exceptions. Some large aircraft must have the redundancy of two *combination recorders* (each of which incorporates both an FDR and CVR). One combination FDR/CVR is located at the aft of the aircraft and another installed near the flight deck. Some helicopter and military aircraft utilize *deployable recorders*, rather than the standard recorders used in commercial aircraft. These deployable recorders can eject from an aircraft and clear the crash area during an accident, and then transmit an emergency signal so that they can be found by search teams.

Flight recorders are not indestructible and are occasionally destroyed in an accident. However, they are built to meet crashworthiness standards specified by the European Organisation for Civil Aviation Equipment (EUROCAE), including the ability to withstand the following:

- impact velocity of 270 knots and a deceleration/crushing distance of 45 centimetres (deceleration of 3400 G);

- penetration force produced by dropping a 227-kilogram (5000 pound) weight from three meters (10 feet);

- crush force of 22.25 kilonewtons (5000 pounds) applied for five minutes; and fire of 1100 degrees Celsius (2012 degrees Fahrenheit) for one hour.[18]

Recorders must also incorporate an *underwater location beacon* (ULB) that, when immersed in fresh or saltwater, automatically activates and radiates an acoustic signal called a *pinger*. The ULB must operate up to a depth to 6096 meters (20 000 feet) for at least 30 days. The ULB's signal can be detected by an underwater receiver, usually with a maximum detection range of two to three kilometres.[19]

Recorders are designed so that they cannot be switched off by passengers or crew. They begin recording before an aircraft moves under its own power and continue until the aircraft is parked.

Did You Know?

The FDR was invented by Australian David Warren whose father was killed in an aviation accident in 1934, when David was only 9 years old. After earning his PhD, and working as a research scientist at the Aeronautical Research Laboratories, Warren completed the first prototype of a data recorder in 1956. In 1960, a Fokker F27 accident in Queensland killed 29 people and investigators were unable to determine what caused the accident. The next year Australia made history when it became the first country to require aircraft to carry flight recorders.[1]

Note

1 Witham, 2005

Accident Investigation Process

After an accident, a formal investigation is launched with the goal of identifying causal and/ or contributing factors that led to the accident. The intent is never to place blame or liability. The recommendations made through an investigation are intended to improve international aviation safety.

Beginning in 1951, ICAO began unifying the way the international community investigates aviation accidents. International standardization is crucial because determining which country has jurisdiction to lead an investigation can become very tricky: an aircraft may be registered in one country, crash in a second country, and carry passengers from several other countries (all of which may want to be involved in the investigation). In addition, without international oversight, the way in which an accident would be investigated around the world – as well as the characteristics of each State's AIA – might vary. Annex 13: Aircraft Accident and Incident Investigation contains SARPs that unify the international community's response to an accident.

Accident Investigation Authorities

Around the globe, each country designates an AIA. To ensure fair and impartial investigations, AIAs must have unrestricted authority over their own conduct, including the

Table 8.1 Accident investigation authorities of ICAO Council States

Australia	Australian Transport Safety Bureau (ATSB) www.atsb.gov.au
Brazil	Centro de Investigação e Prevenção de Acidentes Aeronáuticos (CENIPA) www.cenipa.aer.mil.br
Canada	Transportation Safety Board of Canada (TSB) www.tsb.gc.ca
China	Office of Aviation Safety Civil Aviation Administration of China www.caac.gov.cn/en/
France	Bureau d'Enquêtes et d'Analyses pour la sécurité de l'aviation civile (BEA) www.bea.aero
Germany	Federal Bureau of Aircraft Accidents Investigation (BFU) www.bfu-web.de
Italy	Agenzia Nazionale per la Sicurezza del Volo (ANSV) www.ansv.it
Japan	Japan Transport Safety Board (JTSB) (part of the Ministry of Land, Infrastructure, Transport and Tourism) www.mlit.go.jp/jtsb/english.html
Russian Federation	Interstate Aviation Committee http://mak-iac.org/en/
United Kingdom	Air Accidents Investigation Branch (AAIB) (part of the Department for Transport) www.aaib.gov.uk
United States	National Transportation Safety Board (NTSB) www.ntsb.gov

Source: ICAOa, n.d.

gathering of evidence, issuance of safety recommendations, determination of causes and contributing factors, and completion of the final report. This means that the AIA must be independent of the State's CAA, as the AIA needs to be free to identify shortcomings in regulations that may be the fault of the CAA. The structure and organization of AIAs varies between States.

Accident Investigation Jurisdiction

To determine which State's AIA has jurisdiction over an investigation, ICAO designates authority to the *State of Occurrence* – the country where the accident crash site is located. The AIA in the State of Occurrence is tasked with leading the investigation.

> **Did You Know?**
>
> If an accident occurs in international waters, there is no State of Occurrence. In this case, the State of Registry is responsible for the investigation.

However, there are several stakeholder groups with personal, financial, or reputational interests in the outcome of an investigation. Accidents have tremendous financial and emotional impacts on a variety of groups. The aircraft manufacturer, engine manufacturer, and airline all have a financial interest in the result of the investigation – if an investigation determines the causal factor was linked to one of their organizations, it could devastate its reputation. Likewise, various States (i.e., where the aircraft was registered, where it was built, and where the operator is based) will have interest in an investigation. The regulations within each of these States will have influenced that flight; therefore, they will want to understand whether one of their policies might have played a role in the accident.

To organize this large group of stakeholders, ICAO sets out which States have the right to be included in an investigation (see Figure 8.5).

After an accident, the State of Occurrence is required to notify ICAO quickly, and all the stakeholders listed in Figure 8.5, that an accident has occurred. When notice is received, the States of Registry, Operator, Design, and

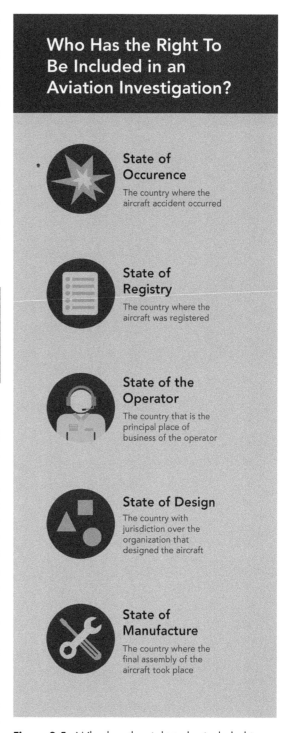

Who Has the Right To Be Included in an Aviation Investigation?

State of Occurence
The country where the aircraft accident occurred

State of Registry
The country where the aircraft was registered

State of the Operator
The country that is the principal place of business of the operator

State of Design
The country with jurisdiction over the organization that designed the aircraft

State of Manufacture
The country where the final assembly of the aircraft took place

Figure 8.5 Who has the right to be included in an aviation investigation

Manufacture must provide the State of Occurrence with any relevant information associated with the aircraft and flight crew involved in the accident or serious incident. These States must also indicate whether they intend to send a representative to be involved in the investigation.

The State of Occurrence's AIA must protect evidence and maintain custody of the evidence for as long as required to conduct the investigation. However, upon request from other stakeholders, the AIA may be required to leave the evidence undisturbed until representatives from other States arrive to witness the process.

The State of Occurrence's AIA may delegate the investigation (or elements of it) to other AIAs through mutual arrangement and consent. Consider that accident investigations can take years to complete and cost millions of dollars – not every country in the world has the resources to take on such a big project. In this case, having another State accept responsibility might make sense to everyone involved. Recall, for example, the case study you read in Chapter 4 – Australia took a lead role in the search for the missing Malaysia Airlines flight MH 370, upon agreement from the Malaysian government, as Australia had greater water-search capabilities.

Accident Investigators

AIAs are staffed with qualified aircraft accident investigators. Accident investigation is a specialized task that requires significant background experience. Before becoming an accident investigator, an individual must have many years of civil or military aviation experience, as a maintenance engineer, aeronautical engineer, or pilot.

8.2   International Society of Air Safety Investigators

The International Society of Air Safety Investigators (ISASI) is an association that exists to share ideas, experiences, and information about aircraft accident investigations. ISASI offers scholarships and student mentoring programmes to support young people who have an interest in aviation safety investigations. See www.isasi.org for information about the organization.

Investigators are typically classified as *operations investigators* or *engineering (technical) investigators*. The former come with experience as a pilot, while the latter have a background in maintenance engineering. To analyse an accident, investigators must have expert-level understanding of both the human operations (i.e., the work of pilots and other aviation personnel) as well as the physical operations of the aircraft structure, systems, and components. Beyond their aviation expertise, investigators must also develop the technical skills to conduct an investigation.

Aircraft accident investigators must be able to...

- ✅ Control the investigation under the leadership of the investigator-in-charge (IIC).

- ✅ Handle unrestricted and immediate access to the accident or incident site and to the aircraft and/or wreckage.

- ✅ Remove parts of the wreckage for examination and analysis.

- ✅ Access the contents of flight recorders.

- ✅ Collect and examine evidence from the bodies of victims.

- ✅ Interview witnesses and survivors with the tact to communicate with people who have been though a traumatic experience.

- ✅ Logically analyse facts.

- ✅ Remain impartial in the recording and analysis of facts and information.

- ✅ Persevere in pursuing lines of inquiry.

- ✅ Communicate and coordinate with owners, operators, and manufacturers of aircraft as well as aviation and air traffic authorities.

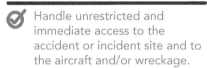

Figure 8.6 Skills of an Accident Investigator

Source: Adapted from ISASI, 2015

To develop their knowledge, skills, and abilities, investigators complete thorough training as detailed in Table 8.2.

Table 8.2 Training for aircraft accident investigators

Phase 1 – Initial classroom training

During phase 1, investigators are trained in administrative arrangements, initial response procedures, and investigative procedures.

The *administrative arrangements* portion of the training includes education on legislation, international agreements (Annex 13), liaising with local and national authorities, investigation manuals and procedures, equipment, and ethics, among other things.

Education in *initial response procedures* includes training in on-call procedures, notification of authorities, securing of records, accident site jurisdiction, investigator safety and psychological stress, and recovery of human remains, among other procedures.

Finally, the *investigation procedures* portion provides training on authority and responsibility, size, scope, and management of investigations.

Phase 2 – On-the-job training

During phase 2, investigators get hands-on practice in procedures and tasks introduced in initial training, and learn more about investigation techniques.

Phase 3 – Basic accident investigation courses

Within the first year on the job, the investigator attends a basic accident investigation course that may cover some or all of the following topics:

- responsibilities of States (as defined in Annex 13);
- accident site considerations;
- investigator equipment and protective clothing;
- examination of wreckage and witness marks;
- recording apparatus;
- witness interview techniques;
- range of in-flight and ground-based recorders;
- how to determine origin and time of fires;
- crashworthiness and survival;
- properties of aircraft structure materials and systems and how they tend to fail;
- aerodynamics and performance;
- examining power plants;
- human performance;
- aviation medicine; and
- report writing.

Phase 4 – Advanced accident investigation courses and additional training

As investigators gain experience, they may enrol in advanced courses to increase their knowledge of special topics related to accidents.

As they may investigate all kinds of accidents, investigators should have basic knowledge of all the major aircraft types operating in their State. Engineering investigators could attend aircraft type courses for maintenance personnel, while operations investigators could attend a pilot's type course that includes introductory training in a flight simulator.

Source: ICAO, 2003, pp. 5–6; ISASI, 2015, p. 9

Accident Investigation

Following an accident, the AIA typically deploys a *go team* to the accident site. This team will be led by the investigator-in-charge (IIC), who will play a leadership role throughout the entire investigation. At the accident site, the investigators will gather evidence, interview witnesses, and document the scene.

Following their time on site, investigators will complete a post-field phase that includes gathering and documenting additional factual information (such as the maintenance history of the aircraft and a timeline of the flight crew's activities in the days before an accident). They will also reconstruct the exact sequence of events that led up to the accident, by analysing flight recorder data, building graphical reconstructions of the aircraft's configuration and manoeuvres, and reviewing a complete history of the aircraft and flight crew.

This extensive data is then analysed to identify the causal and/or contributing factors of the accident. These findings will be written up in the final report through which investigators share their findings. These reports and the safety recommendations therein are widely disseminated to the public and the international aviation community. In exceptional circumstances, if investigators find something that presents an ongoing risk to the aviation industry, they will release preliminary findings in the interest of safety, rather than waiting for the final report to be completed.

Final accident reports follow an ICAO-standardized structure that includes:

- Title
  - name of operator, manufacturer and model of aircraft, nationality, and registration; date and place of the accident
- Synopsis
  - brief description of the accident
- Body
  - *factual information* – history of the flight, injuries to persons, damage to aircraft, other damage, personnel information (background of crew members), aircraft information, meteorological information, navigation aids, communications, aerodrome information, flight recorders, wreckage and impact information, medical and pathological information, fire, survival aspects, tests and research, organizational and management information, and useful investigation techniques
  - *analysis* – evaluation of factual information relevant to the determination of causes and/or contributing factors
  - *conclusions* – listing of the findings, causes, and/or contributing factors discovered through the investigation
  - *safety recommendations* – brief statements of recommendations made for the purpose of accident prevention.[20]

A first draft of the final report must be circulated to stakeholders for comments before being made publicly available. If there is a disagreement over a finding, comments will be added in an appendix to the report describing the dissent.

Did You Know?

Accident reports are in the public domain, and therefore can be found with a quick Internet search of the flight number. As an aviation professional, if you want to learn more about a particular accident, seek out the original report from the AIA in order to consider the information directly from the source (rather than through media reports, which may be biased or distorted).

Consider for a moment that a final report could find a particular State's regulations to be deficient – that State may dispute the finding and adamantly argue to have it removed from the final report. When this situation has occurred in the past, it has sometimes led to a country (other than the State of Occurrence) having its AIA release an independent report from its perspective. Final reports carry a lot of weight and can cause intense and emotional debates before they are finalized.

Once finalized, the report is distributed to any States that were stakeholders in the accident, and to the public. If a safety recommendation is made, States given the recommendations have 90 days to explain what preventative actions they have taken (or are considering) or their reasons for not taking action.

What Causes Aviation Accidents?

With the time and expertise involved in investigating aviation accidents, it is not surprising that a wealth of data exists associated with what causes accidents. However, in the past, it was a challenge to analyse this data because different agencies around the world used a variety of terminology, making it next to impossible to draw conclusions or identify trends on a global scale.

In 1999, the Commercial Aviation Safety Team (CAST) and ICAO established the CAST/ICAO Common Taxonomy Team (CICTT) to standardize the terms used to describe types of accidents. The CICTT was made up of a group of experts including government officials, aviation industry leaders, aircraft and engine manufacturers, pilot associations, regulatory agencies, accident investigation authorities, and ICAO, among others.[21]

The CICTT developed seven operational groupings, within which various *occurrence categories* are clustered, grouping causes of accidents together under standardized terms. This very important work allows the industry to recognize and address the most significant safety risks within specific types of aviation operations. Table 8.3 sets out the occurrence categories within the seven operational groupings.

Table 8.3 CICTT occurrence categories

Operational grouping	Occurrence category	Abbreviation
Airborne	abrupt manoeuvre	AMAN
	airprox/TCAS alert/loss of separation/near mid-air collisions/mid-air collisions	MAC
	controlled flight into/towards terrain	CFIT
	fuel related	FUEL
	glider towing–related events	GTOW
	loss of control – in flight	LOC-I
	loss of lifting conditions en route	LOLI
	low altitude operations	LALT
	navigation errors	NAV
	unintended flight in IMC	UIMC
Aircraft	fire/smoke (non-impact)	F-NI
	system/component failure or malfunction (non-powerplant)	SCF-NP
	system/component failure or malfunction (powerplant)	SCF-PP
Ground operations	evacuation	EVAC
	fire/smoke (post-impact)	F-POST
	ground collision	GCOL
	ground handling	RAMP
	loss of control – ground	LOC-G
	navigation errors	NAV
	runway excursion	RE
	runway incursion	RI
	wildlife	WILD
Miscellaneous	bird	BIRD
	cabin safety events	CABIN
	external load–related occurrences	EXTL
	medical	MED
	other	OTHR

(Continued)

Table 8.3 (Continued)

Operational grouping	Occurrence category	Abbreviation
	security related	SEC
	unknown or undetermined	UNK
Non-aircraft-related	aerodrome	ADRM
	air traffic management/ communication navigation surveillance	ATM
Take-off and landing	abnormal runway contact	ARC
	collision with obstacle(s) during take-off and landing	CTOL
	undershoot/overshoot	USOS
Weather	icing	ICE
	turbulence encounter	TURB
	wind shear or thunderstorm	WSTRW

Source: CICTT, 2013, pp. A-1–A-2

Worldwide Commercial Jet Fleet

Each year Boeing publishes a statistical summary of the most frequently occurring CICTT categories of commercial jet fleet accidents worldwide. The data for 2005 to 2014 are set out in Figure 8.7.

Figure 8.7 clearly shows that loss of control in flight (LOC-I) is the most common type of accident within the commercial jet fleet. This is followed by controlled flight into terrain (CFIT) accidents and landing/approach–related accidents (a broad term that includes three occurrence categories, discussed below).

LOC-I accidents (pronounced *low-key*) result from crew losing the ability to control an aircraft during flight and deviating from the flight path (not associated with a system or component failure). Because it is the deadliest occurrence category, there is an industry focus on how to reduce this type of event. However, there is a broad range of types of LOC-I accidents with many different sequences leading up to these accidents, which has made it very difficult to implement a single solution. One initiative is *upset prevention and recovery training* (UPRT) through which pilots are given the opportunity to experience unusual flight attitudes in a simulator or aerobatic-capable aircraft to practise recovery techniques.

CFIT accidents (pronounced *see-fit*) occur when an aircraft flies into an obstacle, terrain, or the water without any indication that the crew lost control of the flight (anecdotally described as 'flying a perfectly good aircraft into the ground'). CFIT accidents are associated with a pilot's *situational awareness*, which refers to the pilot's mental picture of his or her surroundings. Most CFIT accidents are the result of the pilot losing situational awareness, not understanding that the aircraft is too close to an obstacle or terrain, and flying a

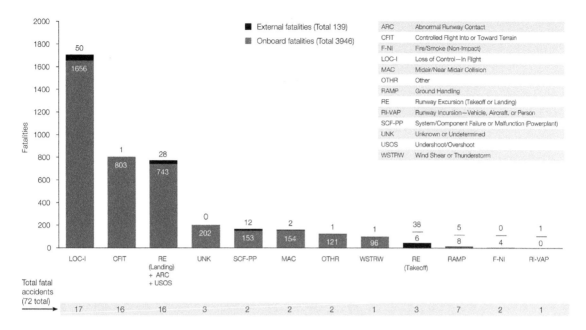

Note: Principal categories as assigned by CAST.
For a complete description of CAST/ICAO Common Taxonomy Team (CICTT) Aviation Occurrence Categories, go to www.intlaviationstandards.org.

Figure 8.7 Fatalities by CICTT aviation occurrence categories, fatal accidents, worldwide commercial jet fleet, 2005 through 2014

Source: Boeing, 2015, p. 22

functional aircraft into that obstacle or terrain. However, as in the case of the Germanwings flight 9525 accident (see Case Study), CFIT accidents may also be the result of a deliberate action. Although CFIT is still the second-deadliest occurrence category, the rate of this type of accident has been reduced over the years with the introduction of *flight management systems* (FMS) and *global positioning systems* (GPS), which assist with the pilot's awareness of their aircraft and the surrounding environment. The CFIT accident rate has also been reduced as a result of *ground proximity warning systems* (GPWS), which alert pilots when they are flying close to terrain and broadcasts a loud auditory warning: 'terrain, terrain, pull up, pull up'.

Landing/approach-related accidents comprise three occurrence categories associated with the landing and/or approach phase of a flight. A *runway excursion* (RE) describes an aircraft overrunning or veering off the side of a runway. Between 2010 and 2014, there were 135 runway excursions in European Aviation Safety Agency (EASA) Aerodromes.[22] *Abnormal runway contact* (ARC) describes an unusually hard landing that causes serious damage or injury, resulting from the pilot's actions (i.e., not associated with a system or component failure). An *undershoot/overshoot* (USOS) is also related to a pilot's actions – a USOS happens when an aircraft was brought down either before (undershoot) or after (overshoot) the runway. Landing/approach–related accidents are primarily associated with poor energy management by the flight crew, meaning that the speed, handling, and/or approach technique was not appropriate for the conditions. Runway contamination (water, ice, or snow on the runway) can also be a contributing factor in these events.

**Case Study: Germanwings Flight 9525 Accident –
Mental Health in Aviation[1]**

On 24 March 2015, Germanwings flight 9525, an Airbus A320, was en route from Barcelona, Spain towards Düsseldorf, Germany with two pilots, four cabin crew, and 144 passengers on board. During the cruise portion of the flight, the aircraft began a steady descent and crashed into the French Alps. There were no survivors of the accident and the aircraft was completely destroyed.

As the AIA in the State of Occurrence, the BEA was responsible for leading the investigation. After collecting the aircraft's flight recorders, including the CVR, investigators listened to the final moments of the flight. They heard the Captain tell the first officer (FO) he was leaving the cockpit for a moment, which is routine for pilots (perhaps to use the lavatory). Seconds after the Captain had left the cockpit, the co-pilot started a descent at a rate of 3 500 feet per minute (350 knots).

The controller tried to contact the flight several times, but got no response from the FO.

After a few minutes, the Captain was heard trying to re-enter the cockpit, calling from the cabin interphone, knocking at the cockpit door, asking for the door to be opened, and eventually making violent blows on the door. The security requirements that led to cockpit doors being reinforced after 9/11 made it impossible for the Captain to access the cockpit.

The GPWS began an auditory warning: 'terrain, terrain, pull up, pull up'. Seconds later, the flight made impact with terrain.

This finding triggered investigators to look more closely into the background of the FO. They found that he had a waiver on his medical certificate because of a severe depressive episode that lasted from 2008 to 2009. It was also found that beginning in December 2014, the co-pilot began showing psychotic depressive symptoms and that he had consulted several doctors and was prescribed antidepressant medications. However, this was not reported to an aviation medical examiner (AME), by the pilot or the medical professionals, and therefore did not impact his medical certificate required for his pilot licence to remain valid.

Neither the airline nor the regulator were ever informed of the FO's mental health conditions. He probably did not report his condition for fear of losing his right to fly, and the associated financial consequences, as there was no insurance covering this condition. At that time, German regulations associated with medical confidentiality were unclear around when a threat to public safety outweighs privacy concerns.

As a result of their investigation, the BEA issued 11 safety recommendations relating to the medical evaluation of pilots with mental health issues, mitigation of the consequences of loss of licence, antidepressant medication and flying status, balance between medical confidentiality and public safety, and promotion of pilot support programmes, among other things.

Note

1 BEA, 2016

In addition to the occurrence categories described above, the CICTT also sets out standard terminology for the phase of flight during which an accident occurs. Standardization of this data offers important insight into when, during a typical flight, accidents are most likely. If you consider the example of a 1.5-hour flight, Figure 8.8 presents the exposure (the percentage of time the aircraft spends in each phase). As expected, the majority of the flight time (57 per cent) is spent in the cruise phase.

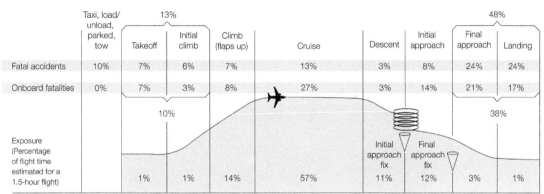

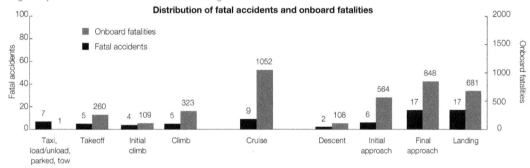

Figure 8.8 Fatal accidents and onboard fatalities by phase of flight, worldwide commercial jet fleet, 2005 thorugh 2014

Source: Boeing, 2015, p. 20

What is evident in these findings is that although the take-off and initial climb represent two per cent of flight time and the final approach and landing represent four per cent of flight time (for a combined total of six per cent of flight time) these two phases of flight represent 61 per cent of non-fatal and 48 per cent of fatal accidents. Clearly, the take-off, initial climb, final approach, and landing phases of flight carry a *disproportionate level of risk* compared to other phases.

It is for this reason that regional airlines – with shorter flights, resulting in a higher proportion of flight time spent in the take-off and landing phases – have a higher exposure to risk than legacy carriers flying long interoceanic flights.

Helicopters – Commercial

Unlike airline operations, rotary-wing flight has an entirely different risk profile. Most accidents and serious incidents occur during manoeuvring, which is not surprising as the manoeuvring stage is the point at which helicopters are near operational limits – such as using almost maximum power (for example, while performing external load operations, power line or pipeline inspections, and agricultural work).[23]

In Europe, for example, between 2005 and 2014, the most common incidents and accidents in helicopter operations include low altitude operations (LALT), loss of control – in flight (LOC-I), collisions during take-off and landing (CTOL), and system or component failure – powerplant (SCF-PP).[24]

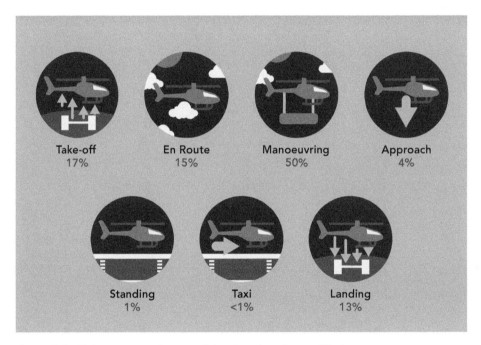

Figure 8.9 Helicopter accidents and incidents by phase of flight
Source: Adapted from EASA, 2015, p. 81

General Aviation

In contrast to the overall accident rate of three accidents per million departures, the GA accident rate is significantly higher. For example, in the United States from 2004 to 2013, there were approximately 65 accidents per million GA flight hours with approximately 70 per cent of these accidents occurring during personal flying.[25] This statistic does not reflect a disregard for safety among these pilots or operators, but might reflect the fact that they may have less training, experience, and/or operational support as compared to airline operations.

Accident risk is statistically linked to a pilot's flight hours with a pilot most likely to be involved in an accident between 40 and 250 hours of flight time, which has been called the 'killing zone'.[26] It was believed that during this period a pilot's confidence outweighs his or her experience. Yet more recent research has determined that the risky zone for pilots may actually peak between 400 and 700 hours of experience and taper off slowly until they reach around 2000 hours of experience.[27] As discussed in Chapter 3, pilots who want to fly commercial airlines typically build experience flying in the GA sector. Therefore, the average airline pilot has more hours of experience than the average GA pilot (including those pilots in the high-risk zones), which may also contribute to the higher rate of accidents within the GA sector.

Figure 8.10 outlines the most common categories of accidents for each aircraft type involved in GA operations in Europe.

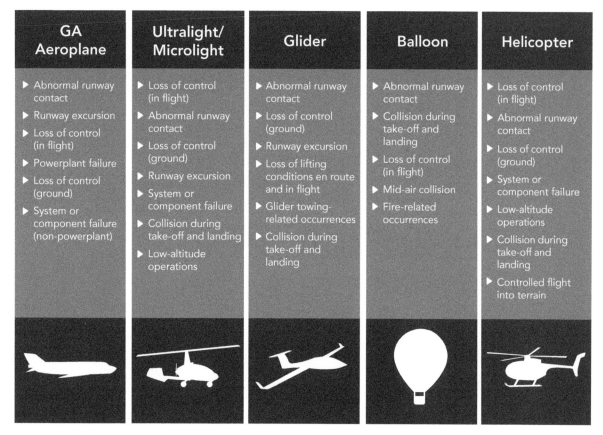

Figure 8.10 Aviation accidents by general aviation aircraft type (in descending order)
Source: Adapted from EASA, 2015, pp. 96–114

Looking specifically at balloon operations, it is worth noting that 61 per cent of the occurrences were during the landing phase (with many reports of hard landings due to aircraft handling errors and collisions with obstacles, such as power lines or objects on the ground).

Air Traffic Management

Of course, not all accidents are caused by aircraft operators; some accidents have causal factors linked to air navigation services. Within the CICTT occurrence categories, such accidents would fall into the non-aircraft-related category air traffic management/communication navigation surveillance (ATM). There are several situations that could lead to an ATM–related accident. Figure 8.11 sets out the most common ATM occurrences along with the severity of the event, as reported by EASA between 2010 and 2014.

It is evident that unauthorized airspace penetration (i.e., an aircraft flying into airspace that it is not authorized to enter) is the most common ATM occurrence, followed by aircraft not following the clearance given to them by an air traffic control officer (aircraft deviation from ATC clearance).

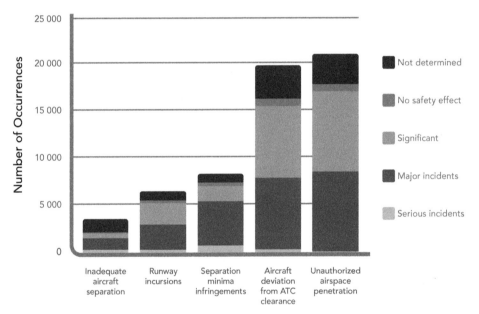

Figure 8.11 ATM-related occurrences

Source: Adapted from EASA, 2015, p. 119

Aerodrome

Another occurrence category not directly associated with aircraft operators is *aerodrome* – within this category are all accidents with a causal factor linked to airport operations. Accidents that occur near or at an airport, but where airport operations did not contribute to the accident, are not included with the category. For example, landing/approach–related accidents are the third-most common occurrence type, but are not included in the aerodrome category when the primary cause of the accident was the aircraft and/or flight crew.

Although some of these categories have already been mentioned, there are a few others directly related to aerodrome operations. Ground collisions (GCOL) refer to events where one aircraft makes contact with another aircraft, vehicle, object, or person during taxi movements. Ground handling (RAMP) events are a particular challenge at aerodromes as the ramp area activities are often conducted by airport contractors rather than the airport employees themselves. RAMP events can be linked to errors during loading, pushback, de-icing, or refuelling; however, the most common event is a collision with a parked aircraft, ground object, or airport vehicle or equipment. EASA reports that in 2012 two fatal RAMP events occurred. The first involved a baggage handler who was killed while loading an aircraft, while the second fatality was a vehicle driver killed by accidentally driving into the wing tip of an aircraft.[28]

The most common categories of aerodrome-related occurrences reported in Europe were loss of control – ground (LOC-G), runway excursion (RE), ground collision (GCOL), and collision with an obstacle during take-off and landing (CTOL).

Human-Caused Accidents

After reviewing the overall aviation accident rate, along with the most common types of accidents, it is important to take a step back to look at the big picture. What is the most common cause of aviation accidents? In fact, 70 to 80 per cent of aviation accidents are primarily caused by human error.[29] While safety interventions in the 1960s and 1970s effectively targeted structural and mechanical improvements, the current plateau in the aviation accident rate is often attributed to a reasonable expectation of human error.

Think of it this way: if an accident occurs and an investigation reveals that a flaw in the aircraft was responsible, a safety recommendation can lead to all aircraft of that type being fixed, thus eliminating the likelihood that this type of accident would happen again. If an investigation determines that a pilot's actions caused an accident, it is much more challenging to 'fix' all pilots.

Although many people outside the aviation industry assume that weather or aircraft failures are the primary cause of accidents, the reality is that they directly cause only a small proportion of occurrences. However, weather or malfunctions are often contributing factors as they can complicate a situation beyond the flight crew's capacity to manage the problem.

The science of human factors has evolved to study the natural limitations associated with human physiology, psychology, and ergonomics. Chapter 9, which looks at safety, will explore human factors along with risk management in the aviation industry.

Conclusion

Although the aviation industry is an impressively safe mode of transportation, accidents do occur; and when they do, they have devastating human, financial, and operational impacts. Safety will always be a top priority for all aviation organizations.

In the rare instance that an accident does occur, the first priorities are locating the crash site and rescuing survivors – SAR professionals undertake these duties. The next task is determining what caused the accident. Accident investigators piece together facts during a thorough and structured investigation process to determine the causes of accidents, with the goal of continually improving the safety of the aviation industry.

These efforts create a constant learning process through which the entire industry learns from mistakes and makes necessary changes to avoid similar tragedies in the future.

Key Points to Remember

1. Perfect safety in the complex and interconnected system of aviation is probably impossible. When accidents happen, the sole objective of the formal investigation is to understand why the accident occurred and to prevent future accidents and incidents (not to place blame or liability).

2. The global accident rate in 2014 was three accidents per million departures. It was highest in the African region and lowest in the Asia and Pacific region. The global accident rate was much higher in the 1960s – about 50 accidents per million departures – but has been plateaued near its current rate since the 1970s. Accidents are expensive and incur both direct and indirect losses.

3. Global standards for search and rescue (SAR) are published by ICAO in Annex 12. The goal of SAR is to reach the accident site as quickly as possible to provide aid to survivors and protect wreckage in order to facilitate the accident investigation.

4. An *accident* is an event that occurs during a flight that results in a fatality, serious aircraft damage, or a lost or missing aircraft. Accidents must be investigated. An *incident* is an unsafe event less severe than an accident – it can be thought of as a near-accident. Incidents might be investigated, depending on the circumstances.

5. Various safety technologies are used to facilitate SAR and accident investigation, including ELTs, which automatically transmit a distress signal when they sense an impact; FDRs, which constantly record aircraft parameters (altitude, airspeed, heading, and so on) to document an aircraft's operations leading up to an accident; and CVRs, which collect the voices of pilots and cockpit sounds (such as alarms, switches, and engine noise) leading up to an accident.

6. ICAO's global SARPs associated with accident investigation are set out in Annex 13.

7. Each State designates an AIA that should be independent of the State's CAA, and have unrestricted authority over its own conduct. The AIA in the State of Occurrence (i.e., the country where an accident happened) has jurisdiction over that

accident. Representatives from other States have a right to be included in the investigation, including the State of Registry, State of the Operator, State of Design, and State of Manufacture.

8. Accident investigators fall into two groups: an *operations investigator* has a background primarily as a pilot; and an *engineering investigator* has a background in maintenance engineering.

9. The Commercial Aviation Safety Team (CAST) and ICAO created the CAST/ICAO Common Taxonomy Team (CICTT) to categorize types of aviation accidents.

   - For the worldwide commercial jet fleet, the most common types of accidents are loss of control – in flight (LOC-I), controlled flight into terrain (CFIT), and landing/approach–related accidents. For commercial jets, the take-off, climb, final approach, and landing phases of flight have a disproportionately higher level of risk than other phases of flight.

   - For commercial helicopters, the majority of accidents occur during the manoeuvring phase of flight. The most common occurrence types include low altitude operations (LALT), loss of control – in flight (LOC-I), and collisions during take-off and landing (CTOL).

   - The general aviation accident rate is higher than that of the commercial jet fleet.

   - Air traffic management (ATM) accidents are most commonly caused by aircraft entering airspace they are not authorized to enter, followed by aircraft deviation from ATC clearance.

   - The most common occurrence categories for aerodrome–related accidents are loss of control – ground, runway excursion, ground collision, and collision with an obstacle during take-off and landing.

10. Seventy to 80 per cent of all accidents are caused by human error.

Table 8.4 Acronym rundown

AAIB	Air Accidents Investigation Branch (United Kingdom's AIA)
ACC	area control centre
ADRM	aerodrome*
ADS-B	automatic dependent surveillance – broadcast
AFI	Africa (regional aviation safety group)
AIA	accident investigation authority

(*Continued*)

Table 8.4 (Continued)

AIR	airborne image recorder
AMAN	abrupt manoeuvre*
AME	aviation medical examiner
ANSV	Agenzia Nazionale per la Sicurezza del Volo (Italy's AIA)
APAC	Asia and Pacific
ARC	abnormal runway contact*
ATM	air traffic management/communication, navigation, and surveillance*
ATSB	Australian Transport Safety Bureau (AIA)
BEA	Bureau d'Enquêtes et d'Analyses pour la sécurité de l'aviation civile (France's AIA)
BFU	Federal Bureau of Aircraft Accidents Investigation (Germany's AIA)
BIRD	bird*
CAA	civil aviation authority
CABIN	cabin safety events*
CAST	Commercial Aviation Safety Team
CENIPA	Centro de Investigação e Prevenção de Acidentes Aeronáuticos (Brazil's AIA)
CFIT	controlled flight into terrain*
CICTT	CAST/ICAO Common Taxonomy Team
CTOL	collision with obstacles during take-off and landing*
CVR	cockpit voice recorder
DLR	data link recorder
EASA	European Aviation Safety Agency
ELT	emergency locator transmitter
EUR	Europe (regional aviation safety group)
EUROCAE	European Organisation for Civil Aviation Equipment
EVAC	evacuation*
EXTL	external load–related occurrences*
FAA	Federal Aviation Administration
FDR	flight data recorder
FMS	flight management system
FO	first officer
F-NI	fire/smoke (non-impact)*
F-POST	fire/smoke (post-impact)*

FUEL	fuel related*
GCOL	ground collision*
GPS	global positioning system
GPWS	ground proximity warning system
GTOW	glider towing–related events*
IAMSAR	International Aeronautical and Maritime Search and Rescue Manual
ICAO	International Civil Aviation Organization
ICE	icing*
IFEN	in-flight entertainment network
IIC	investigator-in-charge
IMO	International Maritime Organization
ISASI	International Society of Air Safety Investigators
JTSB	Japan Transport Safety Board (AIA)
LALT	low altitude operations*
LOC-G	loss of control – ground*
LOC-I	loss of control – in flight*
LOLI	loss of lifting conditions en route*
MAC	airprox/TCAS alert/loss of separation/near mid-air collisions/mid-air collisions*
MED	medical*
MID	Middle East (regional aviation safety group)
NAV	navigation errors*
NTSB	National Transportation Safety Board (United States' AIA)
OTHR	other*
PA	Pan America (regional aviation safety group)
RAMP	ground handling*
RASG	regional aviation safety groups
RCC	rescue coordination centre
RE	runway excursion*
RI	runway incursion*
SAR	search and rescue
SCF-NP	system/component failure or malfunction (non-powerplant)*
SCF-PP	system/component failure or malfunction (powerplant)*
SEC	security-related*

(Continued)

Table 8.4 (Continued)

TSB	Transportation Safety Board of Canada (AIA)
TURB	turbulence encounter*
UIMC	unintended flight in IMC*
ULB	underwater location beacon
UNK	unknown*
UPRT	upset prevention and recovery training
USOS	undershoot/overshoot*
WILD	wildlife*
WSTRW	wind shear or thunderstorm*

* CICTT Accident Occurrence Category

Chapter Review Questions

8.1 Describe the 'true' cost of an aviation accident.

8.2 In what way does an aircraft's black box contribute to international aviation safety? Describe the various types of black boxes.

8.3 List and explain five challenges that aviation investigators face in conducting investigations.

8.4 Why have global accident rates in the commercial jet fleet improved only marginally in the last 35 years?

8.5 There are a variety of stakeholders with an interest in the outcome of an aviation investigation. Name five stakeholder groups and describe their specific interest (e.g., protecting their reputation, improving safety, identifying systematic weaknesses).

8.6 Which CICTT accident categories do you think are the most difficult, and least difficult, to prevent? Do you believe that future technologies have the potential to reduce accidents of a certain category? Explain your answers.

8.7 What is the difference between an accident and an incident? Do some research, if necessary, and provide an example of an accident and an incident that have occurred in your State.

8.8 Depending upon the privacy laws in each State, black box recordings are sometimes made publicly available and other times kept private (shared only with stakeholders in the investigation). As a future aviation professional, which approach do you agree with? Consider the following:

- Can publicly sharing these recordings improve aviation safety or are they voyeuristic privacy violations?

- What is the ideal balance between safety and privacy? Describe the moral and ethical considerations.

- Does your State release recordings after an investigation? (Hint: an Internet search for an accident that occurred in your State will return CVR recordings, or not, depending on privacy laws.)

SWISSAIR FLIGHT 111 ACCIDENT – SEARCHING FOR ANSWERS SCATTERED ACROSS THE OCEAN FLOOR[1]

The Accident

On 2 September 1998, Swissair flight 111 (SR 111) left New York in the United States headed for Geneva, Switzerland. The aircraft was a McDonnell Douglas MD-11 with 215 passengers and 14 crew members on board. Approximately one hour into the flight, the crew noticed a strange smell in the cockpit, which quickly dissipated. The crew decided the air conditioning system was responsible and contacted the Moncton area control centre (ACC) requesting routing to land at a convenient airport. The pilots requested Boston airport – about 300 nautical miles behind them – but after a suggestion from the controller, headed towards Halifax which was only 56 nautical miles from their current location. The crew put on their oxygen masks.

Minutes later, SR 111 was 30 nautical miles from the runway of Halifax Stanfield International Airport when the pilots informed the controller they needed more time to prepare for landing, so the controller instructed SR 111 to turn to a heading of 360 degrees to provide a longer routing for the aircraft to lose altitude. The pilots discussed dumping fuel to reduce the aircraft's weight for landing, and both agreed they had time to do so.

Seconds later, both pilots declared an emergency. They stated that they had begun dumping fuel and had to land immediately. SR 111 declared an emergency again and the controller responded with clearance to dump fuel, but there was no further response from the crew. Observers in the area of St. Margaret's Bay, Nova Scotia, witnessed a large aircraft fly low overhead and described a loud 'clap'. Although SAR teams looked through the night, no survivors were found. The aircraft had impacted the ocean and was destroyed.

The Investigation

As the accident took place in Canadian waters, Canada's Transportation Safety Board (TSB) was responsible for leading the investigation. The TSB had one clue to kick-start the investigation: the crew had told the controller they smelled smoke, which suggested an on-board fire. The investigatory benefit of the aircraft striking water is that the water immediately doused any flames and preserved the state of the wreckage – there was no post-crash fire to destroy evidence. Yet this benefit was countered by the wreckage being scattered across the ocean floor at a depth of 55 meters (180 feet). Although divers recovered the FDR and the CVR, it was discovered that both recorders ceased functioning immediately after the pilots declared an emergency (significantly complicating the investigation). The final six minutes of the flight were lost.

CASE STUDY

This led to an intense 13-month salvage operation that involved more than 4000 people. Ships, including the *Queen of the Netherlands*, which vacuums the ocean floor, combed the seabed and collected everything from tangled wires and cloth to metal scraps.

Tragically, only one of the 229 victims could be identified visually, requiring the expertise of a team of medical examiners. This team included a pathologist and assistant, nurse, Royal Canadian Mounted Police, photographer, dentist, radiologist, X-ray technician, fingerprint technician, and DNA specialists. In total 1370 DNA samples were processed and all 229 victims were identified. The cost of the medical and dental detective work was CAD$800 000, but a greater cost may have been the emotional impact – these professionals later spoke out about the profound, yet intangible, emotional effect this event had on their lives.[2]

Investigators salvaged 98 per cent of the aircraft (by weight) from the ocean. Every item needed to be examined for clues and added to a physical and digital mock-up of the aircraft. A full-size metal frame was constructed, allowing investigators to place identified pieces on the mock aircraft (much like piecing together a jigsaw puzzle). Simultaneously, a digital model was created to rebuild the aircraft electronically. This process allowed investigators to make several key discoveries:

- carpet pieces revealed melted plastic 'drip' marks, indicating that an intense heat in the aircraft attic area had melted the ceiling;

- a charred galley roof; and

- burnt ducts within the overhead attic area above the cockpit.

This narrowed the investigation to the hidden attic area above the cockpit and front galley. A prime suspect was the wiring that runs throughout this area. Over time, with the vibration of flight, wires can develop cracks in their plastic insulation. Experts estimate between 400 and 1500 wire insulation cracks exist per aircraft, depending on the aircraft's age.[3] These cracks can allow electrical current to 'jump' to other wires or the aircraft itself if the wire crack comes into contact with water (such as from condensation). Electrical jumping is called *arcing* and produces an intense heat of around 6600 degrees Celsius (12 000 degrees Fahrenheit). Wiring for the in-flight entertainment network (IFEN) became a suspect as many arcs were found on IFEN wires. Investigators eventually tracked the ignition point down to a few inches. The electrical arc may have sparked the fire, but what fuelled the flames?

The attic area is lined with metallized polyethylene terephthalate (MPET)-covered insulation blankets, which had passed FAA testing and were deemed fireproof. Therefore, TSB investigators were surprised when they salvaged insulation blankets with burn marks. Though insulation is supposed to be fireproof, it quickly caught fire when the FAA retested it as part of the investigation. The FAA also tested end caps of the ventilation system, which were also supposed to be fireproof, and when tested found that they also quickly caught fire. With the end caps burnt away, oxygen flowed freely and fed the flames.

When it was discovered that the fireproof materials had fuelled a fire, the TSB released preliminarily recommendations that these flammable components be removed and replaced in the global fleet in the interest of safety. This was a case where it was clearly necessary to release preliminary recommendations, rather than waiting for the final report when the investigation was complete.

Probable Cause

From the time of the accident until the completion of the final report, the investigation took four years and cost nearly CAD$40 million. Based on the TSB investigation, the chain of events that led up to the accident was as follows:

Condensation caused a cracked IFEN wire to arc within the hidden attic area above the cockpit. The arc ignited the MPET-covered insulation blankets, which fuelled the fire and led to the pilots noticing a whiff of smoke before it was sucked away by recirculation fans. As the pilots didn't know the fire was raging above their heads they didn't consider the event an immediate threat and delayed landing to dump fuel. While completing a checklist, the Captain turned off non-essential power, which stopped the recirculation fans. This caused the fire to be drawn towards the cockpit. The flight recorders recorded a rapid succession of system failures as fire burnt through wires, leading to the autopilot disengaging and causing the CVR and FDR to stop recording (losing the final six minutes of the flight). The plastic cockpit ceiling began to melt and drip down on the pilots. Both pilots declared an emergency. The Captain got out of his seat to fight the fire while the FO piloted the aircraft. The glass cockpit instruments stopped working, and smoke and fire obscured standby instruments, forcing the FO to fly by visual reference (i.e., looking out the window). With a dark sky above and black water below, there was very little visual reference. The aircraft crashed into the ocean five nautical miles southwest of Peggy's Cove, Nova Scotia.

Recommendations

The TSB made 23 recommendations associated with the SR 111 accident. Among other things, the TSB

- identified the risk of wiring becoming cracked;

- suggested that CVRs and FDRs be outfitted with longer duration recordings and their own power supply;

- determined that insulation materials (MPET-covered insulation blankets) were flammable and should be replaced, and that more rigorous flammability test criteria be developed;

- suggested improved firefighting measures (including smoke and fire detection and suppression systems); and

- advised that it should be industry standard for aircraft to land expeditiously anytime smoke is noticed from an unknown source.

Notes

1 TSB, 2003
2 Robb, 1999, p. 241

3 *Nova: Crash of Flight 111*, 2004

CASE STUDY

Case Study Questions

The SR 111 accident stands as an example of the aviation community coming together to solve a complicated and multi-layered accident. Using this case study, and applying what you have learned in this chapter, provide informed responses to the following:

8.9 Consider the cost, complexity, and emotional impact of this accident on the lives of professionals involved in the accident investigation. Some investigators were required to relocate to Canada's east coast away from their loved ones. Put yourself in their shoes: what challenges would these investigators and medical personnel have faced in managing the intense emotional impact of the accident?

8.10 Does this case study help you understand how complex an investigation can be and how challenging it can be to train investigators to handle these events? Which elements of this investigation would require an operations investigator (pilot) and which would need an engineering investigator (maintenance)?

8.11 Canada, as the State of Occurrence, led the investigation of the SR 111 accident. Consider that the State of Occurrence may not always have the strongest connection to the accident, but its AIA has authority over the investigation. What problems might arise from this arrangement? Are there any advantages to it? Can you think of a better way of designating the primary AIA?

8.12 Why is it so important for AIAs to be independent from the government and regulator of their State? Is it possible for bias to creep into an investigation? How can this be managed?

References

ATSB, n.d. *Black box flight recorders: Fact sheet.* [Online] Available at: www.atsb.gov.au/media/4793 913/Black%20Box%20Flight%20Recorders%20Fact%20Sheet.pdf

BEA, 2016. *Final report: Accident on 24 March 2015 at Prads-Haute-Bléone (Alpes-de-Haute-Provence, France) to the Airbus A320-211 registered D-AIPX operated by Germanwings.* Le Bourget: Bureau d'Enquêtes et d'Analyses pour la sécurité de l'aviation civile.

Blumenkron, J., 2017. International safety requirements. In: P. S. Dempsey & R. S. Jakhu, eds. *Routledge Handbook of Public Aviation Law.* New York: Routledge, pp. 9–33.

Boeing, 2015. *Statistical summary of commercial jet airplane accidents: Worldwide operations 1959–2014.* [Online] Available at: www.boeing.com/resources/boeingdotcom/company/about_bca/pdf/stat sum.pdf

BTRE, 2006. *Cost of aviation accidents and incidents: BTRE report 113.* Canberra: Australian Government Department of Transport and Regional Services Bureau of Transport and Regional Economics.

CAST/ICAO, 2014. *About CICTT.* [Online] Available at: www.intlaviationstandards.org/apex/f?p=240:1

CICTT, 2013. *Aviation occurrence categories: Definitions and usage notes.* [Online] Available at: www.intlaviationstandards.org/Documents/OccurrenceCategoryDefinitions.pdf

Craig, P. A., 2001. *The killing zone: How & why pilots die.* New York: McGraw-Hill.

EASA, 2015. *Annual safety review 2014,* Cologne: European Aviation Safety Agency.

IATA, 2017. *2016 fact sheet*. [Online] Available at: www.iata.org/pressroom/facts_figures/fact_sheets/Documents/fact-sheet-safety.pdf

ICAO, 1944. *Convention on Civil Aviation ("Chicago Convention")*. Chicago: International Civil Aviation Organization.

ICAO, 2003. *Training Guidelines for Aircraft Accident Investigators, Cir 298 AN/172*. Montreal: International Civil Aviation Organization.

ICAO, 2004. *Annex 12 to the Convention on International Civil Aviation: Search and rescue, 8th ed.* Montreal: International Civil Aviation Organization.

ICAO, 2010a. *Annex 6 to the Convention on International Civil Aviation: Operation of aircraft, 9th ed.* Montreal: International Civil Aviation Organization.

ICAO, 2010b. *Annex 13 to the Convention on International Civil Aviation: Aircraft accident and incident investigation, 10th ed.* Montreal: International Civil Aviation Organization.

ICAO, 2015. *Safety report*. Montreal: International Civil Aviation Organization.

ICAOa, n.d. *Accident investigation authorities addresses*. [Online] Available at: www.icao.int/safety/AIA/Pages/default.aspx

ICAOb, n.d. *The Convention on International Civil Aviation: Annexes 1 to 18*. Montreal: International Civil Aviation Organization.

IMO Publishing, 2016. *IAMSAR manual*. [Online] Available at: www.imo.org/publications

ISASI, 2015. *Positions on air safety investigation issues, 4th ed.* [Online] Available at: www.isasi.org/Documents/ISASI_Positions_Rev7_Adopted_1May2015.pdf

Knecht, W. R., 2015. *Predicting accident rates from general aviation pilot total flight hours*. Washington, DC: Federal Aviation Administration.

Nova: Crash of Flight 111. 2004. [Film] Directed by G. Glassman, H. Green. s.l.: s.n.

NTSB, 2015. *NTSB US civil aviation accident statistics*. [Online] Available at: www.ntsb.gov/investigations/data/Pages/AviationDataStats.aspx#

Parichat_Pa@Mjets.com, 2015. *Civil aviation search and rescue worldwide*. [Online] Available at: www.ainonline.com/aviation-news/air-transport/2015-11-05/civil-aviation-search-and-rescue-worldwide

Robb, N., 1999. 229 people, 15 000 body parts: pathologists help solve Swissair 111's grisly puzzles. *Canadian Medical Association Journal,* Volume 160, pp. 241–243.

Shappell, S. A. & Wiegmann, D. A., 2000. *The human factors analysis and classification system – HFACS*. Washington, DC: U.S. Department of Transportation: Federal Aviation Administration.

TSB, 2003. *Aviation investigation report: In-flight fire leading to collision with water*. Gatineau: Transportation Safety Board of Canada.

Witham, J. P., 2005. *Black box: David Warren and the creation of the cockpit voice recorder*. s.l.: Lothian Books.

Notes

1 ICAO, 2010b, p. 3–1

2 ICAO, 2015, p. 8

3 IATA, 2017, p. 1

4 BTRE, 2006, p. vii

5 ICAO, 2010b, pp. ATT C-1

6 ICAO, 1944, p. Article 25

7 ICAO, 2004

8 ICAOb, n.d., p. 21

9 Blumenkron, 2017, p. 55

10 IMO Publishing, 2016, p. 2

11 Parichat_Pa@Mjets.com, 2015

12 Parichat_Pa@Mjets.com, 2015

13 ICAOb, n.d., p. 22

14 ICAO, 2010a

15 ATSB, n.d., p. 2
16 ATSB, n.d., p. 2
17 ATSB, n.d., p. 1
18 ATSB, n.d., p. 4
19 ATSB, n.d., p. 5
20 ICAO, 2010b, p. APP-1
21 CAST/ICAO, 2014
22 EASA, 2015, p. 128
23 EASA, 2015, p. 81
24 EASA, 2015, p. 81
25 NTSB, 2015
26 Craig, 2001
27 Knecht, 2015, p. 1
28 EASA, 2015, p. 129
29 Shappell & Wiegmann, 2000, p. 1

1. Human error is the primary cause of _____ per cent of aviation accidents?
    a. 20-30
    b. 40-50
    c. 60-70
    d. 70-80

2. Aviation historically adopted a reactive approach to safety, where safety initiatives would begin after an accident occurred.
    a. True
    b. False

5. Annex 19 outlines SARPs associated with safety management, with the goal of adopting a proactive/predictive approach to safety throughout the global aviation industry.
    a. True
    b. False

3. Two main approaches to understanding why human error contributes to accidents, and how it can be prevented, are the science of human factors and safety management.
    a. True
    b. False

4. Human error generally reflects incompetence, laziness, or a lack-of-effort from aviation professionals.
    a. True
    b. False

Answer Key: 1. d; 2. a; 3. a; 4. b; 5. a.

Learning science suggests that thinking through a few questions before you begin studying new material, even if you answer incorrectly, results in improved learning and retention.
Give it a try!

CHAPTER 9

Safety

CHAPTER OUTCOMES

At the end of this chapter, you will be able to . . .

- Express the similarities and differences between human factors and safety management.

- Describe human factors training for pilots, including crew resource management, line-oriented flight training, and threat and error management.

- Outline several human factors issues that impact all aviation professionals.

- Discuss the origins of the organizational approach to safety, including associated ICAO standards linked to state safety programmes and safety management systems.

- Apply your understanding of human factors and safety management to the Air France flight 447 accident, which illustrates how a technical failure combined with human error can have catastrophic results.

Introduction

In the previous chapter, which discussed the investigation and causes of accidents, it was noted that 70 to 80 per cent of accidents are caused by human error. An initial reaction may be to blame the people involved, asking why professionals make so many mistakes. Historically, accident investigations took a similar approach, and the pilots were typically assigned blame. Unfortunately, blaming and firing pilots involved in an accident did little to prevent similar accidents in the future.

It is important to understand that aviation professionals do not take safety for granted – they are dedicated to their work and have every intention of doing a good job. Yet this makes it difficult to understand the high rate of accidents caused by human error.

The reality is that all aviation professionals have one common limitation – they are all human beings. Although humans have remarkable capacity for critical thinking and creativity, we also have natural and predictable limitations associated with our mental and physical abilities. The science of human factors explores these human limitations and applies that knowledge to the design and operation of aviation systems.

9.1 Human Error and Accident Prevention

Although human error is responsible for the majority of modern aviation accidents, '100 per cent of accidents that are *prevented* are the result of the actions of aviation professionals' (Chris Hadfield, Canadian astronaut).

When exploring why human error is the primary cause of accidents, there are two complementary philosophies:

1. Human error causes accidents and therefore needs to be studied and understood through the scientific discipline of human factors.

2. Human error is a symptom of underlying problems that need to be understood from an organizational approach through safety management, which attempts to proactively identify and eliminate risks.

This chapter will explore these two key elements associated with human error: human factors and safety management.

Human Factors

Historically, the aviation industry had a reactive approach to safety. After a major accident occurred, the investigation would reveal a weakness or concern, which the industry would then correct. If a flaw was identified on a type of airframe or engine, an airworthiness directive would be issued to have that flaw fixed on all airframes or engines of that type. This approach was very successful in reducing accident rates by improving mechanical aspects of aircraft.

If no mechanical or structural malfunction could be identified, accident investigators would turn their attention to the flight crew to determine if someone broke a rule or made a mistake. Historically, the emphasis was on determining responsibility for the accident – what happened, who did it, and when it took place.[1] Investigations did not typically dig into the deeper organizational issues that may have contributed to the accident. Over time, the number of accidents caused by mechanical elements went down at a much quicker rate than the number of human-caused accidents.

Although human factors affect the performance of all aviation professionals, we will begin by exploring the origin of aviation human factors training specifically for airline pilots. After a series of high-profile airline accidents in the 1970s caused by pilot error, increasing attention was given to addressing this issue.

9.2 Consequences of Human Error

Human factors research explores limitations associated with being human, and therefore is not limited to pilots.

A tragic example of how human error can impact the safety of a flight occurred on October 2 1996. Aeroperú flight 603 crashed and fatally injured all 70 persons on board after the pilots lost control when their cockpit instruments failed. The instrument failure was caused by a maintenance engineer forgetting to remove a piece of tape covering the static ports that inform the cockpit instruments.

Within the complex aviation environment, what may seem like a simple human error can have terrible consequences. Therefore, human factors must be considered with respect to all aviation professionals (maintenance personnel, air traffic control officers, regulators, manufacturers, management, and pilots, among others).

Human Factors Training for Pilots

In 1979, NASA in the United States held a workshop called Resource Management on the Flightdeck,[2] intended to bring airlines and researchers together to discuss the prevalence of human error–caused aviation accidents. This workshop had a global impact that led to *crew resource management* (CRM) training, which is now a required component of pilot annual training programmes. CRM instruction teaches pilots to use all available resources, including those on the aircraft and support on the ground, to achieve safe and efficient flight.

CRM training explores human factors issues that relate to pilot errors, with emphasis on communication, leadership, stress, fatigue management, and decision-making. Many airlines involve several employee groups (pilots, cabin crew, maintenance) in annual CRM training to facilitate a culture of teamwork.

Although CRM is an accepted and valued aspect of aviation education, it has its limitations. Some professionals reject the notion of CRM and refer to it as 'psychobabble' or 'charm school'.[3] This lack of acceptance may have been related to the wide variability in quality and content of early CRM training. Initially, airlines did not necessarily agree on what topic areas should be included and how the training should be delivered, which resulted in some low-quality courses.

In modern aviation, the NOTECHS (non-technical skills) framework is commonly used as a standardized structure for CRM training and assessment. There are four NOTECHS categories:[4]

1. Cooperation

   - team building
   - considering others
   - supporting others
   - solving conflict

2. Leadership and Managerial Skills

   - use of authority and assertiveness
   - providing and maintaining standards
   - planning and coordination
   - workload management

3. Situational Awareness

   - awareness of aircraft systems
   - awareness of time
   - problem definition and diagnosis

4. Decision-Making

   - option generation
   - risk assessment and option selection
   - outcome review

Did You Know?

Technical skills (also called *hard skills*) refer to the mental and physical attributes required to operate professionally. These skills can be assessed through traditional written exams and on-the-job assessments (such as in-aircraft flight tests). Non-technical skills (or *soft skills*) relate to human factors issues such as those included within the NOTECHS framework.

One challenge with modern classroom-based CRM training is that it can be very theoretical and seem inapplicable to the real world. Professionals in the classroom may understand the concepts, but be unable to apply them in their professional environment.

Figure 9.1 Evolution of pilot human factors training

To allow pilots to practise encountering in-flight challenges and applying CRM skills, *line-oriented flight training* (LOFT) was developed. As part of LOFT, pilots are placed in a flight simulator and work through scenario-based instruction that incorporates a variety of abnormal or emergency situations that might be encountered on the line (i.e., during the course of their work). The goal of LOFT is to make CRM concepts relevant to the operational flight environment.

However, even with scenario-based LOFT, human factors training can seem artificial. It can come across as overly simplistic or irrelevant to the real-world challenges that are faced by particular companies. Many LOFT scenarios are purchased from training companies and so, as an extreme example, an airline that operates in the heat of the desert, might end up using a LOFT scenario that relates to ice on the runway.

To align human-factors training with the specific operational challenges of a particular company, training can be based on a *line operations safety audit* (LOSA).[5] A LOSA involves expert observers sitting in cockpits during normal operations and making notes of any threats that crew must manage as part of their operations and any errors they make. This data is compiled and analysed to precisely identify the most critical threats to that company's operations. With the LOSA complete, the company develops *threat and error management* (TEM) training. TEM is a combination of CRM and LOFT training, but is specifically tailored to the operational threats and common errors unique to that company's operations.

Case Study: The Benefits of CRM – United Airlines Flight 232[1]

On 19 July 1989, a McDonnell Douglas DC-10 was en route from Denver, Colorado to Philadelphia, Pennsylvania with 285 passengers and 11 crew members on board. About an hour into the flight, a loud bang was heard as the fan blades of the number 2 (tail-mounted) engine failed, separated, and exploded through the engine housing. Fan blades shot through all three lines of the aircraft's hydraulic system, which supported the flight controls. (The three lines existed as redundancies –

it was thought to be impossible for all three to fail.) Without the support of the aircraft's hydraulic system, the flight controls became incredibly heavy and extremely difficult to manoeuvre. This situation, coupled with the loss of one engine, is considered impossible to survive.

The accident was ultimately attributed to poor maintenance inspections that failed to detect a fatigue crack in the engine's fan blades; however, the remarkable aspect of this accident was how well the crew communicated, collaborated, and worked to manage the catastrophic failure.

After completing the engine shutdown checklist, the crew successfully diagnosed the problem, when they saw that the hydraulic system pressure and quantity was at zero. The first officer (FO) told the Captain he could not control the plane as it began a right descending turn. The Captain took control, reduced the number 1 engine thrust, and rolled the aircraft wings level.

The crew decided to make an emergency landing at the airport in Sioux City, Iowa. The crew effectively communicated their situation to ATC, who relayed crucial information to Sioux City where emergency services were prepared well in advance. The pilots also advised the senior flight attendant of the situation and she prepared the cabin for an emergency landing. The flight attendant returned to tell the pilots that a DC-10 check airman (an experienced pilot) was in the cabin as a passenger and that he had offered his assistance. The Captain immediately invited the check airman to the cockpit.

As a team, the three pilots worked tirelessly. They determined that there was no way to move the primary flight control surfaces (ailerons), but that they could make right turns by moving the throttles to vary engine power of the two still functional engines on the wings. The check airman controlled the throttles while the Captain and FO managed the flight controls. Through a series of right turns, they descended and approached the Sioux Gateway Airport. The pilots collectively managed to descend and direct the seemingly uncontrollable flight to the runway.

The aeroplane touched down just to the left of the centerline. The right wingtip made contact with the runway followed by the right main landing gear. The aircraft skidded off the runway and rolled, cartwheeling, before coming to rest.

Although the aircraft was destroyed by the impact and post-impact fire, rescue and firefighting teams began work immediately, as they had been given advance notice from ATC. In total, although 111 people were tragically killed in the accident, 185 people survived. When this scenario was recreated post-accident in a flight simulator, test crews were not able to manage the flight as successfully as the actual pilots had. This accident stands as an example of the potential for teamwork and crew coordination to overcome an impossible situation.

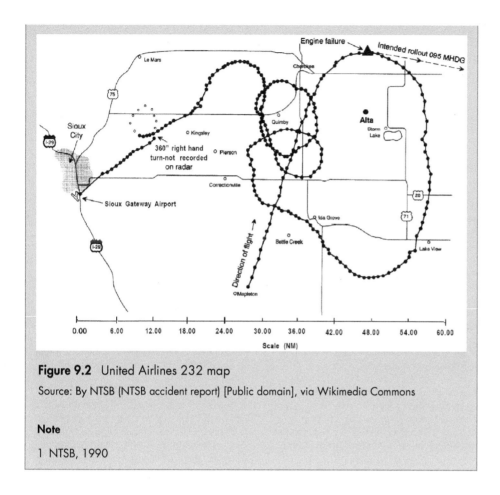

Figure 9.2 United Airlines 232 map

Source: By NTSB (NTSB accident report) [Public domain], via Wikimedia Commons

Note

1 NTSB, 1990

Human Factors Issues

There are, of course, general human factors considerations that apply not just to pilots, but to all aviation personnel:

- Human limitations are natural and predictable. They do not reflect incompetence or lack of effort.

- Often, human factors do not directly cause an accident, but they reduce a profession-al's ability to manage any complications that arise.

- Systems that are used by human operators must be designed to be error-tolerant and user-friendly, and interfaces should be consistent between types.

Those who study human factors have identified several issues of particular importance to aviation operations. These are detailed in Table 9.1.

Table 9.1 Overview of human factors issues

Human factors issue	Description	Application
Culture	Culture impact human behaviour. Aviation professionals are affected by three layers of culture:[1] 1. Organizational – Each company fosters its own culture, setting the expectations for professional behaviours. In aviation companies, the organizational culture can lead to either respect or disrespect for safety and risk avoidance. 2. Industry – The aviation industry itself has a culture, which establishes high-level expectations about what it means to be an aviation professional. The aviation industry has always had a culture of responsibility, punctuality, and pride in the field. However, there are also occasional negative aspects of the aviation culture associated with individuality or questioning procedures. 3. National – People develop different shared cultural values and expectations depending on where they were raised. In a multicultural world, there are challenges in working on a team with people whose cultures take different approaches to expectations, teamwork, and gender roles.	Korean Air Cargo 8509[2] – On December 22 1999, a Korean Air Cargo flight was about to take off in a Boeing 747-2B5F aircraft from London Stansted Airport. The inbound flight crew to Stansted noted in the technical log that the Captain's attitude direction indicator (ADI) instrument was unreliable in a roll. The new outbound crew was made up of four people (Captain, FO, flight engineer, and maintenance engineer). The Captain was a respected and experienced pilot (with 8495 hours flying the B747 aircraft, and 13 490 hours total) who had previously served in the Korean Air Force. The FO was new to flying the B747 (with only 195 hours on this aircraft type, and 1406 hours total). When the aircraft took off, the Captain's ADI remained at a wings-level position (i.e., did not indicate a roll). The Captain began a turn to the left, and the flight engineer called three separate times that there was an issue with the bank. The Captain continued the turn to a bank angle close to 90 degrees, until the aircraft struck the ground. There is no explanation for the Captain's lack of response to bank cues.
Crew collaboration and communication	Within the aviation industry, many professions use teams to extend human capabilities. For example, a Captain and FO working together should be more capable than a single pilot, which is why certain aircraft types require two pilots. Occasionally, issues can arise where crew members do not effectively work together. This can be caused by a steep authority gradient (a strict Captain and a timid FO), poor collaboration (where a Captain wants to handle all flight tasks and the FO does very little), or poor communication (where crew members do not discuss their thoughts or actions, making it impossible to collaborate to accomplish a task).	The FO was making radio calls and supporting the Captain, including scanning his instruments and informing the Captain of any anomalies. Analysis of the CVR revealed that during the taxi before take-off, some of the Captain's remarks to the FO were condescending, such as 'MAKE SURE YOU UNDERSTAND WHAT GROUND CONTROL IS SAYING, BEFORE YOU SPEAK.'[3] A contributing factor to the accident may have been that the FO was intimidated by the Captain and that his national cultural was such that he was reluctant to speak up against an authority figure.

Human factors issue	Description	Application
Mental health/stress	Historically, there has been a reluctance among aviation professionals to discuss and/or treat mental health concerns. This problem is made worse by the fact that professional licences require medical certification, which may cause aviators to lie about or conceal mental health concerns, avoid treatment, or self-diagnose and self-treat to avoid losing their medical certificate. Although severe mental illness is rare, one study of civil aviation pilots determined that 23.7 per cent of those with a heavy workload were suspected to have common mental health disorders (such as anxiety or mood disorders).[4] Mental health concerns are second only to cardiovascular issues as the most common reason for aviation professionals losing their medical certificates.[5] To promote safety, the aviation industry must move to support those suffering from mental health issues and devise treatment plans that allow people to be monitored and return to work when possible.	Germanwings 9525[6] – On March 24 2015, an Airbus A320 with 150 people on board experienced a controlled flight into terrain (CFIT) accident, fatally injuring all on board. This accident was the result of a deliberate action by the FO, who had been suffering from severe depressive episodes and feared losing his medical certificate, and the subsequent financial issues (see Chapter 8 case study on page 285). This tragic accident highlights the need for mental health screening as a component of medical certification, as well as treatment and support options for those who are suffering.
Fatigue	Fatigue is a complicated issue linked to reduced physical and mental performance and alertness. Fatigue is associated with a variety of factors, including sleep requirements, continuous hours awake, and circadian rhythms. Sleep is a requirement for humans. People who sleep even one hour less than required demonstrate lower alertness the following day.[7] After a few nights with less-than-adequate sleep, a person accumulates a 'sleep debt', which further impairs alertness and mental performance.[8] This cumulative sleep debt is a risk for aviation professionals who get less than adequate sleep several days in a row because of a trip or duty schedule. Continuous hours awake can also negatively impact performance. A period of 20–25 hours of wakefulness negatively impacts performance to the same extent as a blood alcohol concentration of 0.10 per cent (a level of intoxication deemed unsafe for driving).[9] Lastly, fatigue can result from jet lag –when a person's natural sleep/wake schedule is out of sync with the day/night schedule of his or her current location. Jet lag is common among crews of long flights, who may spend only a few days in a foreign destination before piloting an aircraft back home.	Colgan Air 3407 – On February 12 2009, an accident in New York illustrated how fatigue and illness can be contributing factors in an accident. The FO of the flight lived nearly 4000 kilometres (2500 miles) away from her airline's hub. She had to commute as a passenger on other flights, before beginning her trip as a pilot. Investigation showed that she had interrupted sleep while commuting, and then napped on a couch in the pilot's rest area before Colgan Air flight 3407 began. Although fatigue does not frequently cause accidents, it can be a contributing factor by impairing pilot performance and reducing capacity to manage workload. (See Chapter 3 case study on page XX.)

(Continued)

Table 9.1 (Continued)

Human factors issue	Description	Application
Situational awareness	Situational awareness refers to a person's mental picture of his or her environment. There are three components to situational awareness: 1. *perception* of the relevant environment (e.g., other aircraft or mountains); 2. *understanding* of the situation (how elements of the environment may impact a flight); and 3. *prediction* of how one's actions may impact a flight.[10] Errors can be made on all three levels – a pilot may not perceive another aircraft during visual flight, a controller may misunderstand a position report, or a maintenance engineer may not be able to predict how his or her actions could impact a flight.	Bashkirian 2937 and DHL 611[11] – On July 1 2002, two aircraft were in the cruise portion of their flights over Germany. Bashkirian Airlines flight 2937 (a Tupolev TU154M) was a charter flight headed to Spain with 60 passengers and nine crew members on board. DHL flight 611 was flying cargo in a Boeing 757-200 from Italy to Belgium, with two crew members on board. Both aircraft had a traffic collision avoidance system (TCAS) installed, which broadcasts an alarm to pilots if the system senses another aircraft on a collision course. The purpose of a TCAS is to avoid mid-air collisions.
Decision-making	Aviation professionals are regularly required to make judgement calls. Often decisions must be made under pressure, in uncertain conditions, with team members, and/or with high levels of risk. Aeronautical decision-making (ADM) is a process for teaching professionals to slow down and carefully consider the information they are presented. They are taught to analyse the cues, generate a hypothesis, and then select an appropriate action. Decision-making and situational awareness are intertwined, as people can only choose the best course of action if they are fully aware of their environment. A variety of biases can negatively impact decision-making, such as people's culture and personality and the organizational climate in which they work.	As the system was relatively new, there was some confusion as to whether pilots should follow directions from ATC or their on-board TCAS. In this accident, the DHL crew complied with the direction from TCAS while the Bashkirian crew were dealing with conflicting instructions – ATC instructed them to descend while their TCAS instructed them to climb. The ATCO repeated the instruction to descend and the Bashkirian crew complied. The aircraft collided with one another and broke apart in the skies over Germany. Following the accident, procedures were changed to ensure that pilots always follow the direction of TCAS alerts.
Workload management	A professional's ability to manage a workload is linked to human attention. There is a limit to the number of things that a person can pay attention to at once.	Eastern Air Lines 401[13] – On December 29 1972, a Lockheed L-1011 crashed into the Florida Everglades in the United States. Of the 176 people on board, 99 people were fatally injured. Investigation of the CVR revealed that the crew had become fixated on an indicator light that did not illuminate. The Captain, FO, and flight engineer all failed to notice the aircraft's slow descent towards the ground. This accident illustrates the importance of prioritizing workload (in this case, the safe flight of the aircraft over minor systems malfunctions) and distributing tasks among flight crew.

Human factors issue	Description	Application
	The challenge is that there are often periods of high workload followed by periods of low workload. Too heavy a workload causes professionals to become overwhelmed and potentially miss crucial information. Too light a workload is also a problem as it requires a lot of effort to pay attention to prolonged tasks with few events. Consider security screeners at an airport – the frequency of identifying a problem may be very low, which makes it difficult to maintain focus. This is called the *vigilance decrement*, which is characterized by very poor performance.[12] Aviation professionals are taught to manage their workload through task prioritization. This requires identifying the highest priority task and focusing on that aspect of the situation. Pilots are trained to prioritize in the order of aviate, navigate, communicate, and finally, manage systems.	

1 Helmreich & Merritt, 2001
2 AAIB, 1996
3 AAIB, 1996, p. 33
4 Feijó, et al., 2012, p. 509
5 Bor, 2007
6 BEA, 2016
7 Carskadon & Dement, 1982

8 Rosekind, et al., 1994
9 Lamond & Dawson, 1999, p. 255
10 Endsley, 1995
11 BFU, 2004
12 Grier, et al., 2003
13 NTSB, 1973

The evolution of human factors training has come a long way in educating professionals about their own limitations in an effort to reduce human error. However, there are situations in which human error is the result of a factor beyond the individual's control. Consider how an organization impacts its employees' behaviour. A company might not provide sufficient training, may push employees to work past their duty time limitations, or could use failing, poorly maintained equipment. Focusing only on a single individual's error can be narrow-sighted as it does not consider the organizational context in which a person works.

Organizational Approach to Safety

The aviation industry's approach to safety management used to be reactive. After an accident occurred, changes would be made to improve the safety of the entire industry. Of course, the flaw with this approach was that an accident must occur before changes were made.

The modern approach is for organizations to integrate a *safety management system* (SMS) into their operations. SMSs take an organizational approach to safety, which facilitates continuous hazard identification and risk assessment to proactively predict how accidents may happen, and eliminate the risk before any harm is done. To identify and eliminate hazards within an organization, it is necessary to consider all aspects of operations, from high-level management decisions to the actions of front-line employees.

The public might believe that a safe transportation system is one in which professionals never make mistakes. However, accomplishing this would require complete and absolute control over every person, aircraft, and element of an organization. This is not considered a realistic approach. Instead, ICAO defines safety as:

> the state in which the possibility of harm to persons or of property damage is reduced to, and maintained at or below, an acceptable level through a continuing process of hazard identification and safety risk management.[6]

The organizational approach to safety assumes that hazards are always present within an organization. The goal is to identify those hazards and eliminate their risk before an accident occurs.

The Origins of the Organizational Approach to Safety

The SMS approach to safety required a shift in the culture of aviation, and this change did not happen quickly. There were two key investigations that led the industry to change its focus from the individuals on the front line to the organization as a whole. These investigations were published within the Mahon Report and the Moshansky Report.

The Mahon Report[7]

On November 28 1979, a DC10 aircraft operated by Air New Zealand Limited left Auckland on a scenic passenger flight over Antarctica, intending to land at Christchurch. The sightseeing flight was conducted in clear weather conditions. Tragically, the aircraft flew into Mount Erebus, (a 12 450-foot mountain) in Antarctica at an altitude of 6000 feet. The aircraft was destroyed and all 257 people on board were killed.

The flight crew had been briefed on a track that would take them safely down the centre of McMurdo Sound. However, the flight plan that was provided to the pilots on the day of the flight (and was programmed into their navigation computer) put them on a collision course with Mount Erebus at any altitude below 12 450 feet.

The final report on the accident, published by the AIA in New Zealand, determined pilot error was the cause of the accident. Specifically, it referred to 'the Captain's decision to make a [visual] descent below the specified minimum safety height while north of McMurdo'.[8] Although the report acknowledged that the crew had followed their navigation computer, which had been changed by another employee shortly before the flight to a route that flew directly into a mountain, the report stated that 'no evidence was found to suggest that they [the crew] had been misled by this error in the flight plan'.[9]

A public inquiry, led by the Honourable Peter Mahon from New Zealand, was conducted. Mahon heard evidence over 75 days and took 3083 pages of notes. The resulting Mahon Report rejected the findings of New Zealand's AIA, which had determined that pilot error

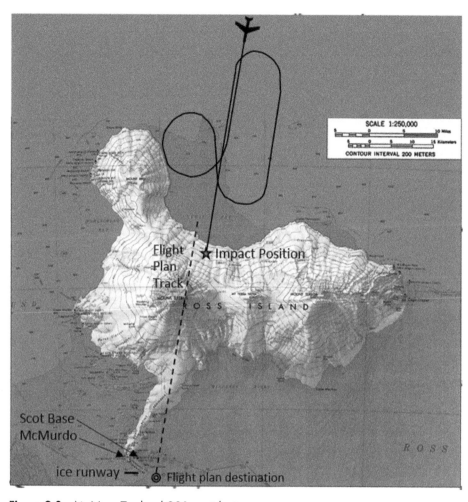

Figure 9.3 Air New Zealand 901 accident map

Source: By ItalianAirForce (Own work) [CC0], via Wikimedia Commons

caused the accident. Mahon acknowledged that the AIA report was written under daunting circumstances and noted that the constraints under which the AIA was working led to a misinterpretation of conclusions by the general public.

The Mahon Report found no fault with the pilots, but concluded:

> the single dominant and effective cause of the disaster was the mistake made by those airline officials who programmed the aircraft to fly directly at Mt. Erebus and omitted to tell the aircrew. That mistake is directly attributable, not so much to the persons who made it, but to the incompetent administrative airline procedures which made the mistake possible.[10]

The accident required the coincidental existence of 10 separate circumstances that combined to make the disaster possible. The chances of the aircraft colliding with the mountain were a million to one.

Airline employees who were interviewed were intent on establishing pilot error as the cause of the accident. At that time, this was a typical response of an airline involved in an accident. The Mahon Report responded that:

> the palpably false sections of evidence which I heard could not have been the result of mistake, or faulty recollection. They originated, I am compelled to say, in a pre-determined plan of deception. They were very clearly part of an attempt to conceal a series of disastrous administrative blunders and so . . . I am forced reluctantly to say that I had to listen to an orchestrated litany of lies.[11]

The Mahon Report was ground-breaking in its allocation of blame to organizational failure. This report was 10 years ahead of its time. It wouldn't be until 1989, when another accident would occur in Canada, that this concept was reiterated within the aviation industry.

The Moshansky Report[12]

On 10 March 1989, Air Ontario flight 1363, in a Fokker F-28, was scheduled to fly from Thunder Bay, Ontario to Winnipeg, Manitoba with a stopover in Dryden, Ontario. The aircraft had 65 passengers and four crew members on board. During the stopover in Dryden, an hour behind schedule, the pilots decided not to have the wings de-iced because their auxiliary power unit (APU) was not working. If they turned off their engines for de-icing, they would have difficulty starting them again.

During take-off, the aircraft failed to gain altitude and crashed into a wooded area about one kilometre (0.62 miles) in front of the runway. There were 21 passengers and three crew members (the Captain, FO, and one flight attendant) fatally injured in the crash and subsequent fire.

Canada's AIA immediately launched an investigation, but it was suspended on March 29 1989, and replaced by a formal Commission of Inquiry to explore the wide-ranging causes and contributing factors that led to the accident. This inquiry was led by the Honourable Virgil P. Moshansky. Although the decision to take off without de-icing could be blamed on the pilots, the Inquiry considered all elements of the aviation system – the pilots, the aircraft, the airport infrastructure, the airline, and the CAA – and how each one contributed to the accident.

Following 166 witness interviews and years of investigation and analysis, the final report made 191 recommendations associated with regulatory reform. It determined:

> the accident at Dryden on 10 March 1989, was not the result of one cause but of a combination of several related factors. Had the system operated effectively, each of the factors might have been identified and corrected before it took on significance. It will be shown that this accident was the result of a failure in the air transportation system.[13]

Figure 9.4 Air ontario memorial sign

Source: By C-FTFC (Own work) [CC BY-SA 4.0 (https://creativecommons.org/licenses/by-sa/4.0)], via Wikimedia Commons

Blaming the entire aviation industry in Canada, the Moshansky Report reinforced the philosophy expressed in the Mahon Report 10 years earlier. Together, they led the international aviation industry to rethink its traditional reactive approach to safety.

Organizational Approach to Human Error

The organizational approach to safety asserts that

- a safe system is not one without risk, but rather one that identifies hazards and works to reduce the associated risk proactively (before an accident occurs); and

- a safe aviation industry cannot be achieved by States or operators working in isolation. It requires a coordinated effort from all stakeholder groups involved with air transportation.

Rather than seeing a pilot's mistake as the singular cause of an accident, this organizational approach considers that mistake just one of a sequence of failures that took place at multiple levels within the organization. James Reason's Swiss cheese model[14] helps to describe the relationship between these levels. Reason proposed that an aviation organization can be thought of as 'layers of Swiss cheese' with each layer representing an element of the company, from front-line employees up to high-level management.

Each layer has holes, which represent weaknesses in the system called *latent failures*. Latent failures can be thought of as accidents waiting to happen – an example would be a wet floor that doesn't immediately cause an accident but has the potential to cause a slip or fall.

Front-line employees can also cause an *active failure*, which is an error that has immediate consequences.

However, as illustrated in Figure 9.5, each layer of an organization represents a safeguard with the potential to prevent an accident. It is only rarely that latent failures in each layer perfectly align with an active failure on the front line – in this case, an accident results.

Building on Reason's work, the human factors analysis and classification system (HFACS) describes human error on four levels: 1) unsafe acts of operators, 2) preconditions for unsafe acts, 3) unsafe supervision, and 4) organizational influences.[15]

1. Unsafe acts of operators include

   • errors (honest mistakes); and

   • violations (intentional rule-breaking).

2. Preconditions for unsafe acts include

   • environmental factors (adverse physical or technological environment);

   • condition of operators (poor mental or physiological state); and

   • personnel factors (crew resource mismanagement and lack of personal readiness).

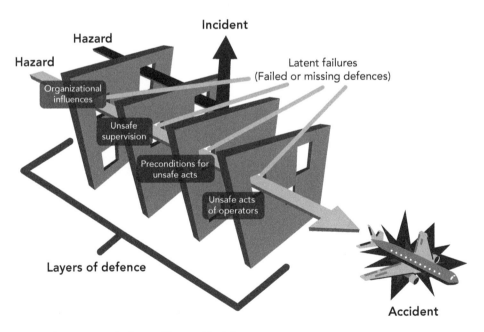

Figure 9.5 Reason's Swiss Cheese Model

Source: Adapted from Reason, 1990; Weigmann & Shappell, 2001

3. Unsafe supervision includes

   • inadequate supervision;

   • planned inappropriate operations;

   • failure to correct problem; and

   • supervisory violations.

4. Organizational influences include

   • resource management;

   • organizational climate; and

   • organizational process.[16]

International Safety Regulations

As the aviation industry shifted towards an organizational approach to safety, ICAO took a leadership role in ensuring uniformity around the globe. In 1997, ICAO launched the Global Aviation Safety Plan (GASP), based on a series of recommendations developed by the Air Navigation Commission (ANC) and industry representatives. The GASP work continued through 2005 when the Industry Safety Strategy Group (ISSG) worked with ICAO to develop a global aviation safety roadmap. This roadmap is the foundation of the global framework for coordinated safety policies, guiding future work of the GASP.[17] Membership in the ISSG includes representatives from Airbus, Boeing, Airports Council International (ACI), Civil Air Navigation Services Organisation (CANSO), International Air Transport Association (IATA), the International Federation of Air Line Pilots' Associations (IFALPA) and the Flight Safety Foundation (FSF).

Did You Know?

There are many aviation industry organizations whose names are written as acronyms (such as ICAO, IATA, ACI, and CANSO among others). With all the acronyms to remember, aviation professionals sometimes refer to these organizations collectively as *alphabet groups*.

The revisions based on the 2005 roadmap were supported by Doc 9859, Safety Management Manual, first published in 2006, which outlines the rationale behind the organizational approach to safety and includes guidance material associated with the development of safety management systems and state safety programmes.[18] The goal of the manual was to harmonize safety activities globally to reduce aviation accidents around the world, particularly where accident rates remain high.[19]

In Doc 9859, a safety management system (SMS) is defined as 'a systematic approach to managing safety, including the necessary organizational structures, accountabilities, policies and procedures'[20] and a state safety programme (SSP) is defined as 'an integrated set of regulations and activities aimed at improving safety'.[21] At that time, safety-related SARPs linked to SSPs and SMSs were incorporated into several existing annexes including Personnel Licensing, Operation of Aircraft, and Air Traffic Services, among others.

In 2010, work began on the development of a new annex to describe safety responsibilities associated with SSPs and SMSs. Annex 19: Safety Management, the first new annex in 30 years, came into force in 2013. On an international basis, it established the organizational approach to safety management, consolidating SSP and SMS standards that were previously distributed through several other annexes. Annex 19 complements and enhances the GASP.

State Safety Programmes[22]

Annex 19 requires States to establish an SSP to promote civil aviation safety domestically. The SSP must include 1) State safety policy and objectives, 2) State safety risk management, 3) State safety assurance, and 4) State safety promotion. As part of its SSP, a State must require domestic service providers to implement an SMS, and the State must maintain oversight. This requirement applies to a variety of service providers, including training organizations, operators of aeroplanes and helicopters, maintenance organizations, aircraft manufacturers, ANSPs, and airport operators.

States must also establish a mandatory incident reporting system to collect information on safety issues and a voluntary incident reporting system to collect data that might not be captured by the mandatory system. Both should be stored in a database to allow for analysis and identification of safety deficiencies.

Voluntary incident reporting systems allow people who have witnessed unsafe practices to report this information without fear of punishment or retaliation from their employer or colleagues (referred to as *non-punitive reporting*). Voluntary reports are de-identified, meaning that reports cannot be traced back to a certain individual or company. This anonymity is important as it makes safety a priority above all else – you can't fix problems that you don't know about. Anonymous non-punitive reports shift the safety culture away from one of blaming and punishing individuals and towards openness and corrective actions. However, the non-punitive aspects of reporting do not excuse illegal or intentional safety violations; these acts are still punishable.

Reporting systems are designed to recognize that even competent professionals occasionally make mistakes, and that these should corrected rather than punished. Examples of international voluntary reporting systems include

- IATA's Safety Trend Evaluation, Analysis and Data Exchange System (STEADES)
  - the largest de-identified database of airline incident reports used to facilitate global analyses

- EUROCONTROL Voluntary ATM Incident Reporting (EVAIR)
  - implemented by EUROCONTROL to collect airline and ANSP voluntary reports to identify safety concerns
- Aviation Safety Reporting System (ASRS)
  - NASA's confidential reporting system that collects aviation safety data from pilots, cabin crew, ATCOs, dispatchers, maintenance personnel, and other aviation professionals

9.3 The Flight Safety Foundation

The Flight Safety Foundation (FSF) is an independent organization that promotes aviation safety around the world. The FSF works to bridge cultural, economic, and political differences within the global aviation community for the common good of safer air travel. FSF membership includes over 1000 organizations and individuals across 150 countries. See flightsafety.org for more details on the FSF.

Safety Management Systems

While an SSP is associated with each State's guidance and oversight of service providers within its borders, a safety management system (SMS) is an internal system, which each organization must develop and implement. The SMS identifies, monitors, and reduces safety risks within each service provider.

It is important to understand that risks are not uniform between companies. The greatest risks within one organization may be linked to utilizing a new type of aircraft, while in another the risk may be linked to high retirement rates and a large percentage of junior staff. An SMS allows each service provider to take a proactive approach to safety, analysing and reducing the risks that have the highest potential to impact its specific operations.

ICAO's *Safety Management Manual* outlines the four pillars of an SMS: 1) safety policy and objectives, 2) safety risk management, 3) safety assurance, and 4) safety promotion.[23] Details on each of these four pillars are set out below.

1. Safety policy and objectives. Service providers must outline the methods they will use to measure and achieve safety objectives.

1.1 *Management commitment and responsibility* – An SMS must be led by senior management within a company. Safety leadership is an issue that must be led from the top down and cannot be delegated to a subordinate.

1.2 *Safety policy statement* – Cultivating a culture of safety within an organization must begin with a clear policy statement reaffirming safety as a core business function that requires continuous improvement.

1.3 *Safety accountabilities* – Service providers must identify a senior level employee as the *accountable executive* with ultimate responsibility for the implementation of the SMS.

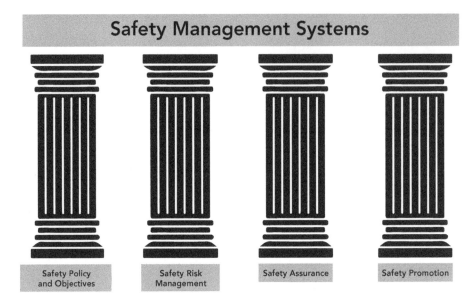

Safety Management Systems

Safety Policy and Objectives | Safety Risk Management | Safety Assurance | Safety Promotion

Figure 9.6 SMS — Four pillars

1.4 *Appointment of key safety personnel* – An organization must hire a safety manager who is responsible for day-to-day implementation and maintenance of the SMS.

1.5 *Coordination of emergency response planning* – A detailed emergency response plan (ERP) must be created that includes coordination with related organization's ERPs.

1.6 *SMS documentation* – The entire SMS process must be thoroughly documented.

2. Safety risk management. This stage can be considered the heart of the SMS. The goal is to proactively identify hazards in order to reduce or eliminate risks before an accident occurs. This is accomplished through hazard identification and risk management.

2.1 *Hazard identification* – A *hazard* is something within a company (e.g., equipment, personnel, procedure) with the potential to cause harm or loss. Hazard identification can be *reactive* (occurring after an unsafe occurrence) or *proactive* and *predictive* (by seeking out safety information and analysing safety trends to determine where accidents may occur). To effectively mitigate risk, hazards must be proactively and predictively identified and documented.

2.2 *Safety risk assessment and mitigation* – After hazards have been identified, a risk analysis must be performed. Risk is the potential *severity* of a hazard multiplied by the *likelihood* that it might occur. Hazards with very high risk must be dealt with immediately while those with lower risk may be monitored and controlled to maintain risk at as low a level as reasonably possible.

3. Safety assurance. To ensure that a service provider's SMS is functioning effectively, the safety assurance process monitors the company to detect deviations that may result in a safety risk.

3.1 *Safety performance monitoring and measurement* – The safety performance of the service operator must be continually assessed and validated to ensure that risks are effectively being controlled. Data sources include mandatory and voluntary incident reporting systems. Mandatory reports are required after serious incidents while voluntary reports, which are confidential and non-punitive, can be submitted by anyone at any time.

3.2 *Change management* – Any change to a service provider's operations must be assessed to determine if it will impact safety risks, and if so, how those potential risks must be managed.

3.3 *Continuous improvement* – The goal of the quality assurance process is continuous improvement of the organization, including the development of corrective actions when system-wide safety deficiencies are identified.

4. Safety promotion. An SMS represents an organizational approach to risk management. Therefore, it requires all employees of a service provider to contribute to the safety of the organization. To accomplish this, employees must be trained in both the organizational approach to safety and the specific requirements of an SMS. Frequent communication about safety initiatives is required to establish a safety culture within the company.

4.1 *Training and education* – The service provider must create and maintain a safety training programme that allows employees to develop competency in completing their SMS duties. The scope and detail of SMS training varies depending on each employee's involvement in the SMS.

4.2 *Safety communication* – The service provider must establish a process for communicating safety information to their employees. Such a process must include ensuring each employee is aware of the SMS, distributing safety-critical information, explaining why and how certain safety actions are being taken, and why safety procedures are in place (or may be changed).

9.4   The Language of Safety

Human factors refers to the scientific study of human limitations, including physiology, psychology, ergonomics, and human–computer interaction. This science has led to several types of pilot training including **crew resource management (CRM), line-oriented flight training (LOFT)**, and **threat and error management (TEM)**, all of which have the same goal: to help professionals recognize their natural human limitations in order to avoid making errors.

A **safety management system** (SMS) is a framework for detecting and eliminating hazards within an organization, including those related to human factors, with the goal of keeping the risk within an organization as low as possible.

Risk is the potential severity of a hazard multiplied by the likelihood that it might occur.

Reason's Swiss cheese model is one way of visualizing the organizational approach to safety, where several levels of an organization are visualized as safeguards with latent failures (accidents waiting to happen) represented by 'holes' in the layers.

Conclusion

Although human error is the primary causal factor of 70 to 80 per cent of aviation accidents, blaming the front-line worker is not an effective approach to corrective action in this industry. Aviation professionals are typically dedicated to their work and committed to the safety of the entire network. However, as human beings, they are subject to natural human factors that can impact performance, including fatigue, difficulty with workload management, and errors in decision-making. To educate professionals about these natural human limitations, the aviation industry has deployed training programmes based on a crew resource management (CRM) approach and a threat and error management (TEM) approach.

Yet even with the most advanced training, not all errors can be prevented. Front-line errors are often preceded by a complex series of shortcomings within an organization, which collectively contribute to an accident. Therefore, in addition to addressing human factors, the industry also requires service providers to maintain safety management systems (SMSs). The intent of an SMS is to assist a company in facilitating a culture of safety, ensuring top-down support for safety, identifying and eliminating hazards, monitoring company-wide safety performance, and training employees in their respective safety roles. Whereas the industry's historic approach to safety management was reactive (i.e., in response to an accident), an SMS approach allows for proactive and predictive safety management.

Within the complex, evolving, and dynamic aviation industry, it could be argued that perfect safety is impossible. However, the aviation industry has continually invested in and deployed the most advanced methodologies to enhance safety. In fact, aviation is a leader in safety – its approaches have been adapted to manage safety in many other industries, including medicine, rail transportation, and nuclear power. Despite the industry's many successes, there is still more to be learned about human factors and safety management in the future. Safety will always be a primary goal within the aviation industry.

Key Points to Remember

1. Between 70 and 80 per cent of aviation accidents have human error as their primary cause, but blaming the individual does not make the industry safer.

2. Over time, the aviation industry has been more successful in reducing the number of accidents caused by mechanical issues than those caused by human error.

3. There are two complementary approaches to understanding why human error contributes to accidents:

   - The human factors approach relates to the scientific study of human limitations that contribute to human error.

   - The safety management approach refers to the process of examining an organization as a whole to determine if company-wide decisions or behaviors are contributing to human error.

4. Human factors training for pilots began after a NASA workshop in 1979 developed crew resource management (CRM) training. CRM evolved to include scenario-based

---

training using simulators, where it is called *line-oriented flight training* (LOFT). If this training is based on a line operation safety audit (i.e., an assessment of the risks within a specific company), it is referred to as *threat and error management* (TEM).

5. Many human factors training courses are based upon the NOTECHS framework, which focuses on developing non-technical skills (or *soft skills*) such as cooperation, leadership and managerial skills, situational awareness, and decision-making. The study of human factors applies not only to pilots but to all aviation professionals, and takes into consideration the skills noted above as well as culture, crew collaboration and communication, mental health/stress, fatigue, and workload management.

6. In general, human factors science suggests that

   - human limitations are natural and predictable, and they don't reflect incompetence or a lack of effort;

   - often, human factors do not directly cause an accident, but may reduce a professional's ability to manage any complications that arise; and

   - systems used by humans must be designed to be error-tolerant and user-friendly, and have consistent interfaces.

7. The organizational approach to safety does not try to create a system without risk. Instead, it assumes risk is always present within organizations and tries to identify and eliminate hazards before an accident occurs. These inherent risks are called *latent failures*. By contrast, an *active failure* refers to a mistake made by a front-line employee. The human factors analysis and classification system (HFACS) describes human error that occurs at four organizational levels: 1) unsafe acts of operators, 2) preconditions for unsafe acts, 3) unsafe supervision, and 4) organizational influences.

8. To facilitate an organizational approach to international aviation safety, ICAO requires States to have a state safety programme (SSP) to promote civil aviation safety domestically. Service providers (such as airlines, ANSPs, and airports, among others) must maintain a safety management systems (SMS). The requirements for SSP and SMSs are set out in Annex 19: Safety Management, and Doc 9859, *Safety Management Manual.*

Table 9.2 Acronym rundown

ACI	Airports Council International
ADI	attitude direction indicator
ADM	aeronautical decision-making

(Continued)

AIA	accident investigation authority
ANC	Air Navigation Commission
ANSP	air navigation service provider
APU	auxiliary power unit
ASRS	Aviation Safety Reporting System
ATC/ATCO	air traffic control / air traffic control officer
CAA	civil aviation authority
CANSO	Civil Air Navigation Services Organization
CFIT	controlled flight into terrain
CRM	crew resource management
CVR	cockpit voice recorder
ERP	emergency response plan
EVAIR	EUROCONTROL Voluntary ATM Incident Reporting
FDR	flight data recorder
FSF	Flight Safety Foundation
FO	first officer
GASP	Global Aviation Safety Plan
GPWS	ground proximity warning system
HFACS	human factors analysis and classification system
IATA	International Air Transport Association
ICAO	International Civil Aviation Organization
IFALPA	International Federation of Air Line Pilots' Associations
ISSG	Industry Safety Strategy Group
LOFT	line-oriented flight training
LOSA	line operations safety audit
NASA	National Aeronautics and Space Administration
NOTECHS	non-technical skills
PF	pilot flying
PNF	pilot not flying
SARPs	standards and recommended practices
SMS	safety management system
SSP	state safety programme
STEADES	IATA Safety Trend Evaluation, Analysis and Data Exchange System
TCAS	traffic collision avoidance system
TEM	threat and error management

Chapter Review Questions

9.1   In your own words, explain the science of human factors. Consider the following:

- What are crew resource management (CRM) and line-oriented flight training (LOFT)? How can they improve aviation safety? What are their limitations?

- What is threat and error management (TEM)? How is it different from CRM? What are its strengths and limitations?

9.2   What is a safety management system (SMS)? How can it save lives?

9.3   Some argue that 'to err is human', meaning that it is human nature to make mistakes. Do you agree with this? Why can it be argued that humans are the riskiest component of aviation?

9.4   Choose an aviation profession, other than pilot, and discuss which three of the following human factors issues have the most potential to cause errors in that role. For these professionals, what strategies can you think of to reduce errors associated with these issues (e.g., vary regulations, add training, new technologies)?

- culture

- crew coordination and communication

- mental health/stress

- fatigue

- situational awareness

- decision-making

- workload management

9.5.   Explain how the Mahon and Moshansky Reports led to improvements in aviation safety. Why were these investigations controversial? Why might aviation companies, at that time, have been unhappy with the findings of these reports?

9.6.   How does Reason's Swiss cheese model apply to SMS and human factors in general? Do some research and provide a real-life example of an accident that occurred in your State. Use the HFACS framework to dissect what went wrong.

9.7   How can manual flight skills be maintained when most flights are automated?

AIR FRANCE 447 – A CASCADING SERIES OF HUMAN ERRORS[1]

In the early morning of 1 June 2009, Air France flight 447 (AF 447) was travelling at a cruising altitude of 35 000 feet above the Atlantic Ocean. The flight was an Airbus A330–203 en route from Rio de Janeiro–Galeão International Airport headed to Paris' Charles de Gaulle Airport. The flight carried 216 passengers, nine cabin crew, and three pilots. As it was an 11-hour flight, the three pilots were scheduled to take turns flying and resting.

The Captain was the most experienced pilot, with nearly 11 000 hours of flight time. The second co-pilot, relief pilot for the Captain, had more than 6500 flying hours. (In this case study, he is referred to as PNF (pilot not flying).) Lastly, the first officer in the right seat was the least experienced with just under 3000 flying hours. (He is referred to as PF (pilot flying) in this case study).

At about 2:00 a.m., AF 447 was over the Atlantic Ocean. The Captain called the second co-pilot (PNF) up from the rest bunk to take over controls so the Captain could rest. The PNF entered the cockpit and was quickly briefed on the flight. The pilots discussed an area of convective turbulence ahead and then the Captain left the cockpit. The first officer in the right seat became the PF, effectively taking the role of Captain, and the second co-pilot in the left seat became the PNF.

Approximately 10 minutes later, the pilots ran into trouble as detailed below:

2:10 a.m.: The autopilot and auto-thrust systems disconnected and the airspeed instruments began displaying unusual readings. The instruments incorrectly displayed a 400-foot descent. The PF said, 'I have the controls', and began manually flying the aircraft. The aeroplane rolled right and the PF, trusting the failed instruments, responded with a nose-up and left input on the controls, causing the aircraft to slow from 275 to 60 knots and the stall warning to sound.

The PNF said, 'we've lost the speeds'. The PF then made rapid high-amplitude control inputs to roll the aircraft left-and-right (the full range of motion of the controls) and a nose-up input that increased the pitch to 11 degrees in 10 seconds.

The pilots performed some checklist items, including turning on wing anti-icing. The PNF said the aircraft was climbing and asked the PF several times to descend. The PF made a slight nose-down input on the controls, but the aircraft continued to climb (from 35 000 feet to more than 37 000 feet).

2:10:36 a.m.: The PNF's airspeed indicator began functioning correctly. The thrust controls were pulled back. The PNF called the Captain to return to the cockpit several times.

The stall warning triggered again and its alarm sounded in the cockpit. The PF continued nose-up control inputs, increasing the aircraft's pitch and reaching an altitude of 38 000 feet and a pitch attitude of 16 degrees. At that point, all airspeed indications were functioning correctly.

2:11:37 a.m.: The PNF said, 'controls to the left' and took over control of the aircraft without any call-out (i.e., the PF did not verbally acknowledge this transfer of control). The PF immediately took back control, also without any call-out, and continued making control inputs.

2:11:42 a.m.: The Captain re-entered the cockpit and sat in the jump seat behind the PF and PNF, who told him that they had lost control. The PNF said he didn't understand what was happening, that they had tried everything. Within a few seconds, all recorded speeds went invalid and the stall warning stopped. The aircraft began falling at 10 000 feet per minute and rolled to the right. The PF made an input fully to the left and nose-up for about 30 seconds.

2:13:32 a.m.: The PF said, 'at level one hundred' meaning they were at an altitude of 10 000 feet. Shortly thereafter, simultaneous inputs by both pilots were made on the controls. The aircraft was quickly losing altitude in a nose-up attitude. The stall warning had been running continuously for three minutes.

The ground proximity warning system (GPWS) began broadcasting 'sink rate' and then 'pull up' to the pilots.

The PNF expressed confusion, saying 'I've got control, haven't I?' before seeming to understand that the PF had been maintaining a nose-up attitude the entire time.

The flight data recorder (FDR) stopped recording at 2:14:28 a.m., with the last readings showing a descent of 10 912 feet per minute and a ground speed of 107 knots on impact with the ocean. The aircraft was destroyed and there were no survivors of the accident. No emergency messages were broadcast by the flight crew.

With the aircraft wreckage on the ocean seabed at a depth of 3900 metres (2.5 miles), the black boxes were not recovered until April 2 2011 (nearly two years after the accident). With the recovery of the FDR and CVR, investigators, investigators began to unravel the mystery of what had happened on board AF 447.

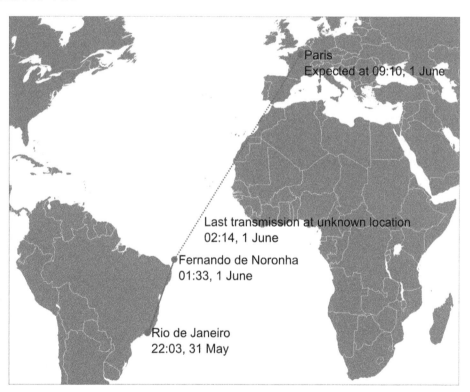

Figure 9.7 Anticipated flight path of air france 447

Source: By Jolly Janner [Public domain], via Wikimedia Commons

Approximately 10 minutes after the Captain left the cockpit, the CVR picked up a loud sound the investigators believed to be ice crystals hitting the aircraft. Unknown to the pilots, the aircraft's pitot tubes (which inform the aircraft's airspeed instruments) had become blocked with ice crystals. This caused the autopilot and auto-thrust systems to disconnect. Incorrect speed information was presented to the pilots. This situation would have permitted continued straight-and-level flight, giving the crew time to troubleshoot the airspeed indicator issue (if incorrect manual flight inputs not been made).

Although all pilots were licensed, trained, and experienced, they were confused by what was happening to their aircraft. Some of their instruments failed while others were functioning correctly. They thought they were flying too fast when they were actually in a stall condition (going too slow).

Investigators determined that the accident was caused by the flight crew's incorrect inputs on the flight controls, following the failure of flight instruments, which destabilized the aircraft. The PNF was late identifying that the PF was making incorrect control inputs. None of the pilots immediately identified or reacted to the stall condition.

This accident illustrates how technical and human factors can combine to cause an accident. The failure of AF 447's instruments, part of a complex and automated modern cockpit, did not cause the accident but it placed the pilots in a confusing situation. The pilots' human limitations led to errors in judgement – perhaps because of stress, fatigue, lack of situational awareness, and a failure to communicate effectively. Yet if the instruments had not failed initially, it is possible that the pilots would have completed the flight without incident.

Note

1 BEA, 2012

Case Study Questions

9.8   With two or more pilots, how might it be possible to communicate and coordinate actions more effectively?

9.9   Complex technical systems can be difficult for humans to diagnose and understand. Consider how challenging it can be to troubleshoot a desktop computer when something goes wrong. Imagine the challenge faced by the AF 447 pilots – they became confused by their instruments and responded inappropriately. How might the human–computer interaction be improved? How can pilots be taught to avoid making the same mistakes in the future?

9.10 Several human limitations impacted the pilots of AF 447. Discuss how the pilots failed to effectively manage their workload, maintain situational awareness, establish leadership of the cockpit, and collaborate and communicate with each other.

9.11 Taking an organizational approach to managing safety, what organizations would you include in your investigation if you were an investigator of the AF 447 accident? Would you include the airline, ANSP, aircraft manufacturer, and/or regulator? Why or why not?

References

AAIB, 1996. *Report on the accident to Boeing 747-2B5F, HL-7451, near London Stansted Airport on 22 December 1999*. London: Air Accidents Investigation Branch.

BEA, 2012. *Accident on 1st June 2009 to the Airbus A330-203 operated by Air France flight AF 447 – Rio de Janeiro – Paris*. Le Bourget: Bureau d'Enquetes et d'Analyses pour la securite de l'aviation civile.

BEA, 2016. *Final report: Accident on 24 March 2015 at Prads-Haute-Bléone (Alpes-de-Haute-Provence, France) to the Airbus A320-211 registered D-AIPX operated by Germanwings*. [Online] Available at: www.bea.aero/uploads/tx_elyextendttnews/BEA2015-0125.en-LR_03.pdf

BFU, 2004. *Investigation report, 1 July 2002, (near) Ueberlingen/Lake of Constance/Germany, Transport Aircraft, Boeing B757-200 and Tupolev TU154M*. Braunschweig: German Federal Bureau of Aircraft Accidents Investigation.

Bor, R., 2007. Psychological factors in airline passenger and crew behaviour: A clinical overview. *Travel Medicine and Infectious Disease,* 5(4), pp. 207–216.

Carskadon, M. A. & Dement, W. C., 1982. Nocturnal determinants of daytime sleepiness. *Sleep,* 5, pp. S73–S81.

CNN.com, 2004. *Swiss air crash controller killed*. [Online] Available at: https://web.archive.org/web/20040226025158/www.cnn.com/2004/WORLD/europe/02/25/swiss.stabbing/index.html

Endsley, M. A., 1995. Toward a theory of situation awareness in dynamic systems. *Human Factors,* 37(1), pp. 32–64.

Feijó, D., Luiz, R. R. & Camara, V. M., 2012. Common mental disorders among civil aviation pilots. *Aviation, Space, and Environmental Medicine,* 83(5), pp. 509–513.

Flin, R. et al., 2003. Development of the NOTECHS system for assessing pilots' CRM skills. *Human Factors and Aerospace Safety,* 3(2), pp. 95–117.

Grier, R. A. et al., 2003. The vigilance decrement reflects limitations in effortful attention, not mindfulness. *Human Factors,* 45(3), pp. 349–359.

Helmreich, R. L. & Merritt, A. C., 2001. *Culture at work in aviation and medicine: National, organizational, and professional influences*. New York: Routledge.

Helmreich, R. L., Merritt, A. C. & Wilhelm, J. A., 1999. The evolution of crew resource management training in commercial aviation. *International Journal of Aviation Psychology,* 9(1), pp. 19–32.

ICAO, 2013a. *Annex 19 to the Convention on International Civil Aviation: Safety management*. Montreal: International Civil Aviation Organization.

ICAO, 2013b. *Safety management manual, Doc 9859, 3rd ed*. Montreal: International Civil Aviation Organization.

ICAO, n.d. *Annex 19 – Safety management*. [Online] Available at: www.icao.int/secretariat/PostalHistory/annex_19_safety_management.htm

Klinect, J., Murray, P., Merritt, A. & Helmreich, R., 2003. *Line operations safety audits (LOSA): Definition and operating characteristics*. Dayton, OH: The Ohio State University, pp. 663–668.

Lamond, N. & Dawson, D., 1999. Quantifying the performance impairment associated with fatigue. *Journal of Sleep Research,* 8(4), pp. 255–262.

Mahon, P. T., 1981. *Royal commission to inquire into and report upon the crash on Mount Erebus, Antarctica, of a DC10 aircraft operated by Air New Zealand Limited*. Wellington: P.D. Hasselberg, Government Printer.

Moshansky, V. P., 1992. *Commission of inquiry into the Air Ontario crash at Dryden, Ontario*. Ottawa: Minister of Supply and Services Canada.

NTSB, 1973. *Aircraft Accident Report: Eastern Air Lines, Inc., L-1011, N3105A, Miami, Florida, December 29, 1972*. Washington, DC: National Transportation Safety Board.

NTSB, 1990. *United Airlines flight 232, McDonnell Douglas DC-10-10, Sioux Gateway Airport, Sioux City, Iowa, July 19, 1989*. Washington, DC: National Transportation Safety Board.

Office of Air Accidents Investigation, 1980. *Aircraft Accident Report No. 79-139 Air New Zealand McDonnell-Douglas DC10-30 ZK-NZP*. Wellington: Ministry of Transport.

Reason, J., 1990. *Human error*. New York: Cambridge University Press.

Rosekind, M. R. et al., 1994. Fatigue in operational settings: Examples from the aviation environment. *Human Factors,* 36(2), pp. 327–338.

Weigmann, D. A. & Shappell, S. A., 2001. *A human error analysis of commercial aviation accidents using the human factors analysis and classification system (HFACS)*. Washington, DC: U.S. Department of Transportation: Federal Aviation Administration.

Notes

1 ICAO, 2013b
2 Helmreich, et al., 1999, p. 1
3 Helmreich, et al., 1999
4 Flin, et al., 2003, p. 100
5 Klinect, et al., 2003
6 ICAO, 2013b, p. 2–1
7 Mahon, P. T., 1981
8 Office of Air Accidents Investigation, 1980, p. 29
9 Office of Air Accidents Investigation, 1980, p. 29
10 Mahon, P. T., 1981, p. 159
11 Mahon, P. T., 1981, p. 150
12 Moshansky, 1992
13 Moshansky, 1992, pp. 5–6
14 Reason, 1990
15 Weigmann & Shappell, 2001
16 Weigmann & Shappell, 2001, p. 4
17 ICAO, n.d.
18 ICAO, 2013b
19 ICAO, n.d.
20 ICAO, 2013b, p. xii
21 ICAO, 2013b, p. xii
22 ICAO, 2013a
23 ICAO, 2013a

1. RPAs flown internationally require a:
   a. Type certificate
   b. Certificate of airworthiness
   c. Certificate of registration with their CAA
   d. All of the above.

2. RPAs have been used within military aviation since World War II.
   a. True
   b. False

5. The term 'drone', describing remotely-piloted aircraft (RPA), is preferred among the vast majority of aviation professionals in this sector.
   a. True
   b. False

3. International aviation regulations have SARPs that govern the domestic use of RPAs.
   a. True
   b. False

4. _____ describes the category of RPA operation within which a pilot must maintain visual contact with their aircraft.
   a. RPS
   b. VLOS
   c. BVLOS
   d. C2

Learning science suggests that thinking through a few questions before you begin studying new material, even if you answer incorrectly, results in improved learning and retention.
Give it a try!

CHAPTER 10

Remotely Piloted Aircraft

Introduction

What the public refers to as a *drone* is referred to by a variety of names within the aviation industry: unmanned aerial vehicle (UAV), unmanned aircraft system (UAS), and remotely piloted aircraft (RPA). ICAO prefers RPA (a term with origins tied to the Vietnam War[1]) and for the sake of consistency, this chapter will describe these aircraft as RPAs.

While various groups may not agree on the preferred name for these aircraft, what everyone does agree on is the fact that they represent a disruptive – and rapidly growing – technology within the traditional aviation system.

10.1 What's in a Name?

There is no international agreement on what to call these devices; however, most RPA professionals dislike the term *drone* and strongly prefer UAV, UAS, or RPA.

The terms *manned* and *unmanned* are commonly used in aviation to distinguish between traditionally piloted aircraft and remotely piloted aircraft. However, some have an aversion to these terms as they are gender-based. For fun, female pilots occasionally share photos on social media of all-female flight crews, joking that these are 'unmanned' flights.

History of Unmanned Aircraft

The history of remotely piloted flight can be traced back to human's earliest experiments with aviation, including a self-propelled flying bird device developed by Archytas (possibly the world's first engineer) around 425 BCE, and the use of kites by the Chinese around 180 CE.[2] In 1783, Joseph-Michel and Jacques-Étienne Montgolfier, French brothers who were paper-makers by trade, achieved lighter-than-air flight and sent up the first aerial passengers (a sheep, duck, and a rooster). Remarkably, this flight, which carried the first passengers of air transport, was also an unmanned flight.

Although far different from the RPA of today, all were aircraft that were remotely piloted.[3]

The general public tends to refer to an RPA as a *drone*, a word that actually means a male honeybee. Unlike the worker bees who collect pollen, the only job of the drone bee is to mate with the queen. Some believe that RPAs are called drones because they 'buzz' like a bee. However, the truth is that in 1935, on a trip to Britain, American Admiral Standley was introduced to the Royal Navy's target drone called the DH 82B Queen Bee. When Standley returned to the United States he wanted to develop something similar, which he called a drone in homage to the Queen Bee. The term stuck[4] and, in fact, *drone* is now being used as a verb as well (i.e., to drone a traditionally piloted aircraft is to convert it to a remotely piloted aircraft).

With respect to RPAs, the military sector contributed far more than just a nickname. The modern history of these aircraft is directly linked to their use in military operations:[5]

- 1849

  - Unmanned lighter-than-air balloons were used as the first aerial bombs during a revolt in Venice against Austrian rule.[6]

- 1916–1920

  - The first experimental military drones were built as *aerial torpedoes* and *flying bombs* to support WWI.

- 1921–1940

  - *Target drones* were built and used for the testing of anti-aircraft weaponry.

  - *Assault drones* were used to carry a bomb into a risky environment without putting a pilot at risk.

  - The Americans experimented with B-17 aircraft, loaded with explosives and flown into targets by radio control.

  - The Germans experimented with piggybacking a drone onto a traditionally piloted aircraft, in order to launch the drone into battle.

- 1941–1950

  - The first unmanned aircraft with mounted cameras were introduced.

  - With the advent of the Cold War, the primary use of drone technology shifted towards reconnaissance. (This is still the case in modern military operations – 90 per cent of drone activities are for gathering information).[7]

- 1951–1970

  - RPAs were used as radar decoys. By fooling radars into believing they were real aircraft, RPAs helped reduce the risk of surface-to-air missiles (SAM) to aircraft with a pilot on board.

- 1971–1990

  - Technology evolved to facilitate long-range high-speed unmanned reconnaissance aircraft, and to recover and reuse RPAs.

  - The first unmanned rotary-wing systems entered service.

- 1991–2000

  - Militaries used RPAs for a variety of missions, replacing traditionally piloted aircraft missions.

  - The goal was to transition to autonomous operations, where RPAs could make decisions on their own. Advancing computing power during this period allowed for initial strides towards autonomy.

On the civilian side, the growth of RPAs faced resistance from pilots who were concerned about their profession becoming obsolete. However, the resistance to RPAs faded following the terrorist attacks of 11 September 2001. Before 9/11, the US army had 30 RPAs and by 2010 they had over 2000.[8] It is generally acknowledged that 2001 was the year that RPA operations began to be considered an industry in itself, and resulted in rapid growth in civilian use.

10.2 The Language of Remotely Piloted Aircraft

Autonomous aircraft are those that can operate without pilot inputs. These can be thought of as self-piloting aircraft.

A **remotely piloted aircraft** (RPA), or **drone**, is an aircraft with no human on board, which is controlled from a **remote pilot station (RPS)** by **a remote pilot-in-command (RPIC)** who manipulates the flight controls. A **command and control link (C2 link)** provides the data connection between the RPS and the RPA.

RPAs are operated by a remote pilot visually watching the aircraft, called **visual line-of-sight (VLOS)**, or through an RPS **beyond visual line-of-sight (BVLOS)**. In the latter type of flying, RPAs must have sophisticated **detect and avoid (DAA)** capabilities to see other aircraft and hazards and ensure collisions do not occur.

RPAs can become airborne in a variety of ways, through **horizontal take-off** and **landing (HTOL)** using a launch platform (similar to an aircraft taking off from a runway) or **vertical take-off and landing (VTOL)** where an RPA becomes airborne vertically (similar to a helicopter).

Small RPAs that incorporate rotary wings (that spin like a helicopter's) can take a variety of configurations and are named depending on the number of rotors: they may referred to as **trirotors/tricopters** (3), **quadrotors/quadcopters** (4), **hexarotors/hexacopters** (6) or **octorotors/octocopters** (8).

The RPA Industry

Many aviation professionals initially dismissed RPAs as fancy remote-controlled (RC) aircraft, with limited use beyond hobby flying (and which presented an occasional annoyance to traditional aircraft pilots). However, as technology advanced, civilian remote pilots began operating devices with operational characteristics (speed and altitude) far beyond traditional hobby craft. For the first time, recreational model-aircraft operators found themselves struggling to abide by regulations, which historically had not applied to their activities.

Regulators were challenged to determine rules associated with how these devices would be piloted, navigated, maintained, and kept secure. Traditional aviation educators begun struggling with how to train RPA operators – were the skills required more aligned with piloting or with air traffic control (ATC)?

Perhaps the biggest challenge with this emerging industry is that aviation is now accessible to the general public, as small RPAs have become available at a wide range of retailers. For the first time in the history of aviation, people with minimal experience or operational understanding can participate in the aviation industry.

A Growing Market

In 2016, the global market for RPAs produced roughly US$9 billion in revenue ($1 billion in the commercial sector and $8 billion in the military sector). This industry is experiencing rapid growth and the annual revenue is expected to grow to $12 billion by 2021.[9]

Looking ahead to 2020, a number of predictions about the RPA market have been made:[10]

- Growth in the market will be driven primarily by recreational sales (expected to quadruple from 2015 sales).

- The civilian sector is anticipated to grow at a rate of 19 per cent annually.

- Public service use of RPAs (e.g., by law enforcement and first responders) is expected to grow.

- The military sector, an early adopter of RPAs, is anticipated to show slower growth moving forward (5 per cent annually).

- Accessability will improve as technology evolves, leading to:

  - the simplification of operations that make RPAs easier to fly;

  - increased affordability as prices drop; and

  - identification of new capabilities.

Growth in the civilan sector is and will continue to be driven by a variety of industries adopting RPAs to support their work (see Figure 10.1). However, the use of these devices is limited by a heavy regulatory burden and even prohibited in some areas around the world. Continued growth in this sector is dependent upon the international community agreeing on regulatory and safety principles that allow commercial RPA operations alongside piloted aircraft.[11]

The manufacturers of RPAs in the emerging commercial sector are distributed internationally. These include

- DJI, China (www.dji.com);

- Aeryon, Canada (www.aeryon.com);

- CybAero, Sweden (www.cybaero.se/en);

- 3D Robotics, United States (www.3dr.com);

- Gryphon Dynamics, Korea (www.gryphondynamics.co.kr); and

- senseFly, based in Switzerland and owned by a French firm (www.sensefly.com).[12]

Certainly, this increasingly popular technology represents a growing market in aviation. Yet, the wide-ranging implications of RPAs on the traditional aviation industry has led to a number of questions and challenges. The remainder of this chapter will explore how RPAs both impact and are impacted by the aspects of aviation discussed in previous chapters of this book.

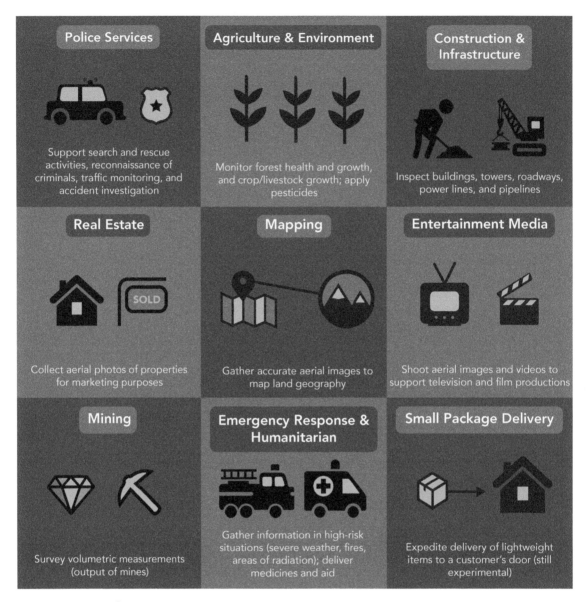

Figure 10.1 Key industries using RPAs

Source: Adapted from Camhi, 2016; Terwilliger, et al., 2017

It should be reiterated that the scope of this book is on *international civil aviation* (and does not explore military nor domestic operations in detail). To provide context for understanding the RPA sector, the discussion that follows will only touch briefly on domestic and military operations, focusing on the international aspects of civil RPA operations. Bear in mind that these international civil operations represent only one portion of a large and rapidly evolving sector.

> ## 10.3 Association for Unmanned Vehicle Systems International
>
> The Association for Unmanned Vehicle Systems International (AUVSI) is a non-profit group for government, industry, and academia. Its mission is to foster, develop, and promote unmanned systems and robotics technologies in both military and civil sectors. They offer student chapters and facilitate student competitions. See www. auvsi.org for details.

International Air Law

RPA technologies can expand the capabilities of the aviation system, and offer a variety of advantages over traditional aviation. However, regulators are faced with the challenge of figuring out how these devices can be safely integrated into the existing aviation network without posing a hazard to piloted flights.[13]

For traditionally piloted aircraft, ICAO has long-standing regulations associated with pilot licensing and medical requirements, aircraft maintenance and certification, radio frequency spectrums, and minimum separation between aircraft, among many others. With the advancement in RPA technologies, regulators must create new rules to accommodate this sector.

The civilian use of RPAs grew out of the recreational model aircraft sector, which has a long history of minimal (to no) regulatory oversight. As RPAs are capable of flight speeds, altitudes, and operations beyond those of model aircraft, civil aviation authorities (CAAs) have moved to create rules for keeping RPA operations safe and separate from other aircraft. This has resulted in variability of regulations between countries (including varying terminology, categorization of RPA types, licensing, and flight rules).

On an international scale, the legal framework for RPA operations stretches back to the Paris Convention, which required pilotless aircraft to have a State's permission before flying over its territory. This concept was amended and included in the Chicago Convention's Article 8, which states:

> *Pilotless aircraft.* No aircraft capable of being flown without a pilot shall be flown without a pilot over the territory of a contracting State without special authorization by that State and in accordance with the terms of such authorization. Each contracting State undertakes to insure [sic] that the flight of such aircraft without a pilot in regions open to civil aircraft shall be so controlled as to obviate danger to civil aircraft.[14]

Whether fully autonomous (operated by artificial intelligence with no human involvement) or controlled remotely from another place, any unmanned aircraft is considered pilotless and this regulation is therefore applicable to it.[15] However, there was little international regulatory work done to build on this premise until the issue was formally considered by the ICAO Assembly in 2005. Subsequent meetings in 2006 determined that a regulatory framework of technical and performance standards would be needed (although only a portion would need to become SARPs). To establish international uniformity, ICAO worked with a joint committee from the European Organisation for Civil Aviation Equipment (EUROCAE) and

RTCA Inc. (formerly know as the Radio Technical Commission for Aeronautics) to develop technical standards.[16]

ICAO's Air Navigation Commission (ANC) approved the establishment of a study group to publish RPA guidance materials, which evolved to become the Remotely Piloted Aircraft Systems Panel (RPASP). The RPASP is the primary coordinator of all ICAO RPAS-related work to ensure global harmonization of regulations.[17] In 2015, they published ICAO's Doc 10019 *Manual on Remotely Piloted Aircraft Systems (RPAS)*, with the intent of standardizing RPA operational practices and terminology with a non-binding guidance document. States use this document to support the development of their domestic regulations, which harmonizes international activities of RPAs while the technology is maturing (with the expectation that SARPs may be needed in the future).

A primary goal of RPA regulation is to protect society from hazards such as mid-air collisions or crashes that damage property or cause injury to persons. As these hazards are not necessarily linked to a specific type of operation, regulations apply to both private and commercial RPA activities.

However, the ICAO guidelines are not all-encompassing. They do not include military aircraft, autonomous unmanned aircraft, or activities associated exclusively with recreational fun flying (as international guidelines for this sector were deemed unnecessary).[18]

Did You Know?

For unmanned aircraft to be widely integrated into the civil aviation industry, they must not negatively impact safety or reduce capacity. ICAO has stated that if this is not possible, it is reasonable to restrict these devices to, for example, visual line-of-sight (VLOS) operations or segregated airspace.[1]

Note

1 ICAO, 2015

Aircraft

The term *aircraft* describes a wide variety of devices capable of flight. If an aircraft is intended to be flown with no pilot on board it is an unmanned aircraft.[19] However, if you look back to Figure 2.3, which lists the ICAO categories of aircraft, you'll see that every category includes aircraft that could be flown unmanned. RPAs vary widely in flight characteristics and size, and include fixed-wing, rotary-wing, and lighter-than-air craft.

In general, unmanned aircraft can be separated into three categories:

1. autonomous aircraft, which fly without any intervention from an operator;

2. model aircraft, which are small unmanned devices primarily used for recreational purposes; and

3. remotely piloted aircraft (RPA).

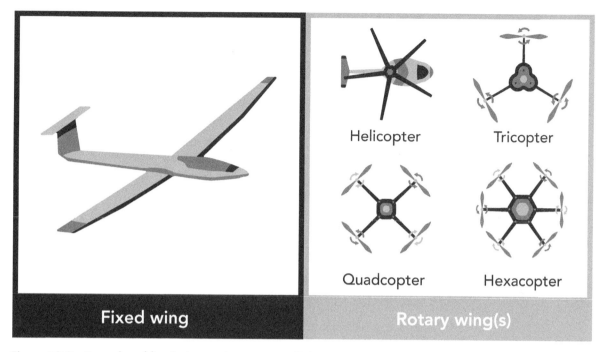

Figure 10.2 Examples of fixed-wing and rotary-wing RPAs

Unmanned aircraft types overlap in two areas: 1) between RPA and model aircraft, where the overlap includes RPAs that are used recreationally and models used for purposes other than recreation (such as professional photography); and 2) between RPA and autonomous aircraft, where the overlap includes RPAs conducting a single autonomous flight segment or an autonomous aircraft conducting an RPA flight segment.

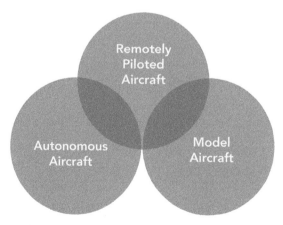

Figure 10.3 Types of unmanned aircraft
Source: Adapted from ICAO, 2015, pp. 1–4

Of these three types of unmanned aircraft, RPAs will have the most significant impact on international civil aviation. The RPA itself is just one component of a remotely piloted aircraft system, as detailed below. The following descriptions are based on ICAO's definitions set out in Doc 10019.

- *Remotely piloted aircraft* (RPA) – an unmanned aircraft piloted from a remote pilot station (RPS).

- *Remote pilot station* (RPS) – the equipment required to pilot an RPA, which can range in complexity from a single handheld device to a station with multiple consoles. An RPS can be permanent/stationary or mobile (such as within a vehicle) and can be either inside or outside. In general, the RPS functions the same way a cockpit of a traditionally piloted aircraft would, and therefore must incorporate equivalent capability to control and manage a flight.[20]

- *Command and control* (C2) *link* – the data uplink and downlink that connects the RPS to the RPA to facilitate flight management. C2 links vary from simple to duplex, radio line-of-sight (RLOS) or beyond radio line-of-sight (BRLOS), the latter which may cause a time delay in communication.

- *Remotely piloted aircraft systems* (RPAS) – the combination of an RPA, RPS, and C2 link, as well as any other components required by the design (e.g., ATC communications and surveillance equipment, navigation tools, launch and recovery equipment, flight control and autopilot, and emergency flight termination systems).[21]

There are a variety of advantages and challenges associated with engineering large-scale aircraft to be remotely piloted.[22] The advantages include

- reduced development costs for accommodation of a human pilot (robotic needs increase but life support, survival, and bathroom facilities are eliminated);

- the fact that crew safety features are unnecessary (e.g., oxygen, seats and restraints, avionics);

- lower costs for small unmanned aircraft than for their full-size counterparts (for flight testing, a scale model remotely piloted version of a traditional aircraft may be developed, as with the F-15 fighter aircraft); and

- operational capabilities that would not be feasible with manned counterparts (for research purposes, a remotely piloted Boeing 720 was deliberately crashed to test the survivability of the passenger aircraft[23]).

On the other hand, the challenges with engineering RPA include

- increased costs for ground support equipment and personnel needs for complex ground control stations;

- increased software costs, which may offset the reduced hardware costs; and

- not-yet-mature technologies necessary for wide-scale use and integration with manned operations (for example, sense and avoid).

The issues noted above relate to larger RPAs (similar in size to traditionally piloted aircraft); however, the largest growth in RPAs is expected in the commercial use of small RPAs that have entirely different operational capabilities than traditionally piloted aircraft.

These smaller unmanned aircraft have two main categories of operations. *Visual line-of-sight* (VLOS) *operations* refer to those in which the remote pilot or a designated observer maintains direct, unaided visual contact with RPA (without binoculars or other assistance). VLOS operations should not occur at night unless methods of mitigating threats have been established. VLOS is used primarily for small model aircraft and some recreational RPAs.

Beyond *visual line-of-sight* (BVLOS) *operations* require a method to detect and avoid (DAA) hazards such as weather, terrain, and obstacles. (Note that in the US, the phrase *sense and avoid* is used rather than *detect and avoid*.) Flight activities must be coordinated with the local ATC unit and be considerate of the class of airspace where operations occur. The C2 link is critically important here, so transaction time must be minimized to avoid any delays. BVLOS operations represent a higher level of complexity and are more common in professional use of RPAS and autonomous aircraft.

Although this organizational framework is helpful, CAAs have moved to further categorize RPA types based on size and weight, which can range dramatically from small RPAs that weigh less than 100 grams (3.5 ounces) to the RQ-4 Global Hawk, which weighs 14 628 kilograms (over 32 000 pounds).[24]

To distinguish between small unmanned aircraft used for recreational purposes and larger devices used for professional activities, the terms *microdrone* or *small unmanned aircraft system* (sUAS) and *macrodrone* or *unmanned aircraft system* (UAS) are sometimes used. This terminology is most common in the United States, as ICAO does not describe RPAs in these terms. However, it is a helpful framework to categorize the large and diverse group of RPAs.

Figure 10.4a RQ-4 Global Hawk (photo)

Source: By Stacey Knott [Public domain], via Wikimedia Commons

Figure 10.4b Small RPA (photo)

Table 10.1 Comparison between microdrones and macrodrones

	sUAS (microdrone)	UAS (macrodrone)
Weight	less than 25 kilograms (55 pounds)	25 kilograms (55 pounds) or more
Type certificate	not required	required
Typical configuration*	battery powered; several propellers; vertical take-off, landing, and hovering capabilities	piston-engine fixed-wing aircraft with traditional propellers, or jet/ducted fan
Primary usage	recreational	transportation or professional services
Origin	evolved from recreational remote-controlled hobby devices	evolved from drones used for military purposes
Regulatory impact	challenging to regulate as their operations differ from traditional aviation operations (close to the ground, buildings, airports, and people)	fits into the regulatory framework that exists for traditionally piloted aircraft
Typical operators	novices with limited background, training, or understanding of the aviation industry	experienced aviation professionals with substantial understanding of the aviation industry
Purchase decision	chosen because of unique capabilities not possible with a traditionally piloted aircraft	chosen through comparison to a similar traditionally piloted airplane or helicopter
Operations*	Can take off and land from anywhere. Regulatory challenges include issues of trespassing, privacy, injury to persons or damage to property, noise, and disruptions to traditionally piloted aircraft and airports. Battery power allows flights from 15 to 30 minutes. Can reach heights of several thousand feet. Command and control uses frequency-hopping spread-spectrum modulation schemes on 2.4 GHz band and 5.7 GHz for video. Almost all have automatic return-to-home capability to return to launch point and land automatically.	Usually use aerodromes for take-off and landing. Operational characteristics vary significantly between types, yet are comparable to traditionally piloted aircraft. Controlled through a sophisticated RPS with advanced C2 link. The challenge is to ensure safe separation from traditionally piloted aircraft.

* Note that these descriptions refer to the most common types; however, variations exist and will continue to evolve in the future.
Source: Adapted from Perritt & Sprague, 2017

The reason for differentiating between sUAS and UAS is that increased regulatory oversight is required as unmanned aircraft become larger and when their use switches from recreational to professional. Table 10.1 sets out some of differences between the two types

In addition to the sUAS/UAS classification popularized in the United States, various other domestic frameworks are in place around the world. For example:

- The Civil Aviation Safety Authority (CASA) in Australia classifies devices as *large* (more than 150 kilograms (330 pounds) fixed-wing, more than 100 kilograms (220 pounds) rotary-wing), *small* (between 100/150 kilograms and 100 grams (between 330/220 pounds and 3.5 ounces)), and *micro* (less than 100 grams (3.5 ounces)). CASA requires large devices to have an airworthiness certificates, exempts small devices from airworthiness certification, and does not regulate micro UAVs at all.[25]

- China requires operators of devices that weigh more than 7 kilograms (15.4 pounds) to have a licence (equivalent to a pilot's licence), and those over 116 kilograms (255 pounds) require certification for both the RPA and the operator.[26]

- In Europe, the European Aviation Safety Agency (EASA) is responsible for regulating drones over 150 kilograms (330 pounds) while individual CAAs regulate lighter craft (although this may evolve in the future).[27]

- In 2014, India's CAA banned the use of drones for any purpose whatsoever until regulations for certification and operation were established.

Did You Know?

Just as a traditional aircraft has a pilot-in-command, an unmanned aircraft has a remote pilot-in-command (RPIC) whose responsibility it is to ensure that the aircraft is operated in compliance with the rules of the air.

Certification, Registration, and Maintenance

As discussed in Chapter 2, aircraft flown by on-board pilots require type certificates and airworthiness certificates, and must be registered with their CAA. While unmanned aircraft operated domestically would follow the rules of their CAA, RPAs flown internationally must abide by the following guidelines:

- *Type Certificates* – RPAs must be issued a type certificate (TC) if they are to be operated internationally. This certification would cover all the requirements for traditionally piloted aircraft (such as structures, materials, systems, propulsion and fuel, flight testing) as well as the C2 link and RPS. Instructions for continuing airworthiness (ICA) – such as maintenance schedules – and an operational flight manual must also be included in the type design approval.[28]

- *Certificate of Airworthiness* – Any aircraft that operates internationally requires a valid certificate of airworthiness (CofA). CofAs are issued by the State of Registry, though maintaining airworthiness requires cooperation between the States of Design, Manufacture, Registry, and the Operator. Any RPA flown internationally requires a CofA that demonstrates it is safe for the planned operations. As with traditionally piloted aircraft, if an RPA is damaged or not maintained, the CofA becomes invalid. If a problem is identified on an RPA type, an airworthiness directive can be issued to ensure all operators of that type fix the issue and maintain safe operations.[29] In addition, RPAs must posses the equipment necessary to meet the requirements of the airspace in which it operates (for example, VFR or IFR).[30]

- *Registration* – RPAs must be registered with their CAA. However, only the RPA component of the RPAS requires registration. A remote pilot station (RPS) does not require independent registration despite the fact that, in most cases, the RPS and RPA components are integral to one another.

- *Maintenance* – RPAS operators must maintain a maintenance control manual (MCM) that describes the maintenance programme, which includes detailed maintenance procedures including procedures for completing and signing maintenance releases, approved by their State of Registry. RPAS operators must retain detailed maintenance records for at least one year post-maintenance release that includes total time in service, compliance with airworthiness directives, modifications/repairs, time since overhaul, and compliance with maintenance programme.[31]

Figure 10.5 RPA pilots in a remote pilot station (photo)

Source: By SSGT REYNALDO RAMON, USAF [Public domain], via Wikimedia Commons

RPAS Operations

RPA operations vary based on whether they are conducted domestically or internationally.

Domestic Operations

Domestically, as RPA technology matures, it is unlikely that all piloted flights will become obsolete. However, RPAs may begin to dominate some sectors within general aviation. For example, piloted rotary-wing aircraft are more expensive to operate and are slower than their fixed-wing counterparts. RPAs, as they are more affordable and convenient, may increasingly be used for tasks previously conducted by helicopter (see Figure 10.1). RPAs also have the advantage of being able to fly into hazardous situations too risky for humans (such as bad weather or hostile/violent conditions).

Another broad market for RPAs are individual hobbyists, people who want to fly exclusively for recreational purposes. Although there has been a long tradition of enthusiasts flying model aircraft (also called remote-controlled (RC) aircraft), this type of flying was generally conducted in a group setting and in cooperation with local airports and ATC. The recreational use of RPAs is more challenging for regulatory bodies as it is typically an individual activity, often by people with little understanding of aviation, and there have been instances where these flights have posed a hazard to traditionally piloted aircraft and airport operations.

International Operations

In terms of regulation, the type of RPAs that are most impactful are the larger devices capable of crossing international borders. The RPA, RPS, and C2 link associated with these operations all present operational and safety challenges that require globally harmonized regulations.

For traditionally piloted aircraft, an operation is considered international if a flight crosses international borders. However, for unmanned aircraft the situation is more complicated. For example, an RPA may be operating in one State but piloted from another, an RPS may be situated on a ship in international waters, or an RPA may move into and operate in a State that is not the State of the Operator. These situations create a variety of challenges. Annex 6 distinguishes between commercial air transport and general aviation, but this distinction is not relevant for RPAS operations. Instead, regulatory distinctions for international operations must be based on the size and complexity of each operation, as a traditional understanding of aircraft operations does not apply to RPAS.[32] A variety of operational issues, along with guidance from ICAO's Doc 10019 are outlined in Table 10.2.

Table 10.2 International RPAS operations and licensing

Operational considerations

Visual and instrument flight rules (VFR and IFR)	These flight rules present some challenges for RPA operations and capabilities. Consider that a RPA flying IFR in visual conditions may encounter VFR traffic (which may have the right of way) – the remote pilot must be able to identify VFR traffic and respond appropriately. Likewise, an RPA flying VFR must be capable of complying with visibility minima, recognizing traffic, and determining if another aircraft has the right of way and taking appropriate action.
Airspace	RPAs must comply with the airspace requirements and may be prohibited from operating in certain areas (such as overpopulated areas) as determined by the CAA.
Visual line-of-sight (VLOS)	During VLOS, the remote pilot must maintain unaided visual contact with the RPA. If visual contact cannot be maintained, the RPA is beyond visual line-of-sight (BVLOS) and the minimum equipment requirements increase significantly to ensure safety.
Nationality and registration marks	As some RPAs are too small to accommodate markings, the State of Registry shall determine the location and measurement of nationality and registration marks on a case-by-case basis.
Operator responsibilities	Operators are responsible for complying with all requirements of the State of the Operator. They must ensure that employees know the laws and regulations associated with their duties. Similar to the sterile cockpit rule established for on-board pilots (see Chapter 3), RPAS flight crew may perform only flight-related activities during critical phases of the flight to avoid distraction and ensure safety.
RPAS operator certificate (ROC)	Comparable to an air operator certificate (AOC), an ROC is required for RPAS operators. An ROC grants the operator the authority to conduct operations as specified by the ROC. To receive an ROC, an operator must demonstrate organization, control and supervision, training, ground handling, and maintenance appropriate to the size and complexity of its organization.
Documents	The Chicago Convention requires aircraft with pilots to carry documentation on board for international travel. The size and configuration of RPAS may make this impractical, so other approaches are required to make documents available to remote pilots, maintenance, and inspectors.

RPAS pilot licensing (not yet in effect)

Licensing requirements	Requirements are expected to be the same as for traditional pilots, regarding rules of the air, RPAS general knowledge, regulations, flight and human performance, navigation, meteorology, operations, principles of flight, and radiotelephony.Licensing may involve the accumulation of experience (operational hours), a written examination, and a hands-on skills test.
Categories of pilot license	While there are several categories of pilot licence (e.g., private, commercial, airline transport) for traditionally piloted aircraft, a single remote pilot licence is anticipated. The licence will grant the privilege of acting as an RPIC for the RPAS category and type indicated (or as an RPA observer).Licensing regulations will include a process for adding ratings and endorsements for different types of RPAS.A student remote pilot permit will be issued (allowing the holder to operate under supervision of an instructor).
Application	Licensing will apply only to RPAS, not to model aircraft used exclusively for recreation (although recreational users will still need to comply with domestic regulations).

RPAS pilot licensing (not yet in effect)	
Issuing authority	Licences for remote pilots will be issued and validated by the CAA in the State where the RPS is located.
Minimum age	18
Medical requirements	Class 3 medical assessment (as is currently required for ATCOs).
ATCO licensing	
ATCO license	It is not expected that ATC licensing will be impacted by RPAS, although additional training on RPA operations should be required when these devices are allowed within controlled airspace.

Source: Table created with information from ICAO, 2015

Navigation

The existing global navigation system, constantly evolving to increase capacity and the efficiency of travel, is based on a robust network of regulations and technologies. RPAS operations have had a disruptive effect on this system. When a new type of user is added to existing airspace operations, the first challenge is to minimize risk to all users.

Annex 2 of the Chicago Convention specifies RPA right-of-way rules, specifying that remote pilots must ensure the RPA avoids passing over, under, or in front of other aircraft unless well clear and avoids wake turbulence.[33]

RPAS must comply with existing air traffic management procedures of airspace. To accomplish this, remote pilots must plan their flights considering airspace organization, density of traffic, and how piloted flights approach and depart airports (to avoid disruption). To ensure the safety of aircraft with pilots on board, ATCOs may specify a section of airspace where RPAs may operate with the instruction to 'remain well clear' (RWC). At all times, the priority is collision avoidance (CA) between aircraft.

RPA operations occur in both controlled and uncontrolled airspace. To enter controlled airspace, RPAs require sophisticated equipment to detect and avoid (DAA) other aircraft and communication technologies (similar to those required for traditionally piloted aircraft).[34] An RPA operating in uncontrolled airspace must be designed to be detectable and conspicuous, so that it can be identified by pilots of manned aircraft, other remote pilots, and ATCOs. This is a significant challenge, as RPAs are often very small. To enhance detection, RPAs may incorporate a transponder, strobe or anti-collision light, or ADS-B.[35] (See Chapter 4 for discussion of ADS-B.)

The C2 link is a crucial component for RPA integration within the global navigation system. The C2 link includes both an uplink to control the RPA, and a downlink to gather sensor data from the RPA about traffic, weather, and visual information. The C2 link should also transmit information to the remote pilot about the 'health' of the RPA (i.e., alert the operator to any system malfunctions) and relay ATC voice and data communications between the RPA and RPS.

Configuration and protection of the C2 link is crucial, as the loss of it results in a critical situation for an RPA with a high likelihood of an accident. The C2 link can be lost for many reasons including ground clutter or terrain, meteorological conditions, equipment failure, human error, or unintentional interference (such as a television broadcast). A lost C2 link procedure must be established for use of a new emergency transponder code (i.e., one that is not used by traditional pilots for emergencies) to alert others to the lost C2 link.[36]

Did You Know?

In some RPA operations, about 10 per cent of flights experience some loss of the C2 link (although it is usually re-established within a few seconds). The frequency with which this issue occurs has resulted in many devices being programmed to automatically fly to a lost C2 link *loiter point*. This is a predetermined location where the RPA can fly safely in a small pattern until the link is re-established.[1] Other RPAs automatically return to home if the C2 link is lost.

Note

1 Neubauer, et al., 2015, p. 16

Of course, the safe integration of RPAs also requires additional training for ATCOs. Controllers working sectors with RPA activities should become familiar with the characteristics of these aircraft (speed, manoeuvring capabilities, wake turbulence, endurance, and impact of bank angle on C2 and communications link). Additionally, RPAS will require different air traffic management procedures because of the lack of an on-board pilot. These procedures may include unique RPA type designators for flight planning, new standard phraseology, separation standards and right-of-way rules, and emergency procedures. Many of these issues are still being sorted out. For example, until RPAS aircraft type designators are formalized, the code ZZZZ is used in flight plans to indicate an aircraft is remotely piloted.[37]

Airports and Operating Sites

Unlike traditional aircraft that operate predictably from airport to airport, RPAs may be operated from almost any location (depending on the RPA's configuration and launch requirements), including established aerodromes. The reality is that each airport is unique, as are the capabilities of different types of RPAs, so their operations out of airports needs to be assessed on a case-by-case basis in cooperation with airport operators.

When an RPA is operated out of an aerodrome, the RPIC must manage wide-ranging issues that impact the flight; he or she must consider regulations, density of aircraft operations, ground operations, C2 link, payload, wake turbulence of other aircraft, take-off and landing performance, infrastructure requirements, and availability of emergency recovery areas.

For RPAS operations to be conducted at an airport, these activites should generate revenue so as not to strip resources from the airport. There are several potential revenue streams for airports hosting RPAS operations, such as

- rental fees for space (hangars, ramp, office, communication centre);

- fuel (depending on type of RPA);

- landing fees and tie-down fees; and

- additional support services.[38]

Integration of RPAs in Victorville, California

Southern California Logistics Airport in Victorville, California hosts remotely piloted aircraft (MQ-1 Predators) flown by the California Air National Guard. Airport operations integrate RPA and traditionally piloted aircraft, both controlled by the tower.
    RPAS operators and ATC collaborated to create procedures for safe operations at the airport, including the segregation of traditionally piloted and remotely piloted operations. When an RPA will be taking off, a NOTAM (notice to airmen) is issued and all other aircraft movements stop until the RPA is airborne. When landing, the RPA will stay at a predetermined holding point until all aircraft are clear, at which point it will approach the airport and land.[1]

Note

1 Neubauer, et al., 2015

Each State's CAA must determine whether RPAs can be safety integrated into aerodrome operations without sacrificing the safety or capacity of piloted flights. Regulators will have to consider

- whether aerodrome signs can be detected remotely;

- how RPAs can detect and avoid collisions while manoeuvring;

- whether an RPA observer will be required to monitor RPAs and ensure conflict avoidance; and

- if new airport infrastructure will be required (landing aids, launch/recovery aids or firefighting equipment).[39]

Where the infrastructure requirements would exceed an aerodrome's capabilities, one possible solution is for States to establish dedicated aerodromes for RPAS operations or for these activities to occur at launch sites not associated with established airports.[40]

For RPA launches at locations other than aerodromes, a different set of considerations are important. In these cases, the RPIC must consider the condition of the take-off area (e.g., Is it muddy? Damp? Is there sufficient clearance?), height of obstructions along the take-off and landing path, emergency recovery area, whether ATC communications are required, C2 link, payload, and potential overflight traffic.[41]

Did You Know?

Some airports use small RPAs to scare birds away from flight paths, in an effort to avoid bird strikes with traditionally piloted aircraft.

Other Considerations

Security

For traditionally piloted aircraft, security is an important consideration that involves safeguards both on the ground and in the air. For example, international SARPs require cockpit doors to be reinforced and secured to prevent unlawful access to the flight controls. However, the 'cockpits' of RPAS are within remote pilot stations (RPS) on the ground. These stations, therefore, may have a higher exposure to security threats as it can be challenging to secure RPSs – they aren't protected by the sophisticated security screening technologies of airports and are often designed to be mobile.

RPAS professionals must carefully consider how all elements of their systems can be secured, including the RPA, RPS, and C2 link. The RPA must be stored in a way that ensures security and prevents unlawful access or tampering with the aircraft. Access to RPS must be controlled through regulations equal to those in place to protect the commercial aviation industry. This can be accomplished with innovative identification technologies, such as biometrics (fingerprint or retina scanners) to ensure that only qualified professionals can access the station to control the RPA. Like traditional pilots, remote pilots should be subject to minimum background security checks. Lastly, C2 links must be protected from hacking, spoofing, and malicious hijacking.[42]

Weather

Hazardous weather conditions pose a threat to RPAs, just as they do to other aircraft. Icing, wind shear, and turbulence can disrupt RPAS operations and lead to loss of the RPA itself. Although VLOS operations will occur only in visual weather conditions, BVLOS flights require RPAs to incorporate sensors that gauge weather conditions (such as icing detection).[43]

Environment

There are no emissions concerns unique to RPAS, as they use the same types of fuel as piloted flights. It is not anticipated that future RPAs will evolve in a manner that will negatively impact the environment. In fact, the opposite may be true. As RPAs do not carry people, the lighter weight results in less fuel burnt per trip. In addition, researchers are exploring new, more environmentally friendly power sources including solar, hydrogen fuel cells, and biofuels for RPAs.[44]

Accidents

Accident rates for remotely piloted aircraft have been high, which is not entirely surprising as the emerging field has not had the opportunity to benefit from the lessons learned through the history of traditionally piloted aircraft. When RPAS accidents occur, they are often related to failures of hardware, lack of sufficient training, and issues with the human–machine interface.[45]

Annex 13: Aircraft Accident and Incident Investigation requires accident and incidents involving unmanned aircraft to be investigated. However, only RPAs with a type certificate or those operated with a remote operator's certificate are covered by this Standard (i.e., incidents involving small recreational RPAs need not be investigated). As with traditionally piloted aircraft, the State of Occurrence is responsible for investigating RPAS accidents and incidents. In the future, it is expected that ICAO will require RPAs operated by BVLOS to install flight data recorders to assist in investigations.[46]

Human Factors

Remote pilots must manage on-board systems, draw information from (often) multiple displays, and make control inputs just as traditional pilots do, yet they lack much of the sensory information that often guides pilots in these actions.[47] Remote pilots can't feel their aircraft's movement, turn their head to check their wings for icing, smell smoke, or hear changes in engine sounds. This data needs to be delivered to the remote pilot through technologies, such as cameras that support peripheral vision, auditory alarms signalling engine failure or smoke, and instruments that continually monitor and report on the health of the RPAS.[48]

Another challenge is the lack of standardization of remote pilot control interfaces – RPSs may vary significantly in their configuration and control input design. For piloted flight, engineers (taking into account human factors principles) standardized a variety of cockpit control and avionic configurations to make it easier for pilots to transition between aircraft during World War II. RPAS do not have this long history to fall back on. It is important that, as RPSs evolve, consideration is given to standardization between systems to support human performance.[49]

Safety Management Systems

To keep their ROC, operators must maintain and document their safety management system (SMS), the requirements for which are based on provisions for traditional operators in Annex 19: Safety Management. At a minimum, the SMS must include a process of identifying hazards that impact operations and methods of proactively managing risk before an accident occurs.[50]

Figure 10.6 An RPA operated VLOS (photo)

Source: By U.S. Navy photo by Mass Communication Specialist 2nd Class Nancy C. diBenedetto, Public Domain, via Wikimedia Commons

One element of risk that might not be immediately evident is fatigue management. Although remote pilots are not physically on board their aircraft, they are subject to fatigue over the course of long or challenging shifts, just as traditional pilots would be. Therefore, remote flight- and duty-time limitations must be in place to minimize the risk of fatigue.[51]

Conclusion

The RPAS industry is still in its infancy. Many regulatory challenges and technological issues must be overcome before these systems can be fully integrated into the traditional aviation system. However, both regulators and industry professionals are working to craft solutions to the many challenges. It is anticipated that in the coming years remotely piloted aircraft will represent a larger segment of international aviation.

The evolution of these devices will be aligned with advances in a variety of parallel technologies, including

- computing power;
- aerial robotics;
- microminiaturization (and handheld electronics);
- materials and manufacturing; and
- power storage.

As these technologies continue to evolve, so will the capabilities of RPAs. It is unlikely that RPAs will ever fully replace aircraft with on-board pilots. Instead we can expect them to expand the range of aviation activities. For example, it is feasible that future RPAS technology might allow airline flights with a single pilot (with emergency support available as needed through a C2 link to a RPS).

At the time of writing this book, international SARPs for RPAS do not yet exist (although ICAO guidance material is available). A variety of approaches to licensing, certifying, and controlling RPAs are in place through CAAs around the world. Although traditional recreational users of model aircraft may be resistant to regulatory oversight, this attitude must evolve as the capabilities of RPAs increase. Whether RPAs are large or small, flown VLOS or BVLOS, the international aviation community will eventually require standardized international regulations to ensure that operations are conducted safely. The ongoing challenge is to capitalize on the opportunities presented without presenting new risks to traditional aviation activities.

Key Points to Remember

1. Although the general public tends to use the term *drone*, aviation professionals call these devices *unmanned aerial vehicles* (UAVs), *unmanned aerial systems* (UASs), or *remotely piloted aircraft* (RPA), the latter being ICAO's preferred term.

2. RPAs have been used, and have evolved significantly, in military applications as far back as WWII. RPAs have been used as aerial torpedoes, target drones, assault drones, radar decoys, and for reconnaissance purposes.

3. The RPA industry is a growing market, with sales in the civilian recreational sector predicted to quadruple between the years 2015 and 2020. Advances in technology will drive this growth, making RPAs easier to fly and more affordable.

4. ICAO published Doc 10019, *Manual on Remotely Piloted Aircraft Systems* in 2015 with guidance material to help CAAs safely regulate RPAs. International Standards for RPAs have not yet been developed, but are expected in the future.

5. There are three categories of unmanned aircraft: 1) autonomous aircraft that fly independent of a human operator, 2) small model aircraft used for recreation, and 3) remotely piloted aircraft (RPA). An RPA can be operated through visual line-of-sight (VLOS), meaning the pilot maintains unaided visual contact with the RPA, or beyond visual line-of-sight (BVLOS), meaning the pilot cannot see the aircraft. For BVLOS, methods are required to control and communicate with the RPA, and to detect and avoid traffic (among other systems).

6. RPAs are expected to have the most significant impact on international civil aviation. A remotely piloted aircraft system (RPAS) is made up of three components:

   - *remotely piloted aircraft* (RPA) – an unmanned aircraft piloted from a remote pilot station;

- *remote pilot station* (RPS) – the equipment required to pilot a RPA; the RPS functions the same way a cockpit of a traditional aircraft would; and

- *command and control* (C2) *link* – the data link that connects the RPS to the RPA to facilitate flight management.

7. RPAs range significantly in size and capability. Smaller RPAs are often preferred for recreational purposes or commercial activities such as photography, police surveillance, or forestry management. Larger RPAs may have many of the same operational capabilities as traditionally piloted aircraft, along with certain unique characteristics.

8. RPAs flown internationally require a type certificate, a certificate of airworthiness, and an adequate maintenance programme, and must be registered with a CAA. International operations for professional purposes must abide by international aviation laws associated with air rules, airspace, and documentation, and the operator must have an RPAS operator certificate. RPAS must also comply with existing air traffic management procedures (including possessing the required equipment and capabilities for each category of airspace). Finally, RPA operators must maintain a safety management system (SMS) to identify hazards and manage risks before accidents occur. Domestic operations for recreational purposes, by contrast, have little regulatory oversight.

9. Although not yet in effect, it is anticipated that RPAS pilots will eventually be required to hold a valid licence along with any required ratings and endorsements.

10. Air traffic control officers (ATCOs) working in airspace with RPA operations may require additional training on RPA operating characteristics.

11. An RPA may or may not need an airport for take-off, depending on its type. For operations at airports, special consideration must be given to ensure safe separation from all other aircraft. When RPAs take off from other sites, the remote pilot must carefully consider take-off conditions, obstructions, ATC communications, and whether overflying traffic could pose a conflict.

12. To facilitate the integration of RPAS into the current international aviation system, certain unique characteristics of RPAS must be taken into considerations:

- the entire RPAS must be secured against unauthorized use;

- RPAs require sensors to detect weather conditions that might pose a threat to the flight;

- environmentally friendly fuel sources (solar power and biofuels) may be more feasible for RPAs than traditionally piloted aircraft;

- RPAS accident and incidents are typically related to failure of hardware, lack of training, or issues with the human–machine interface; and

- RPAs must collect and transmit sensory data to the remote pilot, whenever possible, to assist with system awareness and management.

Table 10.3 Acronym rundown

ANC	Air Navigation Commission
AOC	air operator certificate
ATC	air traffic control
ATCO	air traffic control officer
AUVSI	Association for Unmanned Vehicle Systems International
BRLOS	beyond radio line-of-sight
BVLOS	beyond visual line-of-sight
C2 link	command and control link
CA	collision avoidance
CAA	civil aviation authority
CASA	Civil Aviation Safety Authority (Australia)
CofA	certificate of airworthiness
DAA	detect and avoid
EASA	European Aviation Safety Agency
EUROCAE	European Organization for Civil Aviation Equipment
GPS	global positioning system
HTOL	horizontal take-off and landing
ICA	instructions for continuing airworthiness
ICAO	International Civil Aviation Organization
IFR	instrument flight rules
MCM	maintenance control manual
RC	remote-controlled
RLOS	radio line-of-sight
ROC	RPAS operator certificate
RPA	remotely piloted aircraft
RPAS	remotely piloted aircraft system
RPASP	Remotely Piloted Aircraft Systems Panel
RPIC	remote pilot-in-command
RPS	remote pilot station
RWC	remain well clear
SAM	surface-to-air missile
SARPs	standards and recommended practices
SMS	safety management systems
sUAS	small unmanned aircraft system

(Continued)

Table 10.3 (Continued)

TC	type certificate
UAS	unmanned aircraft system
UAV	unmanned aerial vehicle
VFR	visual flight rules
VLOS	visual line-of-sight
VTOL	vertical take-off and landing

Chapter Review Questions

10.1 Define three important terms specific to RPA operations.

10.2 Explain three reasons why RPAs flown internationally require different operational standards and considerations than those flown domestically.

10.3 Explain three challenges for remote pilots operating out of an airport. How might they deal with each challenge?

10.4 There have been various types of aircraft flying without on-board pilots for over 100 years. Why do you think RPAs are an increasingly relevant topic in aviation today?

10.5 Do you think domestic recreational RPA activities should be regulated? Provide two arguments for and two arguments against such regulation. Does the size and weight of the RPA influence your decision?

10.6 How has ICAO dealt with RPAs international aviation thus far? Do you believe it is likely that a new annex with RPA-specific SARPs will be published in the future?

10.7 Identify an industry in your State where RPAs offer obvious advantages over piloted aircraft. Describe another industry where piloted aircraft offer advantages over RPAs. Justify your responses.

10.8 Considering both small and large RPA operations in your State, explain the impact of any three of the following considerations. Of the three, which consideration presents the greatest challenges?

- security

- weather

- environment

- accidents

- accident investigation

- human factors

- safety management systems

RPA Incidents

The first nine chapters in this book have profiled a major aviation accident linked to the chapter topic, but RPAS operations are based on an emerging technology and therefore (thankfully), there has not been a major RPA accident in civil aviation. There have been some incidents, however, each which raise interesting questions. Consider the following examples:

- In 2016, an Airbus A320 operated by British Airways was believed to have struck an RPA on its approach to land at London's Heathrow Airport, making international headlines as the first incident of its kind. There was no damage to the aircraft. After having received international attention, the situation turned out to have been simply a strike with a plastic bag rather than an RPA.[1]

- In 2015, an out-of-control RPA crashed on the lawn of the White House (the home of the President of the United States). The operator was not charged as he was not in control of the RPA at the time.[2]

- That same year, an RPA carrying more than six pounds (2.7 kilograms) of methamphetamine (illegal drugs) crashed into the parking lot of a supermarket in Tijuana, Mexico. It is believed that the drugs were intended to be smuggled over the border into the US.[3]

- Also in 2015, an RPA with a built-in camera and a bottle containing an unknown substance (emitting radiation) landed on the roof of the Japanese Prime Minister's office.[4]

- In 2014, an Australian triathlete sustained a head injury when she was hit with an RPA being used to film the competition. The photographer operating the RPA claimed an attacker had wrestled control away from him, which led to the loss of control.[5]

- In 2013, a small RPA flew within a few feet of the German Chancellor at a campaign rally before crashing into the stage at her feet. The RPA was operated as a protest and raised concerns about the threat of weaponized devices.[6]

Notes

1 Hughes & Halkon, 2016
2 Forrest, 2015
3 McVicker, 2015
4 Associated Press in Tokyo, 2015
5 BBC News, 2014
6 Gallagher, 2013

CASE STUDY

Case Study Questions

Thinking about the above incidents, and applying what you have learned in this chapter, discuss the following:

10.9 The recreational use of RPAs by operators with limited aviation experience can occasionally result in the devices entering airspace where they pose a risk to aircraft with pilots on board. To reduce this risk, some have suggested manufacturers should be required to install GPS technology that recognizes restricted areas and prevents the drone from entering. This strategy has been called *geo-fencing*. What do you think? Would this be a practical solution? How might it impact the cost of RPA technologies? How would geo-fences be created, maintained, and enforced? Might geo-fences be susceptible to illegal and malicious hacking?

10.10   The international aviation community has been working to establish licensing requirements for RPA operators; however, the wide variability in size, type, and usage of RPAs makes this challenging. A recreational user operating a very small device requires far different qualifications that a professional RPIC flying an aircraft with similar characteristics to a traditional aircraft. How do you think regulatory oversight of both groups can be practically accomplished? Are type and class ratings on a licence sufficient to distinguish the different skill sets required?

10.11   Some debate exists about whether the skills required of a remote pilot are more aligned with the competencies of a traditional pilot or those of an ATCO. Consider that pilots generally use sensory information (visual, auditory, kinesthetic, and olfactory cues) to help them understand their flying environment while ATCOs must develop the skills to visualize three-dimensional situations based on information presented on a two-dimensional display. What do you think? If you had to choose one, would you argue that the skills required for remote pilots are more aligned with traditional pilots or those of ATCOs?

10.12   Security is an ongoing challenge associated with RPAS. These systems can be exploited by criminals seeking profit (drug smugglers) as well as terrorists with political agendas (i.e, using RPAs as weapons). How can the international aviation community ensure the security of these systems? Should the burden of maintaining security fall more on operators, manufacturers, CAAs, or ICAO?

References

Associated Press in Tokyo, 2015. *Drone 'containing radiation' lands on roof of Japanese PM's office.* [Online] Available at: www.theguardian.com/world/2015/apr/22/drone-with-radiation-sign-lands-on-roof-of-japanese-prime-ministers-office

Barnhart, R. K., 2012. The future of unmanned aircraft systems. In: R. K. Barnhart, S. B. Hottman, D. M. Marshall & E. Shappee, eds. *Introduction to Unmanned Aircraft Systems.* London: CRC Press, pp. 181–192.

BBC News, 2014. *Australian triathlete injured after drone crash.* [Online] Available at: www.bbc.com/news/technology-26921504

Camhi, J., 2016. *Here are the technologies that are making drones safer and accelerating adoption.* [Online] Available at: www.businessinsider.com/the-drones-report-market-forecasts-key-players-and-use-cases-and-regulatory-barriers-to-the-proliferation-of-drones-2016-3

Dalamagkidis, K., Piegl, L. A. & Valavanis, K. P., 2012. *On integrating unmanned aircraft systems into the national airspace system: Issues, challenges, operational restrictions, certification, and recommendations.* 2nd ed. London: Springer.

Forrest, C., 2015. *12 drone disasters that show why the FAA hates drones.* [Online] Available at: www.techrepublic.com/article/12-drone-disasters-that-show-why-the-faa-hates-drones/

Gallagher, S., 2013. *German chancellor's drone "attack" shows the threat of weaponized UAVs.* [Online] Available at: https://arstechnica.com/information-technology/2013/09/german-chancellors-drone-attack-shows-the-threat-of-weaponized-uavs/

Holman, B., 2009. *Airminded: The first air bomb: Venice, 15 July 1849.* [Online] Available at: https://airminded.org/2009/08/22/the-first-air-bomb-venice-15-july-1849/

Hughes, C. & Halkon, R., 2016. *Drone crashes into British Airways passenger jet as it comes in to land at Heathrow Airport.* [Online] Available at: www.mirror.co.uk/news/uk-news/drone-crashes-british-airways-passenger-7776727

ICAO, 1944. *Convention on Civil Aviation ("Chicago Convention").* Chicago: International Civil Aviation Organization.

ICAO, 2015. *Manual on remotely piloted aircraft systems (RPAS), Doc 10019.* Montreal: International Civil Aviation Organization.

Jarnot, C., 2012. History. In: R. K. Barnhart, S. B. Hottman, D. M. Marshall & E. Shappee, eds. *Introduction to Unmanned Aircraft Systems.* London: CRC Press, pp. 1–16.

McVicker, L., 2015. *Drone carrying meth crashes near San Ysidro port of entry.* [Online] Available at: www.nbcsandiego.com/news/local/Drone-Carrying-Meth-Crashes-Near-San-Ysidro-Port-of-Entry-289353601.html

Merlin, P., 2013. *Crash course: Lessons learned from accidents involving remotely piloted and autonomous aircraft: NASA Aeronautics Book Series.* s.l.:Military Bookshop.

Neubauer, K., Fleet, D., Grosoli, F. & Verstynen, H., 2015. *Unmanned aircraft systems (UAS) at airports: A primer (ACRP report 144).* Washington, DC: The National Academies Press.

Perritt, H. H. & Sprague, E. O., 2017. *Domesticating drones: The technology, law, economics of unmanned aircraft.* London: Routledge.

Terwilliger, B., Ison, D., Robbins, J. & Vincenzi, D., 2017. *Small unmanned aircraft systems guide: Exploring designs, operations, regulations, & economics.* Newcastle, Washington: Aviation Supplies & Academics, Inc.

Zimmer, B., 2013. *The flight of 'drone' from bees to planes.* [Online] Available at: www.wsj.com/articles/SB10001424127887324110404578625803736954968

Notes

1 Dalamagkidis, et al., 2012
2 Dalamagkidis, et al., 2012, p. 12
3 Jarnot, 2012
4 Zimmer, 2013
5 Adapted from Barnhart, 2012; Jarnot, 2012; Terwilliger, et al., 2017
6 Holman, 2009, p. para. 1
7 Jarnot, 2012, p. 10
8 Jarnot, 2012, p. 15
9 Camhi, 2016, p. para. 4
10 Camhi, 2016
11 Barnhart, 2012
12 Camhi, 2016
13 ICAO, 2015
14 ICAO, 1944, p. Article 8
15 ICAO, 2015
16 ICAO, 2015
17 ICAO, 2015
18 ICAO, 2015
19 ICAO, 2015
20 ICAO, 2015
21 ICAO, 2015
22 Merlin, 2013
23 Neubauer, et al., 2015, p. 6

24 Neubauer, et al., 2015, p. 6

25 Perritt & Sprague, 2017, p. 7.3.1

26 Perritt & Sprague, 2017, p. 7.3.4

27 Perritt & Sprague, 2017, p. 7.3.5

28 ICAO, 2015

29 ICAO, 2015

30 ICAO, 2015

31 ICAO, 2015

32 ICAO, 2015

33 ICAO, 2015, p. 2–1

34 ICAO, 2015

35 ICAO, 2015

36 ICAO, 2015

37 ICAO, 2015

38 Neubauer, et al., 2015

39 ICAO, 2015

40 ICAO, 2015

41 ICAO, 2015

42 ICAO, 2015

43 ICAO, 2015

44 Neubauer, et al., 2015

45 Merlin, 2013

46 ICAO, 2015

47 Merlin, 2013

48 Merlin, 2013

49 ICAO, 2015

50 ICAO, 2015

51 ICAO, 2015

Index

References from Figures are indexed with bold page numbers

flight control surfaces 35, **36**, 162

flight controls 33, 97, 299, 322, 328, 334, 344

flight crews 70, 80, 85, 103, 110, 227, 267, 271, 275, 281, 294, 304, 306, 321; and airline marketing 85; all-female 326; inbound 302; junior 85; senior 85; skilled 250

flight data recorders 27, 98, 142–3, **263**, 264, 282, 284, 287, 289, 318, 321

flight deck 33, 84, 85, 264

flight dispatchers 9, 86

flight engineers 9, 47, 80, 302, 304

flight hours 55, 70–71, 78

flight information centres 10, 118

flight information regions 10, 115, 119, 126, 138–9, 226, 246

flight information services 128

flight instructors 73, **83**

flight management computers 130, 139

flight management system 124, 130–31, 139, 275, 284

flight manuals 98, 337

flight navigators 80

flight operations 68, 89, 162, 227, 260

flight paths 27, 116, 130, 133–312, 136, 240, 248, 264, 274, 344; direct 134, 243; unique 130

flight planning 112–13, 342

flight plans 43, 111, 113–14, 117, 124, 202, 306, 342

flight progress strip **113**, 139

flight recorders 27, 62, 263, 269, 271, 289

flight rules 113, 331, 340

flight safety 127, 260

Flight Safety Foundation 311, 313, 318

flight simulators 270, 299

flight times 51, 70, 78, 104–5, 277, 320

FMC 130, 139; see also flight management computer

FMS 124, 130–31, 139, 275, 284; see also flight management system

FOD 158, 160–62, 164, 170; see also foreign object debris

foreign object debris 158, 160–2, 164, 170–5

foreign States 7–8, 261

FOs 69, 72, 81, 97, 99–100, 102–5, 246, 248, 276, 284, 300, 302, 318, 320; see also first officers

fossil fuel 232

four-stroke internal combustion engine **37**

Fourth Freedom Agreement 7–8

FPS 113, 139; see also flight progress strip

Freedoms of the Air 4–5, 7, **8**, 19, 24; Eighth Freedom Agreement 7–8; Fifth Freedom Agreement 7–8; First Freedom Agreement 6, 8; Fourth Freedom Agreement 7–8; Ninth Freedom Agreement 7–8; Second Freedom Agreement 7–8; Seventh Freedom Agreement 7–8; Sixth Freedom Agreement 7–8; Third Freedom Agreement 8

FSF 311, 313, 318; see also Flight Safety Foundation

fuel 37–9, 44–5, 85, 89, 98, 100, 143, 148–6, 236, 239, 244, 273, 285, 343; and air mixture 37; dumping 287; fossil 232

full-body scanners 186

fuselages 25, 33, 36, 62, 179–80

Future Air Navigation Systems 135, 139

GA 37, 46, 59, 61, 67–8, 76–7, 80, 99–101, 148, 151, 157, 278, 339; see also general aviation

GAMA 48–9, 59; see also General Aviation Manufacturers Association

gas turbine engine **38**

GASP 12, 311, 318; see also Global Aviation Safety Plan

general aviation 37, 46, 59, 61, 67–8, 76–7, 80, 99–101, 148, 151, 157, 278, 339; accident rate 283; aircraft 48, 70; aircraft shipments 49; aircraft type 279

General Aviation Manufacturers Association 48–9, 59

Germanwings 93, 302

Germanwings Flight 9525 275–6

GHG emissions 234–6

glass cockpits 47

gliders 32–3, 57–8, 279; single-pilot 57, 279

global accident rates 255, **256**, 257, 282, 286

global airlines profits and losses 87

Global Aviation Safety Plan 12, 311, 318

Global business jets 48

global navigation satellite systems 130–31, 135, 138–9

Global Positioning System 125, 130–31, 135, 139, 275, 285, 349

GMT 152, 171; see also Greenwich Mean Time

GNSS 130–31, 135, 138–9; see also global navigation satellite systems

GPS 125, 130–31, 135, 139, 275, 285, 349; see also Global Positioning System